Coastal Wetlands of the World
Geology, Ecology, Distribution and Applications

Salt marshes and mangrove forests, the intertidal wetlands of the world's coastlines, provide key ecological services to all areas of the globe, and are vital sinks and sources in carbon budgets. They are crucial indicators of both modern-day anthropogenic impacts on climate and ecosystems, and paleoecological changes during much of Earth's history.

This cutting-edge, richly illustrated book introduces the essential elements of coastal wetlands and their applications. It unites geological and oceanographic approaches in an accessible way, providing scientific names for key plant and animal species. The book opens by introducing coastal oceanography, the physical features of wetlands, their ecology and human impacts upon them, giving students from all fields the necessary background for wetlands studies. It then presents detailed case studies from all areas of the world, with extensive illustrations, presenting students with a broad, global-scale picture of wetlands geomorphology and biodiversity. The final chapters discuss some unique applications of coastal wetlands, including geological monitoring, uses in biotechnology and agriculture and various experimental mesocosms.

This is ideal as supplementary reading to support students on a wide range of Earth and Life science courses, from environmental science, ecology, and paleoecology to geomorphology and geography. Providing citations to a variety of more specialist articles, it will also be a valuable interdisciplinary reference for researchers.

David Scott is a Professor in the Earth Sciences Department of Dalhousie University, where he teaches micropaleontology and Quaternary geology. Other positions held include VP on the Cushman Foundation Board, membership of the Geological Society of America and Paleontological Society, and serving as associate editor for the Canadian Journal of Earth Sciences. He has written over 130 refereed papers, has edited three NATO volumes on coastal geomorphology and paleontological subjects, and is also the co-author of *Monitoring in Coastal Environments Using Foraminifera and Thecamoebian Indicators* (with Franco Medioli and Charles Schafer, CUP 2001). Professor Scott has conducted field work in most major marshes of North America, and several wetlands in South America and Europe, in addition to participating in Ocean Drilling Program studies in the Indian Ocean.

Jennifer Frail-Gauthier is a PhD Candidate in earth sciences and biology at Dalhousie University, and her research topic is small food webs in salt-marsh ecosystems, specifically foraminifera, which form an important part of this book. Her PhD focus is on experimental approaches to salt marshes for studies in ecology, biology, geology, restoration and other human impacts. Ms Frail-Gauthier is a science writing tutor and teaches third-year applied coastal ecology, which focusses on the various coastal ecosystems of the world, including geology, ecology and anthropogenic impacts, and also teaches various biology courses. She has received several major scholarships and awards during her graduate studies, and also holds a Teaching Excellence Award and a University Medal from Dalhousie University.

Petra Mudie is an Adjunct of the Graduate Studies Faculty at Dalhousie University, Adjunct Professor of the Memorial University of Newfoundland, and a Scientist Emeritus with Geological Survey of Canada Atlantic. Her previous work includes heading up a halophyte research laboratory at the Scripps Institute of Oceanography, including surveys of coastal wetlands from Canada to Mexico, working on environmental marine geology for the Canadian Government until 2001, and subsequently leading an international programme, studying palynological records of the history of climate and sea-level change in the Black-Sea–Eastern-Mediterranean Corridor. She is the author of over 80 papers in science journals. Dr Mudie and Professor Scott have collaborated on salt-marsh and Arctic paleoenvironmental studies for nearly 40 years, co-supervising many graduate students.

'This is a major new contribution to the study of salt marshes and mangroves forests. Uniquely comprehensive, the book provides extraordinary coverage of coastal wetlands from the Arctic to the tropics with superb case study examples from Africa, Europe, Asia, and both Americas.

Importantly, this innovative volume covers not only the physical, ecological and human interventions controlling the development, loss and future of coastal wetlands but also provides the reader with modern approaches to geological monitoring, conservation of plant biodiversity, and experimental methods. The readability of the book, with supporting graphics and informative photographs, makes it accessible to readers at all levels.'

– Professor Curtis J. Richardson, *Director, Duke University Wetland Center*

Coastal Wetlands of the World

Geology, Ecology, Distribution and Applications

DAVID B. SCOTT,

JENNIFER FRAIL-GAUTHIER

AND

PETRA J. MUDIE

Department of Earth Sciences, Dalhousie University, Halifax, Nova Scotia, Canada

CAMBRIDGE
UNIVERSITY PRESS

NORTHWEST MISSOURI STATE
UNIVERSITY LIBRARY
MARYVILLE, MO 64468

CAMBRIDGE
UNIVERSITY PRESS

University Printing House, Cambridge CB2 8BS, United Kingdom

Published in the United States of America by Cambridge University Press, New York

Cambridge University Press is part of the University of Cambridge.

It furthers the University's mission by disseminating knowledge in the pursuit of education, learning and research at the highest international levels of excellence.

www.cambridge.org
Information on this title: www.cambridge.org/9781107056015

© David B. Scott, Jennifer Frail-Gauthier and Petra J. Mudie 2014

First published 2014

A catalogue record for this publication is available from the British Library

Library of Congress Cataloguing in Publication data
Scott, D. B.
Coastal wetlands of the world : geology, ecology, distribution and applications / David B. Scott, Jennifer Frail-Gauthier, Petra J. Mudie.
pages cm
ISBN 978-1-107-05601-5 (hardback)
1. Wetland ecology.   2. Wetland conservation.   3. Coasts.   4. Coastal ecology.
I. Frail-Gauthier, Jennifer.   II. Mudie, Petra J.   III. Title.
QH541.5.M3S36   2014
577.68–dc23
2013035761

ISBN 978-1-107-05601-5 Hardback
ISBN 978-1-107-62825-0 Paperback

Additional resources for this publication at www.cambridge.org/coastalwetlands

Contents

Colour plate section is found between pages 178 and 179

Preface

Coastal Wetlands of the World follows the book by Scott, Medioli and Schafer (2001) on *Monitoring in Coastal Environments*. We are motivated to write this new book based on concern about the status of mangroves and salt marshes all over the world, from pole to pole, and by the fact that few students have the chance to look at our changing shorelines from both a geological and an ecological perspective. Coastal wetlands are being destroyed and degraded at alarming rates, and only a fraction remains. These wetlands protect us from storm buffering and have extremely high primary production, making them important storehouses of carbon and energy, habitats that nurture juvenile stages of commercially important fishes and that filter our waste water – yet we continue to damage them faster than we can preserve them. In some areas, less than a third of natural wetlands remain along the coast, and very few are entirely unaffected by *direct* human impacts. Furthermore, all our coastal wetlands are changing in response to *indirect* human impacts: global warming, sea level rise and increasing numbers of severe coastal storms. These impacts are further magnified in the Arctic, where the pace of climate warming is four times faster than other places on Earth, and where disappearing sea ice is encouraging rapid expansion of oil and gas exploration, with the associated risks of long-lasting pollution damage. Arctic people say that 'The Earth is faster now' – and it appears that traditional methods of coastal living are no longer viable. It is likely that circumpolar regions are already irreversibly changed – and the spill-over impacts on global air and ocean systems is already being felt by people in crowded cities of warm temperate regions.

We take an interdisciplinary approach to *Coastal Wetlands of the World* – there literally is something for everyone between the covers of this book. It was initially written for under-graduate students, focussing on classical studies that are the baselines for evaluating recent changes, but it soon became clear that more detail was needed to guide readers towards the proliferation of new scientific literature. As a result, we have included innumerable up-to-date references that will also help graduate students, naturalists and coastal-resource managers obtain a fresh view of tidal wetlands research across a wide spectrum of disciplines. Geologists, ecologists, conservationists, environmentalists, archeologists, historians and social scientists can all learn something new and clearly understand the issues at hand, for any area of the world. The book's global focus and ample illustrations are also intended to draw the student beyond their familiarity with a limited neighbourhood marshland toward a much bigger picture of wetlands geomorphology and biodiversity on a global scale.

Why yet another book covering coastal wetlands and ecosystems? Our literature search of the most widely used texts showed a large imbalance in coverage of the world's continents, despite the shrinking size of our Internet-linked Global Village. We have attempted to fill in large gaps for under-reported regions of Mexico, South America, Africa, Eastern Europe and China, and we provide the only systematic and focussed coverage of global tidal wetlands. Most other wetlands books are broken into vari-authored chapters and/or report on either marshes or mangroves, presenting a somewhat schizophrenic perspective to the reader, as though the world has sharp boundaries. In contrast, our readers are provided a seamless virtual tour from the northern tip of continental North America to the southern tip of New Zealand. From geology to biology to ecology to human impacts, we introduce wetlands from a generic stand point (Chapters 1–6). We then dive into information about coastal wetlands across all continents, giving specific historical case studies, and earmarking new research and paradigm shifts in traditional concepts about drivers of coastal climate changes. The last section of our book focusses on unique applications of coastal wetlands studies, including a chapter on paleoseismology, paleoclimate and forecasting (updated and much expanded in the range of proxies from *Monitoring in Coastal Environments*), and an outline of how coastal wetlands are used as experimental mesocosms to better understand and replace what is lost. We are the first to cover both traditional knowledge and cutting-edge subcellular and genetic knowledge of the potential for salt-tolerant plants to combat crises of soil salinization in agricultural crops. The development of new salt-tolerant crops is a major part of the new Green Revolution needed to feed the world's rapidly expanding human population – simultaneously representing major carbon credits and conserving our fast-dwindling global freshwater resources. Education is the first step in the coastal crisis facing everyone 'living on the edge' – the more that can be taught about tidal wetlands, the more our global population can see the dire need to save what remains and wisely restore what we have destroyed.

We are indebted to many people and organizations who have helped in the writing of this book, answered multiple questions about places less familiar to us and provided illustrative materials. Invaluable help with diagrams comes from Rob Gauthier, Alexandre Pelletier-Michaud, Gary Grant and Matthew Chedrawe, and we are indebted to Ken Wallace for photo-compilations and design. The extended family of Petra Mudie have provided photo-coverage from all continents where there are tidal wetlands (thanks to Anita and Hilton Whittle, Helen Pease and Peter Mudie) and we sincerely thank all those who graciously provided other beautiful photos of wetlands and animals, as acknowledged in the figure captions. Finally, we are most grateful to Laura Clark and others at Cambridge University Press, who provided encouragement, guidance, and answered no less than 100 questions to help get this book from our heads into a beautiful printed volume.

List of acronyms and abbreviations

Organizations

AGEDI	Abu Dhabi Global Environmental Data Initiative
ASEAN	Association of Southeast Asian Nations
CC	Creative Commons (free-to-use images from Wikipedia or Flickr)
CIMI	Canada-Iraq-marshlands Initiative
CITES	Convention on International Trade in Endangered Species
COSEWIC	Committee on the Status of Endangered Wildlife in Canada
CNES	Centre National d'Etudes Spatiales
CONABIO	Comisión Nacional para el Conocimiento y Uso de la Biodiversidad
COSEWIC	Committee on the Status of Endangered Wildlife in Canada
DFO	Department of Fisheries and Oceans (Canada)
FAO	Food and Agriculture Organization of the United Nations
GSCA	Geological Society of Canada Atlantic
IPCC	Intergovernmental Panel on Climate Change
IRD	l'Institut de recherche pour le développement
IRRI	International Rice Research Institute
MEA	Millennium Ecosystems Assessment
MODIS	Moderate Resolution Imaging Spectroradiometer
NASA	National Aeronautics and Space Administration
NEIC	National Earthquake Information Center
NOAA	National Oceanic and Atmospheric Administration
RIS	Ramsar Information Sheet
SRTM	Shuttle Radar Topography Mission
UNEP	United Nations Environment Programme
UNESCO	United Nations Educational, Scientific and Cultural Organization
USFWS	United States Fish and Wildlife Service
USA	United States of America
USGS	United States Geological Survey
WCMC	World Conservation Monitoring Centre
WRI	World Resources Institute

Standard and other abbreviations or notations

‰	parts per million
$\delta^{13}C$	delta carbon-13 value in mille, which is the ratio between stable carbon isotopes (C-12 and C-13) relative to PeeDee Belmnite $\times$ 1000
ASL	above sea level
BCE	before the Common/Christian Era
BP	before present
cal yr BP	calendar years before present
BTEX	benzene, toluene, ethylbenzene and xylenes
CAT	storm/hurricane category
CE	Common/Christian Era
CHC	chlorinated hydrocarbons
DDT	dichlorodiphenyltrichloroethane
DNA	deoxyribonucleic acid
EHW	extreme high water (highest tide line)
GW	gigawatt
IP_{25}	ice proxy with 25 carbons
LIA	Little Ice Age
LGM	last glacial maximum
M_w	unit for earthquake magnitude
MHW	mean high water
MHHW	mean higher after first high water (see Box 2.1 for more details)
MIS	marine isotope stages
MLW	mean low water
MSL	mean sea level
MSX	*Haplosporidium nelsoni* shellfish disease
MW	megawatt
PAH	polycyclic aromatic hydrocarbon
PLF	pingo-like feature
psu	practical salinity unit
RSL	relative sea level
sp.	singular species; plural = spp.
VP	vice president

1 Introduction: what is covered in this coastal wetlands book?

Coastal wetlands, including tidal salt marshes and mangrove swamps, are environmentally stressful and variable habitats, and yet are teeming with life. Their biological productivity exceeds that of coral reefs and matches that of tropical rainforests. These wetlands provide resident plants and animals with shelter, food and continuous renewal of nutrients on each tidal cycle. These coastal wetlands are also vital to neighbouring ecological communities and have important values to humans, serving as natural carbon-capture systems, as sources of oceanic 'blue carbon', as filters of sediment or nutrient-loaded flood water and as buffers against storm tides and rising sea levels. Concern about the twentieth-century destruction of many wetlands in North America, Europe, Australia and New Zealand, and the degradation of wetlands worldwide led to the 1971 Ramsar Convention on Wetlands, held on the shore of the Caspian Sea in Iran (see Box 1.1 The Ramsar Convention). The Convention provides foundations for planning of 'wise use' for all wetlands; preservation of wetlands with international importance for ecology, biodiversity or hydrology; and co-operative protection of internationally shared species. Coupled with this landmark environmental agreement, the United Nations designated 2 February as 'World Wetlands Day', bracketing it with programmes to raise awareness of the strong link between global freshwater supplies and wetland resources. These two themes 'Water' and 'Wetlands' highlight global efforts to promote understanding that without coastal wetland conservation, there will not be enough water for sustainable development, human health and, ultimately, the survival of humankind.

This book is a university-level text that covers the major features of coastal salt marshes and mangrove swamps, introducing the reader to their ecology and geology, and showing how natural and man-made changes have impacted the wetlands on time scales of tens to thousands of years. Professionals can also make use of updated climate and sea level change perspectives provided in this multidisciplinary and globally covered book. Although the book takes a global approach (Figure 1.1), many examples are drawn from North American coastal wetlands (Chapter 7) because they are the best known and least altered by pressures of burgeoning coastal human populations. The book is unique in integrating geological and ecological case histories of changes in salt marshes on a global scale and in providing insight to various applications of coastal wetland studies, including laboratory and field experiments with salt marsh mesocosms.

Chapter 2 covers the physical aspects – geological, oceanic and climatic conditions – of coastal wetland formation, morphology and evolution, explaining what marshes do, and the influence of tides, storms and other climate impacts. This chapter defines the specific characteristics of coastal wetlands, which are all marked by some degree of tidal inundation, as opposed to other types of wetlands included in the Ramsar Convention, but not discussed in this book, such as inland salt lakes, freshwater marshlands and peat-moss bogs. The

Box 1.1   **The Ramsar Convention: international wetlands conservation**

These white storks nesting on the roof of this house in Croatia spent the winter in the Caspian Sea. To protect these iconic birds and other endangered migratory waterfowl, the Ramsar Convention on Wetlands Conservation was signed on 2 February, 1971, at the village of Ramsar on the south shore of the Caspian Sea. This convention is one of the oldest global intergovernmental agreements on the environment and provides a framework for coastal global wetland protection on several levels. The Convention's Mission is 'the conservation and wise use of all [types of] wetlands through local, regional and national actions and international cooperation, as a contribution towards achieving sustainable development throughout the world'. The convention is now co-signed by over 150 nations on all continents – where about 300 coastal wetlands sites are identified for protection (Millennium Ecosystems Assessment, 2005). Details of the Convention, the Ramsar definitions of 'wetlands', the ratifying countries and their wetland preserves are given by Mitsch and Gosselink (2007). Maps of the Ramsar wetlands distributions and links to the United Nations initiative on world water resources co-operation can be found at http://www.ramsar.org/pdf/wwd/13/Leaflet.pdf.

Box Figure 1.1

importance of tides is explained further in Chapter 3, which describes how the frequency and depth of saltwater flooding controls the shoreline zonation and the plants that grow in tidal wetlands. This chapter also explains how plants interact with the marsh sediments to shape the development of the marshland and its tidal channels, and how marsh plants are uniquely adapted to the stresses of the constantly shifting marsh environment of variable salinity, oxygen and nutrients. Following these introductory sections, Chapter 4 explores the inter-action between marshland animals and plants, the zonations of the animals and their adaptations to stressors. The general ecology of salt marshes and mangroves is given here, emphasizing their immense production and complicated energy flow. Chapter 5

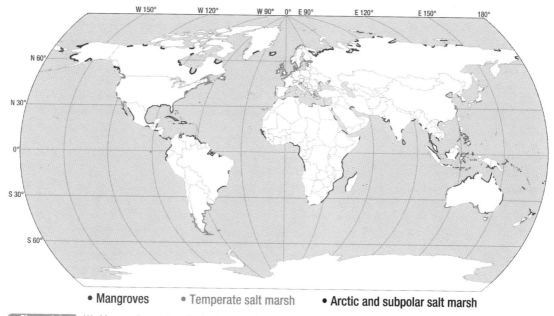

• Mangroves   • Temperate salt marsh   • Arctic and subpolar salt marsh

 Figure 1.1   World map of coastal wetlands. Green marks areas of temperate climate salt marshes; blue indicates polar marshes; red shows areas of tropical mangroves. See also colour plates section.

introduces the kinds of natural catastrophes that change the wetlands, and it gives examples of the human interventions which affect the natural ocean–atmosphere-controlled systems – often endangering modern coastal communities.

The main focus of this book is the unique perspective it presents for coastal marshes on a global scale, including several (e.g. the Beaufort Sea Arctic coast of Canada, the Tigris–Euphrates Delta in Iraq, and The Gambia in tropical West Africa) that have rarely, if ever, been included in texts on coastal wetlands. Chapter 6 introduces the broad layout of coastal wetland distributions that is governed primarily by the nature of the coastline (including steepness, sediment supply and wave energy), the climate (hence latitudinal zonation and regional variations in marsh vegetation zones) and the ocean circulation, which redistributes heat from tropical to polar regions. The chapter also covers the principles of paleogeography and how this drives wetland biodiversity and is a basis for classifying salt marsh vegetation types for comparison on a global basis.

Following this general introductory outline of global-scale patterns that govern the distribution of all coastal wetlands, various specific features are then illustrated with a selection of regional examples. Chapter 7 presents examples from North America, where the largest and least altered salt marshes and coastal wetlands have survived since European colonization. Chapter 8 gives contrasting examples for the Pacific and Atlantic coasts of Central and South America, where the natural landscape is less conducive to development of extensive coastal wetlands outside of the tropical regions where mangroves dominate. In Chapter 9, we present information for selected marsh and mangrove areas in Africa, where pristine marshes are now being replaced by rapidly growing populations and industrialization. Chapter 10 provides contrasts from Europe and Asia, where little of the coastal wetland has survived centuries to millennia of modification by humans, who have reclaimed the land for agriculture and industrial uses. Chapter 11 covers the coastal marshes of Australia and New Zealand that have some unusual features because of their long geographical isolation and relatively recent colonization by Europeans.

The last section of the book introduces some innovative applications of coastal wetland studies. Chapter 12 describes the use of proxy data (such as microfossils, wood and peat layers) in sediment cores for monitoring earthquakes and the history of past earthquakes and tsunamis, as well as deciphering geological records of climate change, hurricanes and tropical cyclones. This chapter also outlines the importance of these coastal wetland archives for forecasting the potential impact of future climate warming, sea level changes and extreme weather events on coastal communities. Chapter 13 outlines the importance of sustaining salt marsh and mangrove-swamp biological diversity as storehouses of genetic information on physiological adaptations (ionomic databases) for coping with the growing problem of salt stress in modern farmlands which face impending shortages of freshwater for irrigation. Chapter 14 describes what can be learned from wetland mesocosms, which are used to examine, on a small scale, the impacts of various environmental threats and predicted problems. Chapter 15 concludes this book with a summary of the main themes discussed and some directions required for future research and protection of global coastal wetlands.

2   Physical aspects: geological, oceanic and climatic conditions

---

Key points

Coastal wetlands exist on the edge of all continents except Antarctica; saltwater wetlands are covered by seawater daily or periodically; tidal range and period (once or twice/day) control the amount of submergence and oxygen supply to marsh biota; tidal vegetation includes halophytic salt marshes, mangroves, seagrass beds and brackish water reed-swamps; salt marshes dominate cooler regions (>30° latitude) and mangroves occupy equatorial regions; local variations follow salt gradients measured as parts per mille or dimensionless psu; coastal wetlands develop in wave-sheltered sites: estuaries, deltas, glaciated fiords, barrier lagoons and tectonic down-faults; sediment supply must compensate for erosion by storm tides (cyclones, hurricanes, typhoons); space is also needed for growth on emerging or submerging coasts; global-warming impacts and human population growth are increasing risks of flooding and erosion.

2.1 What are coastal wetlands (saltwater wetlands)?

The Ramsar Convention defines *wetlands* as areas of marsh, fen or peatland with water which is static or flowing, fresh, brackish or salt, including areas of marine water not deeper than 6 m. Coastal wetlands, however, exist only at the interface of the land and sea, occurring on shorelines marked by some degree of tidal inundation, i.e. the tidal marshes and mangrove swamps, as classified in Mitsch and Gosselink (2007), who provide a dictionary of the many terms used to classify wetlands. All saltwater wetlands are characterized by the presence of brackish or saline water derived from the mixing of marine and fresh waters – including salt marshes, mangroves, seagrass beds and brackish water reed-swamps. Of this suite of intertidal habitats, the best known is the salt marsh, which is the most accessible saltwater wetland and the focus of many classical ecological studies which began over 100 years ago, starting in 1903 in North America (Ganong, 1903) and 1917 in Wales (Doody, 2008).

The strict definition of salt marsh is an intertidal area that is influenced largely by marine tidal cycles. The vertical range of a marsh is governed by the tides, which define the exposure time for the marsh vegetation occurring within different elevation ranges between mean sea level and the highest tides (see Box 2.1 Important tidal reference

Important tidal reference points

Everyone who has visited the seashore knows that seawater moves up and down with tides. However, the precise meaning of mean sea level (MSL) and its upper and lower limits are not so well known. The graph illustrates the main concepts used to set an official reference level (geodetic datum) for measuring long-term changes in mean seawater level. Note that for navigation (nautical/hydrographic) charts, a different reference (chart datum) is used, which is based on the lower low tide (= low spring tide) as its reference point. This provides maximum safety for boaters in shallow waters, but it requires adjustment if used for mapping of tide levels in coastal wetlands.

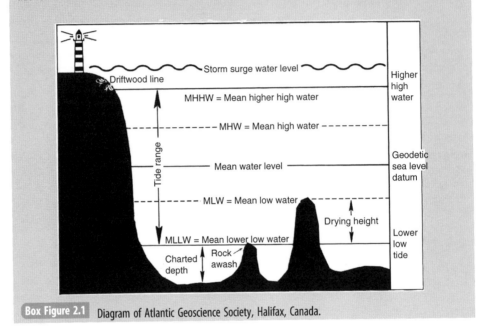

Box Figure 2.1   Diagram of Atlantic Geoscience Society, Halifax, Canada.

points). The tidal range is largely controlled by the periodicity and water levels of each marsh area caused by lunar tidal cycles. Ocean tides can range from 30 cm in semi-enclosed microtidal seas such as the Mediterranean, to large inlets or embayments that have macrotidal heights greater than the global average of 2–3 m. Prime examples of macrotidal regimes are found in eastern Canada: the Bay of Fundy in Nova Scotia (see Section 7.3.2), and Ungava Bay in northern Quebec, where the tides are semi-diurnal (two cycles a day) and tidal amplitude is about 16 m on a 12 h cycle. The opposite end of the tidal scale in North America includes the western Arctic Coast (0.5 m microtides) and areas such as Chesapeake Bay and the Gulf of Mexico, where relatively narrow entrances to the open water of the Atlantic Ocean restrict the range to less than 1 m just once a day (diurnal tides). Regardless of the range magnitude and periodicity, all saltwater marshes require tides to supply nutrients for estuarine and nearshore communities and to provide oxygen needed by the aquatic plants and invertebrate organisms living in the wetlands and channels.

2.2 Where are they found?

Overall, it is estimated that total area of coastal salt marshes in the world today is ~45 000 km^2 (Greenberg *et al.*, 2006), with about half of that (19 382 km^2) being in Canada and the USA (Mendelsohn and Mckee, 2000) and the other large areas being in China and Korea. The largest area (~50%) of tropical tidal wetlands (mangrove forests/swamps) is found in Southeast Asia, with other large areas being in South America and Africa (Valiela *et al.*, 2009). Mangroves are found in 114 countries, covering a total area of about 181 077 km^2 (Spalding *et al.*, 2010) – more than as much as the salt marshes. The long history of agricultural use of marshland in Eurasia means that there are few, if any, undisturbed coastal salt marshes remaining in Europe, North Africa, the Middle East or eastern Asia. In later sections we will present some case histories outlining the original features of these marshlands and the factors that have led to their present-day reduction to pockets or narrow bands along the coast. It should be noted that although our studies include the sparsely vegetated saltflats around the mean higher high water (spring tide) line, we do not cover salt marshes inland of the coastal zone; some of these inland salt marshes contain the same salt-tolerant (= halophytic) plants as the coastal wetlands, but they are not subject to tidal influences. Interested readers can consult Chapman's (1974) book on salt marshes and salt deserts of the world for the most comprehensive account of salt desert vegetation.

Latitudinal differences in climate are the primary factors governing the distribution of coastal wetlands on a global scale. Details of the importance of geographic location, both past and present, are given in Chapter 6, but here we introduce the broad principles. The vegetation of saltwater wetlands responds to climate by adapting to temperature and salinity fluctuations. Annual rainfall and summer temperatures largely determine the range of salinity within the coastal wetland. In cool temperate regions, soil and water salinities may range from 0 to 20‰ over one tidal cycle (for comparison, the salinity range of open ocean water is about 30−38‰ = 30–38 psu; Box 2.2). In contrast, hypersaline conditions can raise the water salinity to 50–60 psu in hot or cold arid regions. High atmospheric temperatures also cause excess evaporation, which increases salinity, and damages or kills plants less resistant to high salt content. However, most coastal wetland plants are adapted to this harsh and highly variable regime. The most widespread tidal marsh types are the temperate salt marshes, which are dominated by tall, salt-tolerant (halophytic) grasses and herbs with succulent stems or leaves (Figure 2.1). These wetlands predominantly occupy the latitudes 45° N and 45° S, which include the mildest of the global climatic regions. In polar regions, a low diversity of salt marsh grasses, sedges and annual herbs survive the short cool summer growing season and tolerate winter temperatures as low as –50 °C. At the other end of the spectrum are mangrove trees and shrubs, which are largely confined between the tropical latitudes from 25° N to S because they cannot tolerate winter temperatures below 16 °C or more than three continuous days of frost.

Some of the largest continents, which occur in both Northern and Southern Hemispheres, have the most extensive mangroves because they straddle the tropical

Box 2.2	Measuring and defining saltiness

Salinity in seawater and soil is measured by the amount of dissolved salt present relative to the weight of the water. During most of the twentieth century, salinity was usually reported as either the amount of salt in a kilogram of water (as mg/kg) or as parts per mille, denoted as ‰. This per mille notation, ‰, is most commonly used by geologists. However, today, by oceanographic convention, salinity is measured as the electrical conductivity of a sample relative to a standard solution of potassium chloride. This number is a ratio, so it has no units of dimension and is expressed as psu, meaning practical salinity unit. Estuaries are often classified by the salinity of the water, which depends on the amount of freshwater river discharge relative to seawater inflow, as depicted in the sketch below. Where river flow is high, the estuary is either vertically stratified, with brackish water (5–25 psu) at the surface and marine water (>30 psu below), or horizontally stratified with oligohaline (0.5–<5 psu) water at the inner end, mesohaline or polyhaline (>5–18 psu) water in the centre, and euhaline (>30–40 psu) water at the entrance.

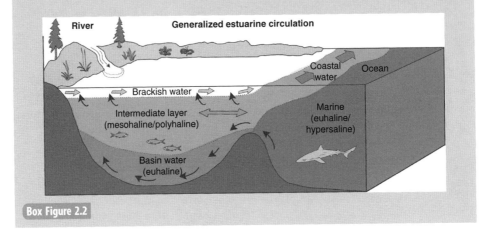

Box Figure 2.2

latitudes: South America (mainly east coast), the Latin American Caribbean and Mexican coasts, Africa, India, Malaysia, Indonesia, Australia (east and west coasts above 10° S). Southern Australia has the most southerly mangroves at ~40° S. The Caribbean Islands, including Cuba, also have extensive marshes and some mangroves, but these are isolated from each other by expanses of ocean. Bermuda and the Japanese island of Okinawa have the most northerly mangroves. There is a small, isolated tidal marshland on the most northeastern part of Hokkaido Island (Japan) – this is the only salt marsh in either Hokkaido or the main island of Honshu.

2.3 How are salt marshes formed?

Salt marshes occur on low-energy shorelines in temperate and high latitudes. The survival of a salt marsh depends primarily on the rate of sea level rise and changes in the tidal range

Figure 2.1 Range of coastal wetland vegetation. Left: temperate region *Spartina* grass-dominated salt marsh, southeastern Canada (photo by P. J. Mudie). Right: tropical riverine mangrove *Rhizophora* in Belize (photo by Helen Pease).

relative to movement of the shoreline, which can be stable, submerging or emerging (Scott *et al.*, 2001; McFadden *et al.*, 2007; Barnhardt, 2009). Figure 2.2 illustrates the changes in shorelines that are expected to accompany long-term changes in sea level associated with glacial and interglacial cycles. When the sea level rise is sudden and large, as associated with tectonic subsidence, then the salt marshes will be rapidly drowned and buried by wave-transported or flood sediments, as discussed later in Chapter 12. Commonly, shorelines with the largest salt marshes consist of mudflats or sand flats (known also as tidal flats), which are sustained primarily by sediment deposits from inflowing rivers and streams. Where sediment supply is sufficient, then accommodation space for horizontal growth of the marsh along the shore becomes the determining factor for salt marsh survival and growth.

Salt marshes typically occur in wave-sheltered environments, such as estuaries, large fiords (= glaciated valleys) and protected bays, areas behind the embankments (levees) of delta tributary channels, and on the leeward side of barrier islands and sand spits (Doody, 2008; Figure 2.3). Less commonly, salt marshes occur on the sheltered shores of submerged tectonic down-faults, such as the Bay of Fundy and San Francisco Bay. On wave-exposed rocky or gravelly shorelines, salt marshes are usually restricted to a few deep-rooted plants near the high tide line, as commonly seen on the wave-swept, glaciated shores of southern Alaska. In the tropics and subtropics, the salt marshes are largely replaced by salt-tolerant mangrove trees (= mangal vegetation) that shade out the low-growing salt marsh herbs and small shrubs.

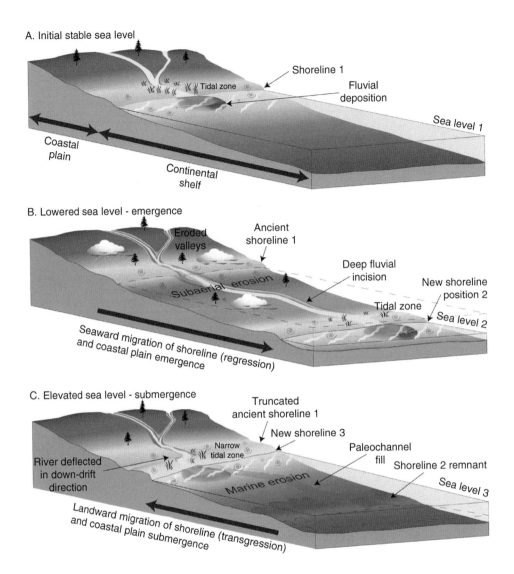

A. Initial stable sea level

Shoreline 1

Tidal zone

Fluvial deposition

Sea level 1

Coastal plain

Continental shelf

B. Lowered sea level - emergence

Eroded valleys

Ancient shoreline 1

Deep fluvial incision

New shoreline position 2

Subaerial erosion

Tidal zone

Sea level 2

Seaward migration of shoreline (regression) and coastal plain emergence

C. Elevated sea level - submergence

Truncated ancient shoreline 1

New shoreline 3

Narrow tidal zone

Paleochannel fill

Shoreline 2 remnant

River deflected in down-drift direction

Marine erosion

Sea level 3

Landward migration of shoreline (transgression) and coastal plain submergence

Figure 2.2 Emergent versus submergent coastline features. A: Salt marsh formed on a stable coastline with no major sea level change; B: Salt marsh retreats seaward on an emerging coastline associated with a falling sea level (= regression), leaving behind saline coastal plain soils and marine shell deposits that are eroded by rain and/or wind; C: Narrow salt marshes form on submerging shores where sea level is continually rising during a marine transgression (modified from Barnhardt, 2009, Fig. 3.8).

Most natural salt marshes occupy extensive areas of low coastal topography, which are preferred areas for human settlement and the reason many salt marshes have been degraded and lost. Deltaic marshes are associated with large rivers such as the Nile Delta, the Danube Delta in Romania and Ukraine, the Mississippi Delta in the USA, the Ganges and Mekong Deltas in Southeast Asia, and the Yangtze and Yellow Deltas in China. These large

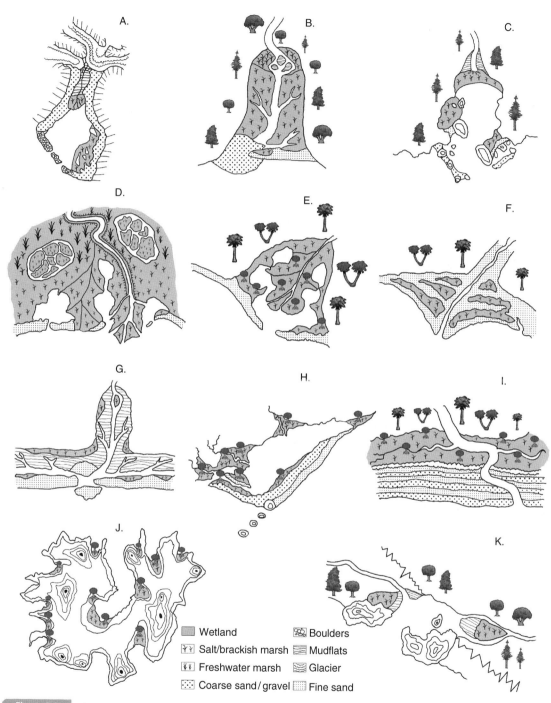

Figure 2.3 Sketches of geomorphological features supporting tidal marshes (grey shade) in areas of sediment deposition. Row 1. A: Glaciated drowned valleys (= fiords); B: estuaries; C: bays. Row 2. Deltas built from terrigenous deposits; D: river-dominated; E and F are wave-dominated beach-barrier and cuspate types, respectively. Row 3. Lagoons behind sand bars; G: barrier lagoon; H: tombola lagoon; I: beach bar sequence; Row 4: structural lagoons of tectonic origin; J: volcanic origin; K: a fault graben bay.

deltas are also the most densely populated (about 0.5 billion people) and intensively cultivated, and their wetlands are at greatest risk of future destruction (IPCC, 2007; Syvitski *et al.*, 2009; see Chapter 5).

Most commonly, salt marshes occur at the heads of estuaries in areas where there is high sedimentation and little wave action (Figure 2.3). Smaller salt marshes are found at the heads or the entrances of steep-sided glacial fiords or lochs where sediment fans or glacial outwash provide a muddy substrate behind gravel or boulder moraines. Back-barrier marshes form on the landward side of coastal sand dunes and are sensitive to the reshaping of the seaward sand barriers. These marshes are common along much of the eastern coast of the United States and the Frisian Islands off Germany. Large, shallow, coastal embayments can contain salt marshes: examples include Morecambe Bay and Portsmouth in Britain, and the Minas Basin and Chignecto Bay at the head of the Bay of Fundy in eastern Canada. Salt marshes often develop in tidal lagoons, such as the Venice Lagoon in Italy and other parts of the Mediterranean Sea (Perez-Ruzafa *et al.*, 2007), and there is a wide array of coastal lagoons in western Mexico (Phleger and Ayala-Castañares, 1969; Gonzales-Zamorano *et al.*, 2012). Many of these lagoons and their saltwater wetlands have been designated as protected Ramsar sites (Christian and Mazzilli, 2007; Santamaria-Gallegos *et al.*, 2011).

2.4 Physical conditions that shape wetlands

Wetlands cannot develop on an open coast because the tidal currents and energetic ocean waves wash away the fine sediments required to maintain marshes. Additionally, most marsh plants cannot grow on hard rock, so even if inlets are protected, there may not be enough fine sediment supply where the base substrate is erosion-resistant rock; also, sediment does not adhere to bare rock. Note however, that in the tropics, larger, sturdier plant forms like shrubby mangroves may grow on old coral reefs.

In the coastal environments suitable for marshes to form, they develop in specific vertical zones controlled by tidal cycles, with the entire marsh occurring between mean sea level (MSL) and mean higher high water (MHHW). There are three basic zonations that occur in marshes. In normal tidal areas (with tidal ranges about 2–3 m), the water level generally fluctuates between MSL and MHHW. High marsh develops in the upper ~20% of the tidal range, the middle marsh forms in the next lowest ~20% of the shoreline, while the low marsh usually occupies the lower ~60% of the marsh down to MSL (Box 2.3 Salt marsh tidal zones). Below MSL, there is no perennial plant growth and bare mudflats extend to the low water mark (MLW); below here, however, perennial seagrass meadows may be found down to the limit of sufficient sunlight.

The higher the elevation within the marsh, the more stresses the salt marsh plants and animals sustain because of greater exposure (i.e. more desiccation and extreme temperatures) unless there is either a continuous freshwater source or high year-round precipitation. The diurnal fluctuations of physical and chemical parameters in a marsh are shown for Southern California during the 1960s (Phleger and Bradshaw, 1966; Figure 2.4). This record

Box 2.3	Salt marsh tidal zones

In Southern California, where the daily tidal range is about 1.7 m and the maximum spring tide range is 2.4 m, many high marsh plants are continuously exposed to the dry climate conditions (less than 30 cm precipitation per year) for two to three weeks (500 hours). In contrast, the low marsh vegetation is always partly submerged, mostly for more than six hours at a time. Seagrass (*Zostera marina*, commonly called eelgrass) is confined to channels or shallow water below MSL. Baseline tidal data are from Mission Bay marsh in May and June 1964 and 1965 (as reported by Macdonald, 1969).

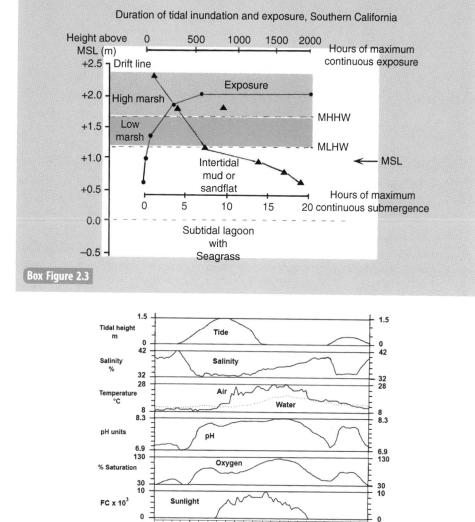

Box Figure 2.3

Figure 2.4 Physical parameters from a 24-h tidal cycle in Mission Bay, California, a warm climatic area. This is one of few complete physico-chemical records from a tidal marsh (after Phleger and Bradshaw, 1966; with permission from AAAS).

is from a warm temperate region with summer drought – in contrast to northern areas, where salinities may be <1 psu with subzero temperatures in the winter months. Fluctuations in all parameters become more extreme with harsher climates, as in Nova Scotia, where the ground is frozen from November to April, and on the Sonoran Desert coast of Mexico, where summer temperatures can reach 49 °C.

The tidal range (the difference in height between MHHW and MLW) also plays an important role in shaping salt marshes because the larger the range, the stronger the currents, which maintain a healthy flow of fresh seawater, but transport large volumes of sediment by scouring mudflats and eroding shorelines. The prime example of this is the Bay of Fundy in Nova Scotia and New Brunswick, Canada, which has the world's highest tides (+16.4 m). These megatides erode vast amounts of sediment from the relatively soft rocks in the Bay of Fundy and deposit these at the head of the bay, building up extensive bare intertidal mudflats seaward of the vegetated tidal marshlands. Another place with tidal ranges similar to the Bay of Fundy is Ungava Bay, northern Quebec, in the Low Arctic region of Canada, but here there is sparse marsh vegetation on the boulder-covered mudflats that creep seaward, pushed by frost heave and gravity, and there is little sediment accretion because the shoreline is formed by ice-scoured, igneous and metamorphic rocks that resist erosion (Eisma, 1997). Some rivers in Western Europe also have megatidal ranges and extensive tidal flats, e.g. the Severn (15 m) and the Seine (~7 m) rivers. The Chang Jiang (Yangtze) Estuary in China and San Sebastian Bay in Tierra del Fuego, Argentina, have tidal ranges of about 9 and 10 m, respectively.

2.5 Impacts of storms and extreme climate events

All salt marshes and other coastal wetlands are ephemeral geological features subject to long-term (millennial-scale) changes in sea levels that accompany glacial and interglacial climate cycles (Box 2.1 Important tidal reference points). Most salt marshes are relatively young, having formed over the past 5000 years, when sea level became almost stable following earlier major tectonic adjustments to postglacial icesheet loading. During these relatively stable sea level conditions, the major components contributing to change are thermal expansion of the ocean (~ 0.5 mm/yr), residual melting of glaciers and icesheets (0.3–0.6 mm/yr) and groundwater depletion (~1.2 mm/yr) from human population growth (Pugh, 2004; Pokhrel et al., 2012). On average, coastal wetlands have been aggrading at rates of about 2–3 mm/yr, and have been keeping pace with the total sea level rise of ~1.8 mm/yr. However, higher sea levels increase the risk of flooding and erosion of the marshes during extreme sea level and coastal flooding events, including tsunamis, storm surges and hurricanes (Table 2.1). *Tsunamis* are rare giant sea waves generated by seismic events like submarine earthquakes or landslides triggered by earthquakes. *Storm surges* are large increases in sea level accompanied by wind-driven waves generated by hurricanes in the Atlantic and other extreme weather events (cyclones or typhoons) in the Indian and Pacific regions. *Cyclones* are storm systems caused by instable atmospheric conditions and they are characterized by a very low pressure centre; *explosive cyclones* are

Table 2.1 Categories and ranges in heights of giant waves associated with abrupt changes in sea level caused by tectonic events (tsunamis), climate-driven events (storms) and combined climate+astronomy-driven events. CAT indicates the category of severity on an index ranging from CAT1 (lowest) to CAT5 (highest).

Wave type	Cause	Example	Wave height and damage	Reference
Tsunami	Earthquake, offshore deep (M = 9.3)	Sumatran earthquake December 2004	20–30 m, 4 km inland; 300 000 deaths; >1 million houses destroyed	Saatcioglu *et al.*, 2005
	Earthquake, onshore (M = 7.7)	Lituya megatsunami July 1958	550 m; saltwater killed spruce forest, washed out wetlands; broke undersea cable at Haines 80 km S	Stover and Coffman, 1993
Storm surge	Ocean heating and strong winds	Hurricane Katrina 29 August, 2005 (CAT3)	4 m waves; 562 km² shoreline eroded; 1800 people died; 300 000 homes destroyed; economic loss totaled >US$100 bn	IPCC, 2007
		Super Typhoon Tip, October 1979 (CAT5)	Honshu Japan; wave heights 'monumental'; winds >280 km/h; 600 mudslides; 22 000 homes flooded; 68 people died	Iacovelli and Vasquez, 1998
		Cyclone Nargis May 2008 (CAT4)	Myanmar; waves 5–7 m, 50 km inland; 138 000 killed; much mangrove loss; saltwater in rice fields	Fritz *et al.*, 2009
Storm tide	Storm surge at highest tides of year (solstices)	Superstorm Sandy Fall 2012 (CAT2)	9.9 m at New York, swept 10 km inland; 47 killed; 1000s of homes lost	Dennison *et al.*, 2012
		Hurricane Juan 28 September, 2003 (CAT2)	9–20 m waves, 2.91 m surge 2 hrs after high tide; winds up to 159 km/h; shoreline property flooded; urban forest lost	Bowyer, 2003

very rapidly developed extra-tropical storms with destructive winds, flooding rains and large ocean waves (Black and Pezza, 2013). *Hurricanes* and *typhoons* are similar devastating cyclones which occur in tropical regions and are also marked by high winds, rain and floods.

The greatest damage to coastal wetlands comes from extreme storm tides which are the sum of the storm surge height and the astronomical tide height. Maximum damage occurs when a storm surge happens at the same time as an extreme high spring tide, like

Figure 2.5   'Superstorm Sandy' 29 October 2012 flooded parts of New York, and Verrazano Bridge (70 m above sea level) at the harbour entrance was closed to traffic because of high winds and waves crashing on the shore (courtesy of Carlos Ayala, SmugMug).

'Superstorm Sandy' (a post-tropical cyclone) which occurred on a full-moon high spring-tide on 29 October, 2012, when the water level was 20% higher than average (Figure 2.5). A new phenomenon in the Arctic Ocean, the 'Great Arctic Cyclone of 2012', has been linked to the unprecedented increase in summer open Arctic Ocean water (Simmonds and Rudeva, 2012).

3

Zonations and plants: development, stressors and adaptations

Key points
Salt marshes and mangroves grow seawards and upwards by sediment accretion resulting from sediment binding by surface algae and roots of pioneer plants; tides transport sediment, nutrients and oxygen to marsh plants twice a day or less, depending on the elevation above MLW; marsh vegetation traps suspended sediment and further raises the marsh; soil salinity stressors increase in the high marsh where there is less regular influence of tidal flow; waterlogging, low oxygen and sediment mobility are the main stressors in low marshes and mudflats; elevational microhabitats have different floras and faunas according to their physiological tolerances of salinity and soil oxygen, resulting in a succession of plant communities; plant adaptations are both structural (e.g. salt glands, creeping roots with air passages, or platform roots with 'lungs') and internal (C_3, C_4 and CAM metabolism) to optimize photosynthesis when alternately submerged and dry; coastal wetlands are thus very productive carbon storage systems; pollen of the different plants marking the marsh zones provides an archive of changes in marsh zonation, salinity and climate over time; pollen of exotic species is used to trace changes in sediment accretion associated with anthropogenic impacts.

3.1 Sediment stabilization and salt marsh development

Algae and halophytic grasses or succulents are the bioengineers in the formation and maturation of a coastal wetland, which is also tightly linked to the baseline coastal geo-morphology. Tidal flats gain elevation relative to MSL by sediment accretion of the mudflats which decreases the rate and duration of tidal flooding and allows pioneer halophytes to colonize the periodically exposed surface. Mud from rivers and streams also increases sediment deposition by fall-out of suspended sediment where freshwater mixes with sea-water over the low-gradient mudflat. Potential colonizing plants arrive on the bare surface as either seeds, propagules (= germinated seedlings) or portions of rhizomes. When conditions are right, germination and establishment of pioneer halophytes begins, and the marsh starts to grow. This colonization is also aided by mats of diatoms and filamentous blue-green algae that bind together small silt and clay particles on the mudflat surface (Figure 3.1). This stabilizes the sediments by preventing resuspension and suppressing wave erosion, thereby enabling colonization by the pioneer halophytes (e.g. *Spartina*, *Salicornia*, *Sarcocornia* spp.), which further trap sediment suspended by the rising tide around their stems and leaves.

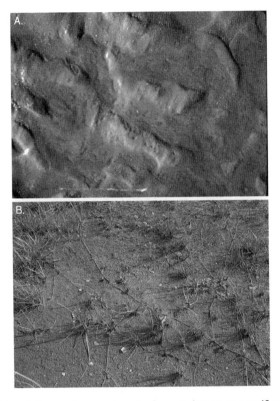

Figure 3.1   Mudflats are stabilized by gelatinous surface-growing microfloras. A: diatoms on a sandflat at Puerto Peñasco, Mexico; B: blue-green algae bind loose sand between stolons ('runners') of colonizing cinquefoil at Taylor Bay, Alaska (photos by P. J. Mudie). See also colour plates section.

These halophytes form low muddy mounds that aggrade vertically, forming depositional terraces, as the subsurface roots and surface runners bind more sediment (Figure 3.2). After the plants develop into these terraces, more sediment deposits, vertically growing the marsh and reducing the length of tidal flooding enough for competitive facultative halophytes with lower tolerance of salinity or water-logging to establish.

The process of salt marsh formation and growth varies depending on latitude, tidal regimes and the geomorphology (see Figure 2.2), drainage basin patterns and local zonation patterns within the marsh (Ibáñez *et al.*, 2013). Two basic models emerge: the 'ramp' model of salt marsh development, where accretion is a direct result of surface elevation and the tidal periods, and the 'creek' model, where drainage channels and tidal creeks play the larger role in development (Pratolongo *et al.*, 2009). Both models influence the maturation of the marsh. Morphological characteristics of colonizing halophytes further affect sediment adhesion and subsequent marsh accretion. Water current velocities are reduced by the stems of tall marsh species that induce hydraulic drag, thereby minimizing sediment resuspension and enabling deposition. Measured concentrations of suspended sediment in the water column decrease from the open water at the marsh edge to the interior because the vegetation canopy traps sediment particles and allows them to settle. In deltas, sediment accretion is

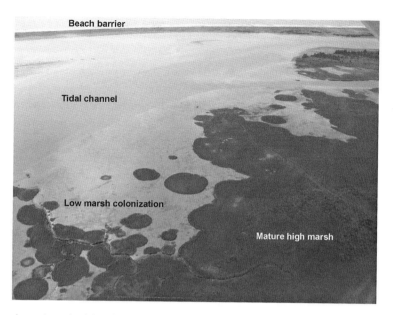

Figure 3.2   Aggrading salt marsh on the delta of the Alsek River at Dry Bay, Alaska, shows the distinctive circles formed by colonizing low marsh perennials (here *Carex lyngbyaea*) that raise the marsh surface and coalesce to form continuous vegetation cover in the high marsh (photo by P. J. Mudie, August 2004).

highest next to the river channels where it builds embankments called levees. An example from the Yangtze Delta in China shows the amount of sediment adhering to marsh grasses (i.e. *Spartina alterniflora* and taller species) decreases with the distance from the sediment source. Here, the tall cordgrass in the low marsh caused over 10% of the total marsh accretion (Yang, 1999). Consequently, accretion is lowest in the higher zones, where tidal inundation is much less frequent than it is in the low marsh.

3.2 Salt marsh zones and physical stressors

Coastal saltwater wetlands differ from other wetlands because of the tidal flow, which floods the shoreline by a variable amount on a daily basis. The amount of tidal inundation, which is ultimately determined by elevation, leads to differences in soil and water salinities, and in soil type and drainage patterns. The high marsh has the least tidal inundation, which leads to either low salinities in areas of high rainfall, soil runoff and river flow, or high salinity in arid climates. In the low marsh zones of daily tidal inundation, soil salinities are essentially the same as the seawater, with little seasonal change from evaporation and salt accumulation. Areas of regular inundation are colonized only by the hardy halophytes, but as elevation increases, the plant diversity becomes greater. This zonation produces various microhabitats in which different species of plants and animals thrive according to their physiological characteristics (e.g. Table 6.2; also Chapter 6). For example, in the temperate

British Isles alone, there are almost 30 different communities of salt marsh vegetation which reflect these specific requirements (Rodwell, 2000).

The zonation of marsh plants is dictated by a species tolerance to both abiotic and biotic factors. Lower marsh zones have stable salinities and moisture levels, and higher nutrients, but the plants require special adaptations to lower soil oxygen levels, as well illustrated in books on peatlands (e.g. Rydin and Jeglum, 2006) and other wetlands books (Keddy, 2010; Mitsch and Gosselink, 2007). In higher marsh zones, plants must tolerate times of low nutrient availability, dryness and more interspecific competition. In all intertidal ecosystems, physical tolerance dictates the lower limit of species occurrence, while competitive inter-actions influence the species in the upper limits. For example, the robust grass *Spartina alterniflora* with internal air passages and salt glands has better adaptations to stresses of the lower zones (salinity, water logging and oxygen depletion), but if transplanted, it can live in any elevation of the marsh. However, *S. alterniflora* is absent in the higher marsh because there it is out-competed by other plants, such as the shorter, fine-leafed *Spartina patens*, which, in contrast, normally cannot survive in the lower marsh zones. A good example of this visible plant zonation is seen in North Atlantic mesotidal marshes of North America, such as those from the Atlantic coast of Nova Scotia or from New England. The low marsh is colonized by a monospecific assemblage of *Spartina alterniflora* (Figure 3.3A). Higher elevations have zones of *Spartina patens* (marsh hay) and succulent glasswort *Sarcocornia perennis*, followed by black rush *Juncus geradii*, then various grasses and shrubs, such as *Scirpus* and *Iva frutescens* (Figure 3.3B, C). These species all have different tolerances to salt and soil drainage that make the different zones along the marsh best suited for each individual.

The highest marsh zone above MHHW is only covered by seawater during the extreme high tides of the summer and winter solstices, and by wave surges during coastal storms. The middle high marsh is the intermittently flooded zone between the low and high marsh zones. Both these marsh zones usually have sandier soil and may include areas of extremely high salinity that prohibit any plant growth. The bare sandy or waterlogged muddy areas in or at the upper edge of the tidal wetlands are called 'salt pans/pannes' or, if extensive, 'saltflats'. The low marsh plants can survive there only in tidal channels and they are stunted in their growth. On the other hand, many annual *Salicornia* species and salt marsh daisies, e.g. *Lasthenia glabrata*, are adapted to these salt pans and the seeds germinate and grow very rapidly after spring rainfall so that they can set seed before the salinity is too high for continued growth as the soil dries in summer.

3.3 Plant adaptations and species diversity

Over the millennia, adaptations to the high salinity, periodic submersion and low-oxygen soils of salt marshes have evolved in many different plant families that today are important sources of genetic diversity for development of salt-tolerant crops (Chapter 13). The most common salt marsh plants are adapted to the high salinity (mostly sodium chloride) either by storing the salt in succulent leaves or stems which are later shed as dry tissue, or by having

Figure 3.3 Chezzetcook Inlet marsh, Nova Scotia. A: low marsh with *Spartina alterniflora* in front and in parts of the channel in the background. B: high marsh, showing in foreground various shrubs, herbs and grasses at the terrestrial transition, zone of dark-coloured black rush (*Juncus gerardii*), then light green *Spartina patens* marsh hay; channel bordered by marsh is in the background (Photos by J. Frail-Gauthier). Bottom: Nova Scotian marshes; C: glasswort (*Sarcocornia perennis*) can dominate a high marsh; it is a succulent perennial herb with petal-less flowers that produce abundant pollen in protruding yellow anthers (white dots); D: *Salicornia europaea* (pickleweed or samphire) is a small succulent annual herb found in bare areas, 'pans', with a surface mat of blue-green algae (dark soil areas) binding the soil and absorbing nitrogen from the air (photos by P. J. Mudie).

salt glands that excrete excess salt (Figure 3.4). Widely distributed succulent halophytes are the glassworts (annual *Salicornia*, perennial *Sarcocornia* herbs, *Arthrocnemon* shrubs) and seablites (*Suaeda* spp.). Salt glands are found in many grasses, including cordgrass (*Spartina* spp.), and salt grasses (*Puccinellia; Distichlis*), and are also found in high marsh sea lavenders (*Limonium* spp.) and orache (*Atriplex*), as well as some mangroves (*Avicennia*). When the soil level is high enough for rainwater, runoff or subsurface seepage to dilute the soil water, then less salt-tolerant plants like plantains (*Plantago* spp.), and various sedges (*Carex* and *Scirpus* spp.) and spikerushes (*Juncus* spp.) can establish themselves. Regional variations on this basic successional theme are given in Chapters 7–11. The specialized root characteristics of the salt marsh plants and mangroves are illustrated in Figure 3.5 and in sections of Chapters 7–10.

 Figure 3.4 Leaves of tidal wetland plants. A: black mangrove has leaves with salt glands and the white flowers of the Avicenniaceae family; B: salt glands on leaves of the cordgrass *Spartina foliosa*; C: leaves of the salt marsh perennial *Frankenia grandifolia* are small, hairy and inrolled to conserve moisture; D: salt grass *Distichlis spicata* has salt glands and small, tightly inrolled leaves; E: red mangrove has succulent leaves and the white flowers of the Rhizophoraceae family; F: seablite has succulent leaves and tiny petal-less flowers (at base of lower leaves); G: succulent leaves of *Batis maritima* (foreground) and succulent grey stems of pickleweed (*Salicornia pacifica*) in a California marsh; H: succulent leaves and fruits of sweet mangrove (*Tricerma phyllanthoides*); I: fleshy leaves of a mangrove shrub *Scaevola taccada* and its white flowers of the family Goodeniaceae. Photos by P. J. Mudie.

Some coastal wetlands plants have photosynthetic adaptations in addition to their physical adaptations, and specialized metabolic pathways give them higher capacities for carbon production than normal C_3 photosynthetic plants (Box 3.1 Comparison of plant photosynthesis and metabolism). Consequently, coastal wetlands are extremely productive habitats, yielding up to twice as much as either well-watered farm fields (Teal and Teal, 1969) or terrestrial forests (Mcleod *et al.*, 2011). Many of the specialized halophytic plants such as cordgrass cannot be grazed by higher animals, but die off and decompose to become food for microorganisms, which in turn become food for fish and birds. The

Figure 3.5 Root systems of salt marsh and mangrove plants. Top left: white roots of salt grass (a) have air passages and include shallow vertical anchor roots and spreading horizontal roots (b); sea lavender has one thick black tap root that grows deep to obtain subsurface low-salinity water. Top right: tropical *Scaevola* mangrove shrub also has a deep anchor root and shallower surface roots absorbing water from rain. Bottom left: black mangrove has abundant pencil-like pneumatophores (foreground) protruding above the mud; red mangroves have branching stilt roots keeping them high in the water (photos by P. J. Mudie). Bottom right: lung roots of the apple mangrove at Tamil, Indonesia (photo by Patrik Nilsson). See also colour plates section.

marshlands thus serve as depositories for a large amount of organic matter, which feeds a broad food chain of organisms from bacteria to mammals and enriches coastal waters when washed out into the sea (Chapter 4). The organic matter stored in salt marshes, mangroves and other coastal ecosystems is called 'blue carbon' and is very important for sequestering carbon (see review by Mcleod *et al.*, 2011). Although the total size of coastal wetlands is small, the amount of carbon stored within them is disproportionately large. This is because they trap sediments and carbon from other environments and bury it where it cannot be easily aerobically metabolized, thereby creating a long-term storehouse for organic carbon that will not be released into the atmosphere as a greenhouse gas. Because the anaerobic decomposition in marsh and mangrove soils is less efficient than the aerobic metabolism found at the surface and in other terrestrial systems, the organic carbon is stored indefinitely until exposed by shoreline erosion or by soil desiccation. It is estimated that salt marshes and mangroves bury and store between 31–34 and 5–87 Tg of carbon per year, respectively (Chmura, 2011; Mcleod *et al.*, 2011; see Chapter 4). The above- and below-ground production of *Spartina alterniflora* is the highest of any salt marsh plant, making it the main contributor to salt marsh productivity as well as carbon sequestration.

Box 3.1	Comparison of plant photosynthesis and metabolism

Most flowering plants and trees of the upper edge of salt marshes have C_3 (Calvin-cycle) photosynthetic pathways in their leaves. The metabolism of C_3 plant leaves is less efficient than the C_4 metabolism found in *Spartina* and many tropical grasses because it uses a slow enzyme system (RuBisCO) in the leaf cells containing chlorophyll (the mesophyll) to trigger the capture of carbon dioxide (CO_2) from leaf openings (stomata, plural; stoma, singular). Relatively large air spaces and multiple stomata also allow water vapour and carbon dioxide to escape easily, making the plants vulnerable to desiccation in hot, dry climates or under water stress. In contrast, C_4 (Hatch–Slack pathway) plants have evolved a more efficient, less wasteful, two-step pathway in which the bundle sheath cells surrounding the food- and water-transporting veins also have chlorophyll and store CO_2 up to a level that switches off the inefficient RuBisCO and turns on another enzyme, PEP carboxylase (or variant, e.g. PEP-PK) that increases the efficiency of the RuBisCO. The PEP carboxylase does not capture CO_2 as a substrate like RuBisCO, but it uses bicarbonate, HCO_3^-, which is formed when CO_2 dissolves in water. The C_4 acids produced from the carboxylation of PEP are transferred to the bundle sheath cells where CO_2 is rereleased and then fixed again through Calvin-cycle processes. C_4 plants have fewer, smaller stomata and more compact leaf cells, so they conserve water vapour better in addition to being more productive of plant foods. The most drought-tolerant plants, including *Salicornia*, are succulents that have a third type of metabolism, CAM (Crassulacean Acid Metabolism), which uses a variation of the C_4 pathway, but separates the reaction in time rather than in tissue space. CAM plants have leaf anatomy more like that shown for C_3 plants, but with few stomata that are sunken into thick wax-coated leaf surfaces like those of the ornamental Jade plant. During the day, the stomata are closed and there is no intake of CO_2 and also no loss of moisture from transpiration. At night, the stomata open, and CO_2 enters the mesophyll cells, where it is fixed by the C_4 enzyme PEP carboxylase to produce the acid malate. The malate is then converted to malic acid, which can be stored in large vacuoles (cavities) within the cytoplasm until the next day. During the day, the malic acid is used as a carbon dioxide source for the C_3 cycle and photosynthesis by the Calvin cycle depletes the leaf of CO_2. At that stage, the malic acid is shunted out of the vacuole to the cell cytoplasm where CO_2 is released and used for C_3 photosynthesis although the stomata are closed. CAM metabolism is therefore the best for conservation of water within plant tissues and in dry or salty environments, where storage of water to dilute salts in the tissues can be critically important.

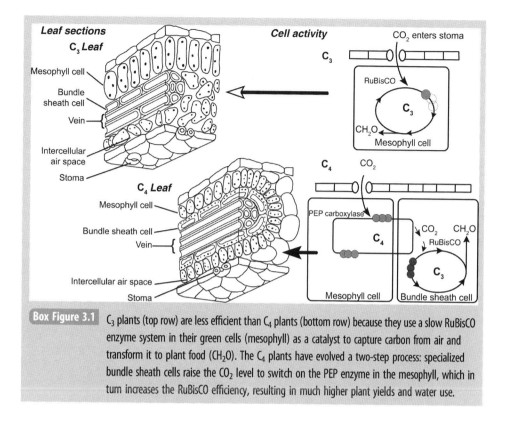

Box Figure 3.1 C_3 plants (top row) are less efficient than C_4 plants (bottom row) because they use a slow RuBisCO enzyme system in their green cells (mesophyll) as a catalyst to capture carbon from air and transform it to plant food (CH_2O). The C_4 plants have evolved a two-step process: specialized bundle sheath cells raise the CO_2 level to switch on the PEP enzyme in the mesophyll, which in turn increases the RuBisCO efficiency, resulting in much higher plant yields and water use.

3.4 Mangrove forest diversity and adaptations

In tropical climatic regions, salt marsh vegetation is usually replaced by mangroves that have a similar range of adaptions to seawater inundation and play essentially the same role as salt marsh plants in stabilizing mudflat sediments. In mangrove swamps, however, the main plant forms are trees and shrubs rather than grasses and herbs. Tropical tidal marshes are characteristically dominated by mangrove vegetation consisting of salt- and flood-tolerant trees and shrubs in the intertidal area. The word mangrove probably comes from the Portuguese word 'mangue' meaning 'tree', combined with the English word 'grove', referring to a group of trees (Mendelsohn and McKee, 2000). Mangrove swamp and tidal forest vegetation is also sometimes called mangal, with the word 'mangrove' being restricted to individual tree/shrub species (Chapman, 1977).

In 2007, the amount of mangrove vegetation in the world was estimated as 230 000 km^2 by Wolanski *et al.* (2009) and as 152 000 km^2 by Spalding *et al.* (2010) – amounts about five and three times greater than the present-day area of coastal salt marshes. However, the mangroves are squeezed into a narrower climatic belt than the salt marshes because these halophytic trees require warm ocean waters and they cannot tolerate frosts more than once a decade. In general, mangroves only flourish where the coldest month is above 20 °C, although a few hardy species can grow where the coldest month is 10 °C. The distribution of warm currents and summer rainfall in the world's coastal regions means that although there are more than 40 mangrove species, only 9 grow in the Americas and Africa, compared to ~20–30 in India and Australia, and 31–40 in Southeast Asia (Field *et al.*, 1998). Many mangroves reproduce by water-buoyant fruits or propagules (reproductive buds) that drop off the trees and are widely dispersed by ocean currents (Figure 3.6; see Tomlinson, 1986, for details); therefore, they have naturally colonized many outer islands of the western Pacific Ocean. However, no mangroves drifted as far as the Hawaiian Islands in the central Pacific Ocean, but red mangroves were first taken there in 1902 by sugarcane farmers to stabilize the mudflats created by soil erosion in their cane fields (Allan, 1998) and today, six kinds of mangroves are naturalized in Hawaii.

Mangrove vegetation includes a wider diversity of plant growth forms than the salt marshes, including forest trees over 30 m tall, various shrubs up to 3 m high, two palms (*Nypa fruticans*; *Raphia farinifera*) and a large bushy fern, *Acrostichum*, in addition to salt-tolerant swamp reeds and climbing vines. There are 27–28 genera in total, but only 17 live in the intertidal part of the mangrove swamps (Duke *et al.*, 1998; Ellison *et al.*, 1999). These mangrove genera belong to 19–20 different families, two of which are almost exclusively mangal plants: Rhizophoraceae with eight species of *Rhizophora*, the red mangrove, four orange mangrove species of *Bruguieria*, two *Ceriops* spp., one *Kandelia* sp. and one non-mangrove species *Carallia brachiata*. The family Pellicieraceae includes just one species *Pelliciera rhizophoreae*, the tea mangrove, which is found only in tropical America. According to Giesen *et al.* (2007), two other exclusively mangrove plant families are the Avicenniaceae, with eight *Avicennia* species (black mangroves), and the Sonneratiaceae, with five *Sonneratia* species (mangrove apples).

All the genera in these four more-or-less exclusively mangrove families share a number of specialized characteristics that adapt them to tidal inundation by seawater and waterlogged mudflat conditions (Duke *et al.*, 1998; Spalding *et al.*, 2010). All have exposed 'breathing' roots above the mud surface: red mangroves have stilt-like roots; the tea mangrove has buttress roots, like those of the Florida swamp cypress or the California redwood; and the black mangroves and mangrove apples have spreading, underground roots from which vertical breathing roots (pneumatophores) grow up to ~25 cm above the mud (Figure 3.5). These specialized root systems stabilize coastal mudflats and at the same time, provide attachment surfaces for shellfish such as oysters and crabs. Other special adaptations include: (1) the ability of these trees to concentrate salts in their leaves so they can extract freshwater from the seawater by reverse osmosis; (2) shedding of salt from leaf hairs (Figure 3.4) and (3) reproduction by vivipary, in which the seeds germinate while still on the trees, forming large, wax-covered, buoyant propagules (= germinated seedlings) that drop off on the high tide and survive for weeks in seawater until landing on a muddy shore and taking root (Figure 3.6).

3.5 Pollen archives of wetland vegetation development

The pollen and peat remains of salt marsh plants and mangroves leave a fossil record in the marsh sediments, which can be used as paleoecological archives of past changes in marsh

Figure 3.6 Seeds and reproductive propagules of mangroves and mangrove associates. Top left: elongated propagules of red mangrove on a mudflat, and one small germinated seedling of black mangrove (circled lower right). Top right: young red mangrove seedlings trap detritus that provides nutrients and helps aggrade the muddy soil. Centre: eelgrass *Zostera* spreads locally by rhizome fragments; seeds are eaten by birds and spread to other inlets. Bottom left: Coconut (*Cocos nucifera*) palm seeds are covered by thick reddish fibres (husk) and a waxy surface layer (a) that traps air and keeps them buoyant for very long distances in surface waves; the seed inside is protected by a waxy dark brown cover and filled with nutritive liquid ('coconut water'). Bottom right: once beached above the high tide, coconuts can germinate rapid and establish as palm seedlings (b). Photos by P. J. Mudie.

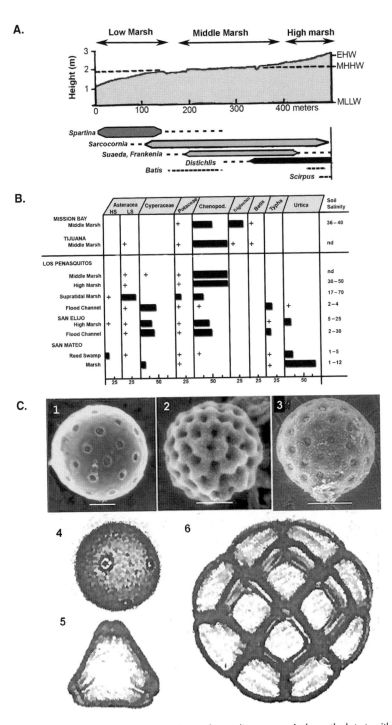

Figure 3.7 Pollen of salt marsh plants is used to trace vegetation history from sediment cores. A: the method starts with field work to record the relation between marsh plants and sea level, as shown here for Mission Bay marsh; B: pollen from surface samples in the marsh shows that different kinds and percentages of pollen provide cryptic 'fingerprints' for marsh vegetation zones in five Southern Californian coastal wetlands with a wide range of soil salinities. C: most of the fossil pollen grains represent the marsh vegetation, but some are from nearby upland forest or farmland. Scanning electron microscope images of three common marsh halophytes: 1, *Chenopodium* (brackish); 2, *Atriplex alaskensis* and 3, *Sarcocornia perennis* (tidal marshes; Mudie *et al.*, 2005a); light microscope photos of three pollen markers for Southern California marshes: 4, *Plantago lanceolata*; 5, *Eucalyptus globulus* and 6, *Acacia baileyana* (Mudie and Byrne, 1980).

vegetation composition, soil salinity, sedimentation rates, tidal exchange and, sometimes, major shifts in climate (Scott *et al.*, 2011). Although the pollen and peat do not provide as sensitive a record of sea level change as foraminifera, the pollen archives are important for understanding how the world's wetland vegetation has altered over time – and how it may change in the future. Examples of these paleoecological pollen studies will be given later for different coastal wetlands of the world. Here we provide an introduction to the method that has been developed so that fossil pollen of salt marsh or mangrove plants can be used in geological studies to backtrack changes in coastal wetland vegetation.

Development of the pollen tracer method starts with a survey of the marsh vegetation zonation according to elevation and tidal submergence, as shown here (Figure 3.7A, B) for the same Southern California marsh studied by Phleger and Bradshaw (1966; Figure 2.5). The fresh pollen of living salt marsh plants is collected and matched with the fossil pollen found in surface mud samples along a profile across the marsh. The soil salinity is also measured at the time of the surface sediment sampling so it can be correlated with the pollen species assemblage. Figure 3.7C shows some of the main pollen types of native salt marsh halophytes and the pollen of introduced plants such as plantains (e.g. *Plantago lanceolata*) and exotic forest species (e.g. *Eucalyptus*, *Acacia*) found in Southern Californian marshes. When the time of first introduction is known for the pollen-producing plants, e.g. *Eucalyptus* trees brought from Australia to California by Europeans around 1850 CE, then the pollen provides a dating horizon for that point in a sediment profile of the salt marsh (Mudie and Byrne, 1980). These historical pollen records are important for accurate dating of marsh sediment accretion rates over the past ~150 years (Scott *et al.*, 2011 and references therein) because this youngest time interval is not measured accurately by the radiocarbon dating method used for older sediments (see Roberts, 1998, for details of this and the pollen dating method).

4

Animals in coastal wetlands: zonation, adaptations and energy flow

Key points

Salt marshes contain abundant and diverse terrestrial and marine animals based on their high primary productivity; animals are classified by trophic functional group and size class; foraminifera are important in the marsh microecosystem and used for reconstructing past sea levels; vertical zonation is based on physical tolerances and biological competition from marine mudflats to terrestrial high marsh; in the tropics, mangrove forests support many large animals, while invertebrates inhabit their subaerial roots; coastal wetlands are highly productive ecosystems with the energy produced by plants and algae supporting large detritus-based food webs; marshes are often nitrogen-limited and are impacted by enrichment from sewage and agricultural runoff; secondary production is concentrated in detritivorous food webs rather than upper-level consumer webs; mangroves are the most carbon-rich forests on Earth.

4.1 Animals that inhabit salt marshes

As seen from Chapter 3, salt marshes and mangroves are extremely productive, which results in a large food web base that can support a high abundance and diversity of species dependent on the vegetation directly or indirectly for food or shelter. Recently, however, there has been a paradigm shift in understanding of the role that tall grasses such as *Spartina alterniflora* play in the feeding structure of a salt marsh (Mann, 2000). New stable isotope studies show that algae (diatoms and algal mats) are more important food sources to the deposit feeders than the decaying grass (Galván *et al.*, 2011), and picophytoplankton can play key trophic roles on intertidal flats (Paterson *et al.*, 2009). Regardless of differing views on salt marsh basal food webs, however, it is clear that salt marsh vegetation is an important driver of the faunal diversities and abundances in temperate regions. The physical environment of coastal wetlands supports a high faunal diversity because it is low-energy and contains many different lateral and vertical zones and microniches for both terrestrial and marine animals. The gradient between the terrestrial high marsh and marine mudflats not only influences plant zonation, but also the animal distributions. Animals must be adapted to a transient environment that changes with the tides and seasons, as well as anthropogenic changes such as habitat reduction and invasive species.

In any square metre of temperate salt marsh, most of the animal diversity is made up of invertebrates such as crabs, burrowing shrimp, gastropods (snails) and hundreds of species

Macrofauna ⟶ **Meiofauna (<1 mm)**

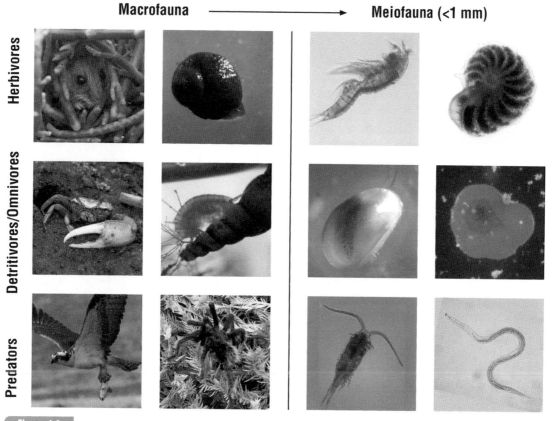

 Examples of macro- and meiofauna represent different trophic levels, from herbivore to carnivore. Top row: salt marsh mouse (photo by P. J. Mudie); *Littorina* periwinkle (photo by Kayla Hamelin); Harpacticoid copepod (photo by Kayla Hamelin); *Elphidium* foraminifer (Malcolm Storey, www.bioimages.org.uk). Middle row: Fiddler crab (photo by Patrick Coin/Flickr); Amphipod (photo by cephalopodcast/Flickr); *Trochammina inflata* foraminifer (photo by J. Frail-Gauthier); ostracod (photo by Kayla Hamelin). Bottom row: Osprey (photo by Dori Chandler); Tarantula (photo by Hester Bell) (photo by P. J. Mudie); Calanoid copepod (photo by Kayla Hamelin); Nematode (photo by Scot Nelson/Flickr).

of marine worms. The assemblages of animals that live in or depend on the salt marsh differ with latitude and climate, but similar functional groups are found in all coastal wetlands, including herbivores, detritivores and predators. The types of animals may also be classified by size (Figure 4.1): macrofauna, which are easily visible (>1 mm; worms, shrimp, fishes, birds, mammals), and meiofauna, which are the size of sand grains (63 microns to 1 mm), including foraminifera (Box 4.1 Tidal wetland foraminifera), ostracods, copepods and nematodes. Salt marsh microfauna (silt to clay size) and microbial processes such as sulfate reduction and methane production are important emerging topics (see Joye *et al.*, 2009), but details are not given in this book. The animals can also be differentiated by their habitats: surface-dwelling organisms living on marsh sediments (e.g. crabs, bivalves, gastropods), and infauna that live under the sediment surface (worms and clams); others move in and out of sediment burrows (fiddler crabs, goby fish). For many of the macrodetritivores, there is a differentiation in feeding mode: filter feeders versus deposit feeders. The filter feeders, such as ribbed mussels and oysters, take organic detritus from the water column only when inundated by the tide. Deposit feeders ingest sediment and excrete inedible organics; this group includes many worms, grass shrimp, amphipods, mud snails (e.g. *Ilynassa obsoletus, Melampus* spp.) and fiddler crabs (*Uca* spp.) in the USA (Bertness, 2007). Fishes in tidal wetlands include residents, species using the salt marsh as a nursery ground before moving to open coastal waters and anadromous fish migrating to spawning grounds. Fish such as mummichogs and killifish (*Fundulus* spp.), minnows (*Cyprinodon variegatus*), silversides and sticklebacks are usually in intertidal areas, but are sometimes trapped in salt ponds at high tide.

Shore birds and specialized salt marsh birds, such as the clapper rail and savannah sparrow, use the marshes for food and protection, raising their young amongst the higher grasses during the summer. They use the abundant fish in the shallows, which are easier to capture than in the open ocean. Any coastal wetland can support over 100 species of birds, of all size classes and functional groups (Figure 4.2), from small song birds, waterfowl, waders and skimmers, to carnivorous birds of prey. Zonation of birds along the marsh gradient depends on their food source, e.g. plant seeds and terrestrial insects or mudflat flies and marine worms. Many types of birds require wetlands and mudflats for feeding during long

Box 4.1	Tidal wetland foraminifera

One gram of surface sediment from the salt marsh can contain many thousands of individual protists in the taxonomic Order commonly called Foraminiferida. These foraminifera recycle organic matter, and are food for many small invertebrates (Lipps and Valentine, 1970; Frail-Gauthier *et al.*, 2011). These single-cell protists are part of the salt marsh microbe population. Emerging details on the importance of other microorganisms (bacteria, microalgae, protists and fungi) for primary productivity, nutrient cycling and soil biogeochemistry are covered in the text edited by Perillo *et al.* (2009). The foraminiferal meiofauna acts as a bridge between the decomposition food web and upper consumer levels. Furthermore, foraminifera have carbonate or organic outer walls (Box Figure 4.1) that leave a fossil record in the marsh sediments, which can be used as paleoenvironmental and paleo-sea level tracers (see Chapter 12) to backtrack past changes in sea level (Scott *et al.*, 2001). Elsewhere, however, the accuracy of paleo-sea level placement depends on tidal regimes (Callard *et al.*, 2011) and baseline sea level data (Barlow *et al.*, 2013). Less is known about mangrove foraminiferal distributions and their value for reconstructing sea level (Phleger and Ayala-Castañares, 1969; Ellison, 2009).

Paleoecological studies start with measurements of sea level elevation, salinity over several seasons and associated foraminifera in a selected marsh such as Chezzetcook Inlet, Nova Scotia (Scott *et al.*, 2001). The change in foraminifera with increasing height and varied salinity is then used to calculate paleo-sea level in radiocarbon-dated sediment cores (Scott and Medioli, 1978): high marsh assemblages can relocate a former sea level within ±10 cm. Box Figure 4.1 illustrates some common, and often abundant, marsh foraminifera that can be used for these paleo-sea level constructions.

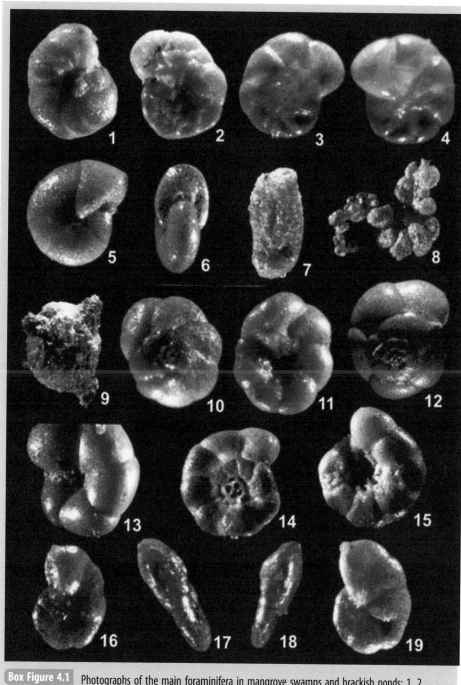

Box Figure 4.1 Photographs of the main foraminifera in mangrove swamps and brackish ponds: 1, 2 *Discorinopsis aguayoi*; 3,4 *Helenina andersoni*; 5,6 *Haplophragmoides wilbertii*; 7 *Miliammina fusca*; 8 *Polysaccammina ipohalina*; 9 *Pseudothurammina limnetis*; 10,11 *Tiphotrocha comprimata*; 12,13 *Trochammina inflata*; 14,15 *Trochammina macrescens*; 16–19 *Trochammina macrescens* f. *polystoma*. Scale is 0.5–1 mm (Plate after Javaux and Scott, 2003, with permission of the Paleontological Society; photos taken with scanning light microscope).

Figure 4.2 Examples of birds commonly found in coastal wetlands. Top: bald eagle (photo by P. J. Mudie); roseate spoonbill (photo by Harold Wagle/Flickr); great blue heron (photo by Dori Chandler). Bottom: greater yellowlegs (photo by J. F. Gauthier); clapper rail (photo by USFWS/Flickr); song sparrow (photo by Dori Chandler). See also colour plates section.

seasonal migrations over thousands of kilometres. For example, marshes in the Bay of Fundy support millions of birds during migratory stops en route to South American wintering areas; the birds depend on an ample supply of invertebrates such as amphipods (*Corophium*), without which they periodically starve.

Reptiles such as snakes and turtles may feed in salt marshes and there are many small mammals such muskrats, weasels, mink and otters that also hunt in salt marshes for crabs, insects, clams, worms, bird eggs and fish. The salt marsh harvest mouse nests in pickleweed marshes. All these smaller animals hide in the vegetation to avoid predation by hawks and eagles. Coyotes or foxes also hunt in these regions for small animals, and deer graze in the high marsh. Other examples are given in the chapters on regional salt marshes.

4.2 Salt marsh zonations, stressors and adaptations

Animals occupy vertical zones within the salt marsh based on biological interactions (predation and competition) and physical tolerances. Because the marsh is a transition between terrestrial land and coastal waters, there is a mixture of land and aquatic animals, dealing with their own set of stressors with every shift of the tide. Terrestrial air-breathing animals such as the coffee bean snail, grasshoppers and dragonflies must migrate up the stems of plants or fly away as the tide rises. Conversely, aquatic animals such as fish and polychaetes are active during high tide, and then either remain in subtidal water or burrow into the sediments during low tide to remain wet and hide from predators on the mudflats. Fish have to adjust to large ranges of salinity (polyhaline species) or remain within high or low-salinity channel areas (stenohaline species). Birds have many adaptations to excessive salinity, mostly through specialized glands in their noses to concentrate, and then remove, the salt. For infaunal animals, the oxygen content of the marsh sediment is low, and the animals must cope with these suboxic conditions. Many worms and crabs create burrows that aerate the sediment and water immediately around their bodies; other animals are simply more tolerant to levels of hypoxia.

Despite being a variable, physically stressful environment, salt marshes support an extraordinarily high abundance of animals, especially meiofauna. Protists, polychaetes, nematodes, ostracods, mites, midges, insect larvae, saltwater oligochaetes, amphipods and isopods can be extremely abundant in surface samples of a salt marsh (Frail-Gauthier *et al.*, 2011). This abundance is supported from the high productivity entering into the consumer system, which, in turn, sustains the migratory and resident birds and fishes. Polar and subarctic animal abundances tend to be lower because of the short (4–6 month) primary production season, but the wetlands are nonetheless important for survival of some endangered animals (see Box 4.2 Animals of polar and subarctic marshes).

Box 4.2	Animals of polar and subarctic marshes

The tundra wetlands of the Mackenzie River Delta on the shores of the Beaufort Sea support many small and large mammals. The large animals (Box Figure 4.2.1) include caribou, wolverines, moose and polar bears plus up to 100 species of small mammals such as lemmings, shrews and voles (Martell and Pearson, 1978). There are also two introduced species: reindeer and dogs. The Mackenzie Delta provides important habitat for waterfowl (ducks, geese, swans) and migratory fish on which the local Inuvialuit people depend for food (Solomon *et al.*, 2000; Solomon, 2005). The Delta is the breeding ground for the lesser snow goose (*Chen caerulescens*) and about half of Canada's tundra geese (*Cygnus columbianus*). The Kendall Island Migratory Bird Sanctuary is the summer feeding ground of >60 000 shorebirds, including the highly endangered Eskimo curlew (*Numenius borealis*; Box Figure 4.2.2; COSEWIC, 2009) and >100 species of migratory birds. More than 5000 beluga whales calve in the adjacent Mackenzie River estuary and barren-ground grizzly bears live on Richards Island. The salt marshes of Hudson Bay are also vital for migratory waterfowl; Henry and Jeffries (2009) outline the complexity of the animal–plant–microbial interactions in these polar coastal wetlands. A new circumpolar marsh stress is the increasing numbers of extreme Arctic weather events that bring rain rather than snow (Hansen *et al.*, 2013). Heavy rain on snow entombs vegetation in ice, thereby preventing winter foraging by herbivores and triggering population fluctuations across the entire vertebrate community.

Box Figure 4.2.1 On the Mackenzie Delta there are several large animals including polar bears (A; photo by Alan Wilson), caribou (B; photo by David Payer) and many smaller animals such as the Arctic fox (C) and Arctic ground squirrel (D; photos by P. J. Mudie).

Box Figure 4.2.2 Arctic salt marshes are vital habitat for some endangered animals. A: the possibly extirpated Eskimo curlew migrates annually from South America to its breeding area in James Bay and Mackenzie Delta; B: Eskimo curlew; painting by J. D. Taylor, from Ontario Government, Ministry of Natural Resources.

4.3 Mangrove animals

Mangroves support complex ecosystems involving thousands of other species of plants, animals, fungi and bacteria (Spalding *et al.*, 2010). The mangrove trees are ecosystem engineers that build and maintain the physical structure of a habitat supporting many herbivores (crustaceans, molluscs and crabs), as well as larger browsers, including the Javan rhinoceros, the Bornean proboscis monkey, and the tiny hutia rodent of the Greater Antilles. The architecture of the mangrove forests also provides shelter, nesting sites and food for an array of birds, from fish-eating egrets, pelicans and kingfishers to insectivorous passerines (Figure 4.3). Mangrove roots support many sedentary epiphytic animals typical of hard-substrate environments, which are important to the cycling of the nutrients within the mangrove system. These include sponges, tunicates, oysters and barnacles (Bertness, 2007). Mobile invertebrates are also common, including many species of crabs, snails and an extraordinary abundance of insects. The soil microbes (bacteria, protists, nematodes and flatworms) play a key role in nutrient recycling, and tree growth is enhanced by nitrogen-fixing bacteria next to their roots. Fish are abundant in number and diversity, using the mangroves as nursery grounds to coastal water (including coral reef) fish stocks, in addition to hosting the amphibious mudskipper. Crocodiles, caimans, crab-eating frogs, snakes, lizards and turtles are part of the mangrove fauna in tropical North and Central America, Asia and/or Africa. Other rare or endangered mangrove animals are listed in Chapters 9 and 10.

Figure 4.3 Mangroves shelter some large animals. A: endangered crocodile (*Crocodilus nilotica*) is protected in a sacred pond at Kachikali near Banjul, The Gambia (photo by Anita Whittle); B: brown pelican in a Ramsar-protected mangrove preserve at Magdalena Bay, Baja California (photo by Nic Slocum).

4.4 Energy flow in coastal wetlands: animal–plant interactions

All is flux, nothing is stationary

Heracleitus

According to Wiegert and Freeman (1990), a salt marsh is one of the few areas where exclusively terrestrial animals co-inhabit with marine animals. They occur together among the halophytic grasses, which are the 'powerhouse' of the salt marsh. Therefore, the ecological energy flow of these systems is complex, encompassing many different sources and sinks of energy and nutrients. It is well-known that salt marshes are one of the most productive (vascular) plant communities on Earth (Odum, 1970; Whittaker, 1975). First, the above-ground production of *Spartina alterniflora* can reach over 3000 $g/(m^2yr)$, with below-ground production either matching or exceeding that number. Second, the base of all salt marsh food webs is the energy used to consume and convert the dead plant matter, recycling it back into the plant systems or passing it up the food chain.

The importance of salt marshes as productive ecosystems is emphasized by John Teal (1962) who found that *Spartina* marshes are one of the most productive ecosystems in the world, with detritus being the dominant food source. Teal showed that almost half of the net production in a salt marsh is removed by the tides to coastal waters, increasing the productivity of those ecosystems and showing how the interconnectivity of coastal wetlands sustains other ecosystems by 'outwelling', similar to the process of deep-water upwelling. Over 90% of the detritus leaving a salt marsh is from *Spartina*, hence the salt marshes apparently support the outwelling hypothesis (Odum and de la Cruz, 1967). However, although Teal's main findings were unquestioned for decades, later studies show the immense variability of salt marsh ecosystems and the fates of their production (Boaden, 1985). This variability includes consumer control, where herbivore grazers of salt marsh grasses play a larger role than the detritivores. For example, Arctic snow geese (Jefferies and Rockwell, 2002), the marsh periwinkle snail, *Littoraria irrorata* (Silliman *et al.*, 2009), and graspid crabs (Bortolus and Iribarne, 1999) can greatly affect the production of the grasses to the point of complete destruction, showing it is not just humans who cause the dieback of salt marsh vegetation.

Common terminology used in studies of salt marsh energy flow is introduced in Table 4.1. Lindeman (1942) studied the energy flow through trophic levels in any given ecosystem, revealing the importance of decomposers in the flow of energy. This work first added to ideas of linear food chains and then led to models for food webs involving many preys and predators for one chain link, as first discussed by Charles Elton in 1927. Since the mid 1940s, qualitative and quantitative use of trophic dynamics has been one major way to study and compare ecosystems.

4.4.1 Primary production

In salt marshes, there are three main areas of processes important to the salt marsh ecosystem nutrient cycling and energy flow: (1) the above-ground stems and leaves of plants, with

Table 4.1 Important ecological terms	
Term	Definition used here
Carbon sink	Reservoir that accumulates and stores carbon-containing compounds
Carbon source	Producer and releaser of carbon-containing compounds
Denitrification	Process facilitated by microorganisms by which nitrogen is reduced to its gaseous version
Detritus	Decomposing plant or animal material
Ecosystem service	Resources and processes supplied by natural ecosystems (i.e. carbon sequestering, food, water, primary production, etc.)
Food web	Representation of the possibly feeding relationships (what eats what) in an ecological community
Net production	Amount produced (mass/area/time) after processes of growth and respiration are subtracted from the total amount of energy produced
Nitrogen fixation	Process where atmospheric nitrogen is converted to a form of ammonia
Primary production	Production of organic compounds from atmospheric or oceanic carbon dioxide and an energy source (i.e. sunlight)
Secondary production	Generation of biomass of consumer organisms in a system
Trophic level	Position an organism occupies in a food chain (food web)

associated fauna; (2) tidal water, including swimming and surface organisms; and (3) soil and its functions of decomposition, sequestration and infaunal microorganism metabolism (Wiegert and Freeman, 1990).

The mostly monospecific temperate-region salt marsh floral communities with a visible abundance of *Spartina* support a very diverse and large food web. In addition, microscopic benthic algae (mostly diatoms, which comprise 75% of the benthic algae biomass) and floating mats of algae (green *Enteromorpha* and brown *Ectocarpus*) are other important primary producers (Wiegert and Freeman, 1990). Phytoplankton in tidal channels also contributes to the primary production. Because *Spartina* grasses are rarely grazed, most of the energy that supports the food web comes from their leaf detritus and the benthic algae living among the grasses and on mudflats. The problem with determining salt marsh production is the lack of a standard method for measuring or estimating primary production (Mann, 2000). Various methods used to measure biomass include total weight versus dry weight, living versus living + dead plants, above-ground and below-ground biomass, seasonal changes and, furthermore the geographic location also changes the amount of production seen in salt marshes, which ranges from 1000 to 400 g/(m²yr) dry weight between Texas and Maine, respectively (Turner, 1976). Other above-ground estimates of *Spartina* production based on amount of carbon vary from 60 gC/(m²yr) in Northern Canada, to 812 gC/(m²yr) in New Mexico. This variation in measurement methods makes precise comparison of marsh production difficult.

Spartina alterniflora below-ground biomass far exceeds that of the shoots (over 60% of net production), giving this species an edge for dealing with the many stressors associated with salt marshes, especially where the soil freezes in winter (Wiegert and Freeman, 1990). Because much biomass is stored below-surface after the leaves die in the winter, the cordgrass can regrow rapidly in spring. This reservoir of below-ground organic matter is relatively constant, providing bacteria and other organisms with a continuous supply of detritus. Most of the oxygen demand in marsh soils is from microbial degradation of the organic carbon (Packham and Willis, 1997). However, the relationship between the organic carbon production of the grasses and the subsequent recycling of carbon by microorganisms is just one path of nutrient fluxes in the salt marsh, and is still under investigation. Locally, a salt marsh is a carbon sink, where all the carbon is 'absorbed' into the ecosystem. In fact, models reveal hundreds of grams of carbon that are not accounted for within the above-ground marsh (Wiegert and Freeman, 1990), so it was proposed that this carbon stays within the marsh system because microorganisms play a crucial role in carbon recycling. Therefore, marshes act as net carbon sinks, taking back and recycling most of the carbon it produces. Although it remains uncertain if salt marshes are an overall carbon source or sink, many authors agree that they are net exporters of caloric energy and nutrients (Bertness, 2007). Lately, their importance as carbon sinks has been emphasized (see reviews by Mcleod, 2011 and Chmura, 2011). This blue carbon (Chapter 3) is buried and stored in the sediments rather than being exported into other coastal systems where the carbon is transferred to the atmosphere as the greenhouse gas, CO_2. However, these carbon sinks can become large carbon sources when anthropogenic impacts, climate warming and sea level rise disrupt the burial and metabolism of the terragrams of stored organic carbon and release them to the atmosphere through shoreline erosion (Jorgenson and Brown, 2005).

Although salt marshes are extremely productive, they are nitrogen-limited ecosystems and are highly susceptible to enrichment by anthropogenic impacts such as fertilizer runoff. One consequence of this excess increased nutrient supply is the proliferation of 'dead zones' at the mouths of the world's rivers – where rapid decomposition of fertilizer-driven algal blooms removes the water oxygen, causing the death of shellfish, fish and mammals. It was believed that coastal salt marshes could solve this problem by capturing and chemically transforming the nutrients from nitrate into nitrogen gas before it reaches the ocean. New studies, however, show that in Massachusetts marshes this comes at the cost of a high rate of tidal wetland loss (Deegan et al., 2012) because the nitrate-enriched cordgrass produces heavier leaf growth and weaker roots that topple the plants into the tidal channels, resulting in bank and marsh erosion.

Other recent studies have shown that although nutrient enrichment increases plant growth and herbivores, there is little effect on the detritivores and other consumers (McFarlin et al., 2008; Johnson and Fleeger, 2009); however, nitrogen loading can cause overall shifts in the ecosystem, accelerating vegetation changes and the availability of marshes as carbon storehouses. Nonetheless, both in impacted and non-impacted systems, bacteria perform nitrogen fixation and denitrification in the salt marsh sediment. These processes mean that some bacteria transform atmospheric nitrogen into useable forms of ammonia, whereas others turn these ammonia compounds back in to gaseous nitrogen, with various steps in between (Mann, 2000). To add to the complications of nitrogen budgets in salt marshes, tidal inundation transports and removes varying amounts and types of nitrogen to and from the marsh. Export of dead organic matter and denitrification results in a net transfer of nitrogen from the system (White and Howes, 1994) and it leaves *Spartina* marshes nitrogen-limited and susceptible to nutrient loading. Comparable complex pathways are involved in the microbial fixation and release of sulfur in marsh soils which controls the acidity (pH) and precipitation or mobilization of nutrients such as iron and toxins, e.g. mercury and arsenic (Tobias and Neubauer, 2009; Stern *et al.*, 2012).

4.4.2 Secondary production: consumers versus detritivores

In addition to the importance of detritus in the salt marsh food web, the role of herbivores must also be considered. In undisturbed areas, salt marsh plants are not readily consumed by herbivores (usually less than 10%) due to their structural armor, low nutritive qualities and defense chemicals (Bertness, 2007) although they are grazed by horses, sheep, goats and cows in temperate regions and by birds, deer and bears in polar regions. Most of the energy from the salt marsh plants is consumed through the detritus system, as shown by the Teal (1962) energy-flow diagram, which is the cornerstone of many salt marsh energy and food web studies (Figure 4.4). In other words, although the macrobenthos (insects, spiders, crabs, snails, bivalves, etc.) account for much of the animal biomass in a salt marsh system, they only consume a small amount of the total food available, indicating that the primary production is mainly going elsewhere within the system (Kuipers *et al.*, 1981) – to detritivores.

The detritus system involves bacteria and fungi converting the decaying plant matter into an 'edible' form, which is then eaten by small invertebrate and protozoan detritivores and other deposit- and filter-feeding animals, and transferred up into the larger food web. This

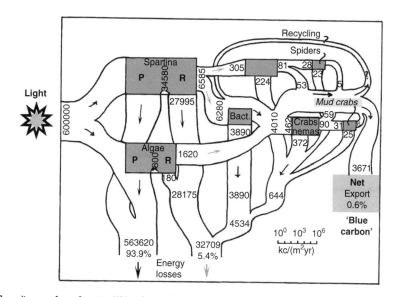

Figure 4.4 Energy-flow diagram for a Georgia, USA, salt marsh (modified from Teal, 1962). Numbers represent the amount of energy as kilocalories (kc/(m^2yr)). Widths of pathways are relative to the amount of energy passing through. Entering as light input, only 6% of this energy enters into the secondary production chain (as 94% is used in high primary production, i.e. photosynthesis, P minus R, respiration). Only 0.6% of the original light energy is exported from the food chain to the coastal ocean as blue carbon.

'small food web' emphasizes the importance of tiny organisms in the energy flow of a salt marsh, consuming up to 80% of all organic material available (Kuipers *et al.*, 1981). High densities of the meiofauna can yield larger production numbers than the same-sized macrobenthic animal. The small food web, then, helps explain the fate of the high amounts of plant detritus, and also provides the food for small macrofauna and other consumers using the salt marsh as a nursery ground. An example of the connectivity between coastal wetlands and nearshore systems is the relationship of *Gammarus* amphipods and the common *Fundulus* mummichog fish. The amphipod eats *Spartina* detritus, and is prey for the mummichog, which then transfers this energy and biomass into other coastal systems when consumed up the food web (Parker *et al.*, 2008).

4.4.3 Mangrove energy flow

Mangrove ecosystems are complex, often highly productive and detrital-based ecosystems, with much of the products of photosynthesis falling into the water as dead and decomposing leaves (Alongi, 2009). Part of this leaf material is eaten directly by larger organisms such as crabs and snails; the rest is attacked by bacteria and fungi to enter the decomposition food web, after which the bacteria are consumed by deposit feeders and other meiofauna (Bertness, 2007). Recent studies show that benthic microalgae are more important direct food sources than the tannin-rich, indigestible mangrove leaves. Like salt marshes, variations in measurement methods means that the total primary production in mangrove communities is debated, though it is agreed they are an important local carbon source (Mann, 2000; Alongi, 2009). Overall, tropical mangrove forests are estimated to be among the most carbon-rich forests in the tropics, accounting for more than 50% of Earth's blue carbon and for 71% of all the carbon stored in ocean sediments (Rovai *et al.*, 2012); however, shrubby subtropical mangroves are not always so productive. The distribution of mangrove forests (see Section 9.3) can also rule the amount of energy exported into other coastal ecosystems. For example, fringing mangroves export approximately 95% of their leaf litter, leaving little for the local detritivore system, whereas backwater (basinal) forests only export 20%. The nitrogen cycle within mangrove ecosystems is more complex than salt marshes and it is regionally variable (Twilley and Rivera-Monroy, 2009), but often there is very little net nitrogen export because mangals are largely a sink for nitrogen imported in tidal water and transformed into organic products by nitrogen fixation (Boto and Robertson, 1990). The herbivorous macrofauna can release nitrogen into the surrounding waters through feces, which helps to support the deposit feeders in the detrital food web and maintain the nitrogen balance of the mangrove system. Water and sediment oxygen levels and rates of sediment deposition also influence the nitrogen, sulfate and trace metal biogeochemistry (iron, manganese) of the ecosystems. Oxygen from mangrove pneumatophores also fixes toxic trace metals from industrial waste.

The greater three-dimensional complexity of mangrove forests compared to the herbaceous salt marshes confers a correspondingly larger amount of ecosystem and food web complexity. The mangrove roots support complex tide-related zonations (see Chapters 8 and 9), including interspecific competition for holdfast space and stimulation of host root production (Alongi, 2009). Like other forests, the canopy space supports diverse insect, bird and mammal species, although epiphytic plant diversity (e.g. orchids and bromeliads) appears to be less important. High rates of photosynthesis reflect not only highly evolved physiological mechanisms, but also energetically efficient linkages among soil nutrient pools, microbes and the trees. At the top end of the consumer food web, predatory tigers, estuarine crocodiles and the dwarf hippopotamus unfortunately are losing the contest for food and shelter to predation by humans.

5

Human intervention causing coastal problems

Key points

The location of coastal wetlands on deltas, estuaries and lagoons make them targets for landscape alteration by dredging, shipping and air industries; land 'reclamation' for agriculture, aquaculture, urban development and tourism has transformed ~30% of the world's wetlands; population growth, rising sea level, dams and soil desiccation increase wetlands flooding from higher water levels and increased storminess; shrinking Arctic sea ice, permafrost and glacier melting increase erosion, adding to greenhouse gases and change ocean–atmosphere circulations pole-to-pole; replacing salt marsh and mangroves by landfill removes natural shoreline protection, but artificial barriers create worse erosion; attempted wetland recolonization often fails because introduced species are invasive; drainage to control mosquitos and tropical diseases changes wetland productivity; pollution from nitrogen loading and oil spills cause long-lasting damage, up to >30 years.

5.1 Human population growth and landscape alteration

Anthropogenic impacts on coastal wetlands include landscape alteration and reclamation of tidal wetlands, accelerated climate warming and sea level rise, spread of alien plant and animal species, construction of dams, draining of tidal wetlands and discharge of pollutants – deliberately or accidentally. Coastal wetlands are particularly vulnerable to the impacts of sea level rise (Cahoon *et al.*, 2006). In the Stern Review of the Economics of Climate Change (Stern, 2007), the costs of future coastal flooding are projected as around US$7.5 to $11 billion per decade for Europe and North America, respectively. Syvitski *et al.* (2009) have shown that 17 of the world's largest deltas (Table 5.1) are critically vulnerable to being flooding and converted to open ocean. In the past decade, 85% of these deltas experienced severe flooding, with a total area of $260\,000\ km^2$ being temporarily submerged. The Syvitski team estimate that the area vulnerable to flooding could increase by 50% under projected values for twentieth-century sea level rise. In contrast, other models predict that increased storminess will transport more sediment into coastal wetlands and enable salt marshes to keep pace with sea level rise (Schuerch *et al.*, 2013).

Marine-based plants and animals tolerate large variations in salinity, subaerial exposure time and tidal creek shifting that occurs naturally in wetlands, but they are vulnerable to human interferences such as dredging, levee and road construction, and filling of marshland.

Table 5.1 Deltas listed by Syvitski *et al.* (2009), as in peril or great peril from lowered twentieth-century aggradation and/accelerated compaction. Aggradation change (mm/yr) is the difference between early twentieth-century and present rates; asl = above sea level.

Country	Delta	Area < 2 m asl (km^2)	Storm surge area (km^2)	Aggradation change (mm/yr)	Relative sea level rise (mm/yr)
In Peril					
Bangladesh	Ganges	6170	10 500	−1	8–18
Myanmar	Irrawaddy	1100	15 000	−0.6	3.4–6.0
Columbia	Magdalena	790	1120	−3	5.3–6.6
Vietnam	Mekong	20 900	9800	−0.4	6
USA	Mississippi	7140	13 500	−0.7	5–25
Nigeria	Niger	350	17 000	−0.7	7–32
Iraq	Tigris–Euphrates	9700	1730	−2	4–5
In Dire Peril					
Thailand	Chao Phraya	1780	800	−0.2	13–150
Mexico	Colorado	700	0	−34	2–5
India	Krishna	250	840	−6.6	~3
Egypt	Nile	9440	0	−1.3	4.8
China	Pearl	3720	1040	−2.25	7.5
Italy	Po	630	0	−3	4–60
France	Rhone	1140	0	−6	2–6
Brazil	San Francisco	80	0	−1.8	3–10
Japan	Tone	410	220	−4	>10
China	Yangtze	7080	6700	−1.1	3–28
China	Yellow	3420	1430	−49	8–23

Natural sedimentation and erosion rates determine if a wetland will be stable or not. However, human interference and land development can accelerate the destruction of salt marshes and mangrove environments. Coastal wetland landscape alteration is widespread, but is greatest in temperate regions of industrialized Europe, USA and Mexico, and in parts of Southeast Asia, where developers have converted coastal wetlands into marinas, airports, industrial parks, holiday resorts, housing tracts and shrimp farms.

Wetlands destruction promotes erosion of the shoreline, triggering landslides and levee breaks that further weaken the coast, and makes it vulnerable to excess damage from storms, earthquakes and tsunamis. Damage results from the following activities: (1) filling or destroying marshlands for development ('reclamation'); (2) changing of the tidal circulation or tidal amplitude by building causeways that trap sediment and constrict channels; (3) cutoff of freshwater supplies by upstream dams and overexploitation by agriculture draw-down of the water table; (4) petroleum industry removal of oil, gas and water from the delta's underlying sediments causes compaction and subsidence; (5) dredging of channels alters the flow and natural water course; and (6) deforestation leads to landscape erosion, and downstream, the washed-in sediment clogs tidal channels and raises marshes above the tidal range. Books by Seabrook (2012), Weis and Butler (2009), Silliman *et al.* (2009), Bertness

(2007), Pilkey and Young (2009) and Teal and Teal (1969) describe the impact of these problems in USA tidal marshes. Classical books by Chapman (1960, 1977) remain the most comprehensive references for salt marshes on a worldwide basis; other authors who take a global approach are Long and Mason (1983), Packham and Willis (1997), Mitsch and Gosselink (2007) and Spalding *et al.* (2010).

In the tropics, many mangrove swamps are being replaced by farmland (mostly rice and oil palm plantations) and aquaculture ponds, particularly in highly populated areas of the Indian Ocean and Southeast Asia. Under natural conditions, the roots of mangrove trees and shrubs stabilize the low marsh and mudflat habitats. Mangroves also tolerate wide ranges of salinity from river flooding and dense mangrove forests provide a buffer against hurricanes and tornadoes, thereby protecting the coastline from erosion. In many regions, however, population growth has led to overuse of mangroves for fire wood and buildings, and forest clearance for farmland or shrimp ponds. Thinned-out fringe forests and unprotected coastlines are more easily eroded, and scattered trees are defoliated or uprooted by tsunamis or cyclones. The loss of the mangroves then reduces the natural habitat for many fish and other mangrove-dependent species in a top-down trophic cascade. Cyclone Nargis (see Table 2.1) provides a good example of the devastating impact of tropical storms on shorelines where mangroves have been removed for rice cultivation. The IPCC (2007) report notes that although the vertical accretion of mangroves is commonly ~5 mm/yr, and thus should keep pace with sea level rise, in fact many mangrove shorelines are subsiding; this report estimates that by 2020, 9–20 million Southeast Asian people per year may be victims of coastal flooding, with further loss of life from starvation due to salinization of storm-flooded rice fields.

5.2 Land reclamation

Communities in many areas are 'reclaiming' marshlands by filling them in for housing or tourism, especially on the southeast coast of the USA. Besides destroying the wetlands, the shoreline is impacted by breakwaters, retaining walls and other structures that contribute to shoreline erosion by changing the natural coastal currents and supply of sediment to the wetlands. All seawalls prevent waves from running up the beach; instead, the wave energy undermines the coastal structures, doing more damage than just leaving the area unprotected. The coastal geomorphologists Pilkey and Theiler (1992) show some dramatic disasters for shorelines around the world where various kinds of barriers, breakwaters and seawalls have been built (see also examples in Pilkey and Young, 2009 and Figure 5.1).

Reclamation of land for agriculture by converting tidal marshland to upland was historically a common practice in Western Europe (see Section 10.1) and eastern North America. Dikes and drainage channels were often built to make this land change and to provide tidal flood protection further inland. For centuries, livestock such as sheep and cattle grazed on the fertile reclaimed wetland. Reclamation for agriculture results in many changes, including shifts in vegetation structure and composition, sedimentation, salinity, water flow, biodiversity loss and nutrient inputs. Later, attempts were made to counteract these problems. For example, in New Zealand, the cordgrass *Spartina anglica* was introduced into the Manawatu River estuary in

 Figure 5.1 Archival photos of storm damage to coastlines (from the Pilkey slide collection). A: New Jersey shore during a Nor'easter storm with high surge waves; the 'protective' seawall had little effect, despite its great expense; B: North

1913 to reclaim the estuarine mudflats for farming. As a result of increased sedimentation, there was a shift in structure from bare tidal flat to pastureland, and the cordgrass also invaded other New Zealand estuaries (Chapter 11). Native plants and animals struggled to survive as rapid-growing non-natives out-competed them. Efforts are now being made to remove these invasive cordgrass species to avoid further damage.

5.3 Accelerated global warming

Large fluctuations in temperature of about 6–8 °C are a natural consequence of glacial–interglacial cycles and have been accompanied by corresponding changes in sea level of up to 125 m (see Scott *et al.*, 2001 Figure 3.11 on p. 41). However, these cyclical temperature and sea level changes occurred on a timescale of thousands of years. In contrast, the accelerated warming of global air temperature during the twentieth century (Figure 5.2) has resulted in an average worldwide atmospheric temperature increase of about 0.6 °C during the past 30 years. In the Arctic region, this temperature increase has been much greater (2–3 °C), partly because of the rapid melting of sea ice and the corresponding increased inflow of relatively warm Atlantic Ocean water. Reduced sea ice in turn lowers the heat-reflecting Arctic Ocean albedo, and open water absorbs more summer heat.

Climate warming affects global sea level in three ways: (1) directly through ocean thermal expansion (a fairly predictable linear event); (2) by transfer of fresh water to the ocean from melting of glaciers and icesheets and (3) by the calving of ice from West Antarctic ice-shelves. Melting of the polar icesheets involves non-linear, poorly predictable events dubbed '800-pound gorillas' by Pilkey and Young (2009) who provide details on historical break-up events that caused jumps in global MSL. Over the past three decades, the average global MSL level has increased by about 30 cm, and the Intergovernmental Panel on Climate Change (IPCC, 2007) projects a further rise of 18 to 59 cm by 2100 (Figure 5.2). Theoretical estimates based on collapse of the Greenland Icesheet and the Antarctic shelves could amount to a total increase of up to 140 cm (Shum, personal communication, March 2013). Another possible underestimated factor may be the absence of volcanic forcing influences in most predictive models of future global sea level rise (Gregory *et al.*, 2013). When added to the models, an additional 5–30 mm/yr can be expected. More than 10 m of sea level rise is possible over time spans of centuries or longer (IPCC, 2007), and geological

---

Caption for Figure 5.1 (cont.)
Myrtle beach, North Carolina, after the 1954 Hurricane Hazel, with damaged houses on the upper beach; C: northern Assateague Island, Maryland, migrated landward when jetties built to protect the marsh after a 1933 hurricane blocked sand supply to the southern barrier island; only the north side was protected at the expense of the beach to the south (lower image) which is over-riding former marshland; D: North Myrtle Beach, showing the futility of retaining walls: ocean currents flow around sea walls and scour out the sediment immediately downstream; E: Capers Island, South Carolina: beach migrating landward over mature marsh and exposing younger marsh, now eroding since the protecting beach moved inland.

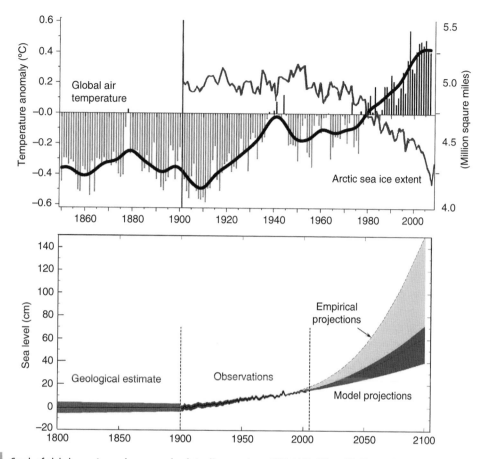

Figure 5.2 Graph of global warming and mean sea level rise (Sources: Ice – IPCC, 2007; RSL – C.K. Shum, Ohio State University, 2013).

records indicate a 6−10 m increase above present MSL during the past ~1 million years (Berger, 2008). Some scientists therefore project a possible global hyperwarming by combining the 'worst scenario' ice melting with accelerated atmospheric and ocean warming as the sea floods continental lowlands and these shallowly flooded seas gain warmth. This global hyperwarming feedback mechanism could take global temperatures well beyond the ~5 °C predicted by some of the 'worst case' models.

5.4 Arctic sea ice tipping point

During the past 30 years, summer sea ice in the Arctic has decreased by over a million km² (Figure 5.2), and models forecast unprecedented amounts of open water after 2011 (Overpeck *et al.*, 2005). The amount of sea level change along the Arctic coastline varies

Figure 5.3 Melting of subsurface permafrost (white cliff area) in the Mackenzie Delta causes massive collapse of the shoreline, resulting in mudslides and new colonization of salt marsh vegetation (green areas on the mud). Note the polygonal cracks in the bluff vegetation, which are typical of tundra soils and aid in release of methane gas on melting of permafrost. Jorgenson and Brown (2005) estimate an average of 0.18 million metric tonnes of carbon are released annually from erosion of the Alaskan-Beaufort sea coast (photo by P. J. Mudie). See also colour plates section.

with the distance from the last great icesheets over Canada and Siberia. Most of these coastlines are still experiencing postglacial rebound, with uplift of up to about 1 m/century, but the western North American coast is subsiding (Forbes and Hansom, 2011). The rising sea level increases the risk of flooding, storm wave damage and sea ice erosion. The most severe flooding risks for shoreline communities are associated with large storm surges, sometimes more than 2 m above MSL (Manson *et al.*, 2005). Storm waves during severe surges can cause rapid coastal retreat of over 10 m in a single event. Further warming, combined with sea level rise, reduced sea ice and melting of permafrost (= soil colder than 0 °C for at least two consecutive years) causes collapse of the shoreline (Figure 5.3) and will probably maintain or increase the high rates of coastal retreat. Shoreline and permafrost collapse in turn release more carbon to the Arctic Ocean and methane to the atmosphere (Ping *et al.*, 2008; Schaefer *et al.*, 2011), adding to global greenhouse gas warming. Melting ice also lowers the salinity of the Arctic Ocean water, changing plankton and fish populations, and their carbon cycles – perhaps triggering a domino effect that could impact the world's climate and ocean circulation (Figure 5.4). Some scientists (Overpeck *et al.*, 2005; Dunton *et al.*, 2006; Carmack and McLauchlin, 2011) think that this irreversible tipping point may already have been reached and that a new Arctic 'normal' is being established. The Arctic climate engine is tightly coupled to the climates of the North Atlantic and Pacific Oceans and these, in turn, could change the ocean circulation and climate as far away as the Antarctic.

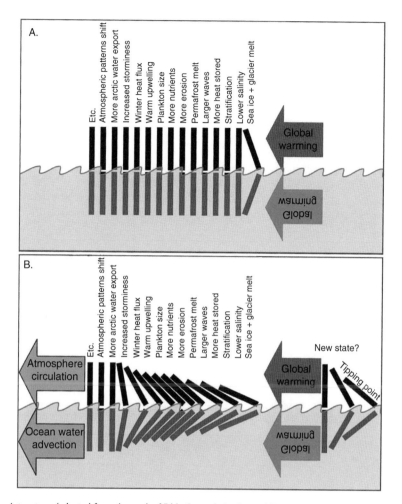

Figure 5.4 Tipping point cartoon (adapted from the work of Eddy Carmack, Institute of Ocean Science, Patricia Bay, Canada, with permission).

The potential impacts of Arctic warming on the extensive freshwater peatland of the circum-Arctic region are discussed in detail by Mitch and Gosselink (2007). Briefly, the vast Siberian wetlands represent a 'ticking time bomb' in the form of methane release from permafrost thaw as temperatures increase to 2.8 °C as deep down as 10 m (Schaefer *et al.*, 2011). The methane release adds atmospheric greenhouse gases, which lead to more changes in the climate, known as the methane 'flywheel' effect. The same sort of 'flywheel' effect has been triggered in the Arctic Ocean by the melting of sea ice and increased melting of glaciers and the Greenland Icesheet. The complex series of events involved is illustrated schematically in Figure 5.4. Overall, it is predicted that by 2022, there will be a ~30−60% decrease in area of arctic-subarctic permafrost, by which time the Arctic permafrost carbon sink will become a carbon source (Schaefer *et al.*, 2011).

5.5 Biological invasions

Spartina alterniflora (salt marsh cordgrass) is a native plant from the eastern seaboard of the United States (Figure 3.3), but it is considered a noxious weed in the Pacific Northwest, where it threatens to displace the western native species *Spartina foliosa*. In eastern USA and other warm, temperate regions, the reed *Phragmites australis* has also invaded coastal wetlands, expanding into lower marshes and becoming a dominant species (Figure 5.5A). *P. australis* is a salt-tolerant grass that spreads rapidly and out-competes many native plants. The loss in biodiversity is not only seen in floral assemblages, but also in many animals such as insects and birds, as their habitat and food resources are altered. Some of the invasive salt marsh species introduce toxins, including *Spartina anglica* (Nehring *et al.*, 2012). Seagrasses can be invaded by aggressive seaweeds, e.g. *Caulerpa taxifolia* and *Caulerpa racemosa* that are spread in ballast water discharged from ships. The seaweeds shade out seagrasses, leading to the collapse of the delicately balanced seagrass food chain. Beaches along infested shorelines are carpeted by the weeds which have a foul odour, making swimming unpleasant (Figure 5.5B). Details of extensive salt marsh plant invasions are given by Silliman *et al.* (2009), and Spalding *et al.* (2010) describe the more limited invasion of mangals by *Nypa* palm. In China, introduction of the fast-growing Indian Ocean mangrove *Sonneratia apetala* to help restore degraded wetland sometimes leads to loss of slower-growing native mangroves (Ren *et al.*, 2009).

Figure 5.5   Invasive plant species in coastal wetlands include salt-tolerant trees and tall grasses that draw down the water table and invade mudflats, and seaweeds that smother seagrasses. A: *Phragmites australis* (cane reed) is a native of European freshwater marshes, but is invading the shores of the Black Sea, destroying mudflats important for migratory waterfowl, as in Saskia liman; B: the seaweed *Caulerpa racemosa* is a recent invasive species introduced into the Black Sea by ships from the Mediterranean; in Odessa, summer beaches are covered by the slimy weed that local people try to remove (photos by P.J. Mudie).

5.6 Wetland drainage and conversion for farming

More than half of the world's population lives within 60 km of the ocean, and in the past, salt marshes were perceived as coastal 'wastelands', more useful when drained for agriculture, flooded for salt production or dredged for boating recreation. Mangrove wetlands were also widely destroyed before their importance for estuarine and nearshore fisheries were well understood. In addition, wartime defoliation of mangals in Vietnam resulted in desiccation of the wetlands and failure of attempted mangrove recolonization. Consequently, 25–35% of the world's mangal resources have been lost by drainage and construction of shrimp farm or salt harvesting ponds, in addition to coastal erosion from tsunamis and storms, and desiccation after deforestation and/or river dam construction.

Today, most Southeast Asian countries have areas in which marsh/mangrove systems are still being converted for aquaculture to farm shrimp, prawns and coastal fish for export. Solomon and Forbes (1999) discuss this problem as it affects the coastlines of low-lying South Pacific island countries north and east of Australia and Papua New Guinea, including the Solomon Islands, Tuvalu, Tonga, the Cook Islands and French Polynesia. This group of 15 states and 7 territories depends heavily on coastal resources, but the region is especially vulnerable to rising sea level, monsoon flooding and wave damage from severe storms and cyclones. Here many mangrove swamps have been channeled and drained for port facilities, airports, housing and tourism. Sea walls are then built to try and protect the reclaimed land, which further changes the shoreline and sediment supply. Mud becomes trapped on the upstream side of the walls and may fill in tidal lagoon channels and mangrove swamps, while the downstream shoreline becomes sediment-starved and vulnerable to erosion. In the Mekong Delta, despite a recent decrease in mangrove loss from wood harvesting and reforestation of mudflats, significant net loss of mangal continues because of the expanding, high-profit shrimp farm industry (Thu and Populus, 2007). Further problems related to application of chemicals and disposal of organic shrimp waste are presently under investigation (Vaiphasa et al., 2007; see Chapter 10 for details).

Drainage of coastal wetlands for mosquito and disease control is widespread, particularly in urban regions of North America and Australia, where insect carriers (vectors) of malaria and West Nile virus are spreading because of accelerated global warming and international travel. Mosquito bites are unpleasant and irritating to humans and wild mammals, so mosquitos have long been a target for avoidance or eradication. In tropical countries, including most of Australia, mosquitos are extremely important because they can carry potentially deadly diseases (Box 5.1 Mosquito life cycle) – including malaria, dengue fever, filariasis, West Nile virus (now also in North America), Ross River virus, Barmah Forest virus (Australia) and viral encephalitis (Dale and Knight, 2008; 2012). These mosquito vectors breed in salt marshes and/or mangrove swamps and lay their eggs wherever there are pockets of standing water to host their larvae ('wrigglers') for a few weeks. When channels are open to daily tidal flushing, most of the larvae will be removed. However, when channels are blocked by construction, the area for mosquito breeding greatly expands; the eggs will remain dormant long after the area dries out, until reflooding triggers hatching.

Most research on mosquito-borne tropical diseases has focussed on a cure for malaria, which is an infection of the liver and blood cells by the malaria parasite (usually *Plasmodium falciparum*). Infected female *Anopheles* mosquitos carry *Plasmodium* sporozoites in their saliva and inject them into people when they bite. Within 30 minutes, the sporozoites infect the person's liver and later they burrow into blood vessels and red blood cells. Once in the blood cells, they divide rapidly (as schizonts), releasing massive numbers of trophozoites into the blood stream that can cause high fever and death. Malaria kills up to one million people each year, mostly children and pregnant women. For decades, doctors have tried to develop a vaccine against the malaria parasite because young children cannot tolerate the medications used to protect adults. However, the vaccination research failed because of the relatively large size of the parasite, and its complex life cycle and high mutation rate. A new discovery, however (Regev-Rudzki *et al.*, 2013), reveals that malaria trophozoites have to 'talk' to each other with DNA message packages in order to ensure long-term survival, which involves sexual reproduction (gametocytes) and transfer back to mosquitos for fertilization in the insect guts. The intercellular trophozoite communication network signals when the parasites should transform into gametocytes for transfer to the guts of blood-sucking mosquitos. This biomedical breakthrough paves the way for new drugs or vaccines to block *Plasmodium* communication networks and prevent the further spread of malaria by mosquitos.

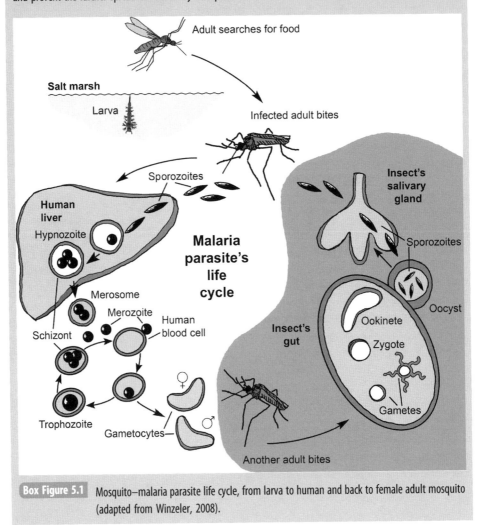

Box Figure 5.1 Mosquito–malaria parasite life cycle, from larva to human and back to female adult mosquito (adapted from Winzeler, 2008).

Earlier in the twentieth century, it was believed that draining of salt marshes would help mitigate mosquito populations. In many locations, particularly in the northeastern USA, residents and government agencies dug deep ditches to drain the marshland (see Figure 7.27). The events and impacts of these actions are detailed by Crain *et al.* (2009). One major result was a depletion of shallow marsh channel habitat for killifish. Furthermore, the killifish is a mosquito larva predator, so the loss of habitat led to higher mosquito populations and less food for wading birds that feed on killifish and eat mosquito larvae. Excavation spoil left on the sides of the drainage ditches led to invasion of woody plant species (*Iva* and *Baccharis*). In other places, insecticides or oil to smother the larvae are applied to the marshlands although mosquitos have a high resistance to chemical warfare and the long-term effects of larvicides on habitats and humans are unknown. In Australia, a method of shallow drainage called runnelling is being applied successfully to salt marshes and is being modified for use in the less-accessible northern mangrove swamps. The long-term solution, however, probably lies in new understanding of how the diseases transmitted by biting insects replicate in humans and their vectors rather than habitat modification.

5.7 Pollution by excess nitrogen and oil spills

In Section 4.4.1, we described the importance of nitrogen loading on salt marsh ecosystems. Many major cities, including Boston, San Francisco, Amsterdam, Rotterdam, Venice and Tokyo, have extended into salt marsh areas where the wetlands are vulnerable to human-induced nitrogen enrichment from sewage, urban runoff, and agricultural and industrial wastes. The increase of nutrients changes the plant species through shifts in competitive advantages. For example, N-enrichment in New England salt marshes results in the spread of tall cordgrass from the lower marsh into the upper marsh zone where it out-shades other species.

Salt marsh ecosystems are also susceptible to oil pollution – usually as the result of accidental spillage from ships and port refineries (Davy *et al.*, 2009). Light oils are highly toxic to marsh plants and animals, whereas heavier crude oils smother plants and small animals. Oil slicks may catch fire, causing decade-long damage to salt marshes in arid regions (Section 10.4) and altering marsh species composition (Pezeshki *et al.*, 2000). Birds are particularly vulnerable because they cannot fly with oil-slicked feathers and they ingest lethal BTEX (= benzene, toluene, ethylbenzene, xylene) products when trying to clean their plumage. Another source of oil pollution is the catastrophic release of light or heavy oil from either oil tanker wreckages (e.g., the *Torrey Canyon* tanker and *Amoco Cadiz* spills in the English Channel, and the *Exxon Valdez* in Prince William Sound, Alaska) or from oil rig blowouts, e.g. the 2010 *Deepwater Horizon* accident in the Gulf of Mexico. Davy *et al.* (2009) report that from 1967 to 2002, about 0.5 million tonnes of oil spilled into the English Channel, and attempted marsh cleanups caused more damage than found in 'uncleaned' marshes 12 years after the oil spill. In Alaska, 36 000 tonnes of crude oil spilled from the damaged *Exxon Valdez* tanker in 1989 that caused US$2 billion of damage, contaminated >2000 km of shoreline and resulted in the death of 2000 sea otters, 302 harbour seals and about 250 000 seabirds within a few days of the spill. Salt marsh fiddler crabs were also badly impacted, but other invertebrates typical of rocky shores (e.g. mussels) were recovering a decade later (2001), despite residual amounts of petroleum under the beach gravel (Graham, 2003). However, 2006 surveys of birds and sea otters show no recovery trend for oil-covered areas (McKnight *et al.*, 2006), and a 30-year recovery period is estimated.

The long-term impact of large oil spills on wetlands vegetation is poorly understood, and differs with vegetation, season and soil type. The impacts of the 2010 *Deepwater Horizon* blowout of $>8 \times 10^5$ m^3 of crude oil into the Gulf of Mexico 80 km offshore also produced some unpredicted results (Silliman *et al.*, 2012; Stokstad, 2013). After three years, the Gulf coastal marsh sediments still contain 1000 times the background level of BTEX toxins, and salt marsh insects (such as fire ants living inside *Spartina* stems) and spiders are still dying, even when they are experimentally protected from direct oil contact. Tar-balls on the beach enclose toxins, removing them from decomposition by oil-eating bacteria. Use of the oil dispersant Corexit to break-up the oil slick inhibits the formation of bacterial clumps ('marine snow'), which helps to remove oil and contaminants from the water column, but the BTEX levels are still increasing in deeper water fish, although blue crab and nearshore fish appear to be recovering.

6

Coastal wetlands worldwide: climatic zonation, ecosystems and biogeography

Key points

All wetland ecosystems provide valuable services to local people and influence global carbon budgets; climate delimits four major ecosystems: arctic and subpolar marshes, temperate marshes, subtropical tidal wetlands and tropical mangrove swamps; wetlands worldwide have specialized plant forms adapted to physical extremes of soil salinity, instability and low oxygen; global biogeographic studies are the tools of wetlands biodiversity studies; primary regional differences in species composition reflect ancient paleogeographic changes; plate tectonics and icesheets isolated tropical 'Old World' from 'New World' biological provinces; coastal wetlands are azonal vegetation types highly influenced by soil conditions; differences in temperature and precipitation regimes shift the ecosystem boundaries north or south on different coasts of continents; seasonal variations in temperature–precipitation conditions drive subregional soil water salinity differences.

In this chapter, general distinctions (both physical and biological features) of major latitudinal and climatic zones are outlined, so the reader can appreciate the broad similarities and dramatic differences in coastal wetlands across the globe (Chapters 7 to 11). We also look at the phytogeographical variations found on different coasts of the same continent and explain why the tropical floras of Atlantic and Pacific regions often are different, even within the same latitudinal range.

6.1 Climate zones and coastal wetland ecosystems

On a worldwide basis, there are four broad latitudinal zones of tidal wetlands (Figure 1.1). These world coastal wetland zones are: arctic and subpolar marshes; temperate zone marshes; subtropical marshes and mangroves; and tropical mangrove swamps. Each of these major ecosystems comprises a set of biological components interacting with their physical environment and providing characteristic wetland services (Figure 6.1) These services are defined by the Ramsar Convention as 'the benefits people obtain from ecosystems, including provisions (food and water), regulating functions (buffering from floods and storms), nutrient cycling, and cultural needs (recreation, spiritual, and esthetic qualities)'. The vital services for human wellbeing include water purification and nitrate detoxification, climate regulation through sequestering and releasing of carbon in the

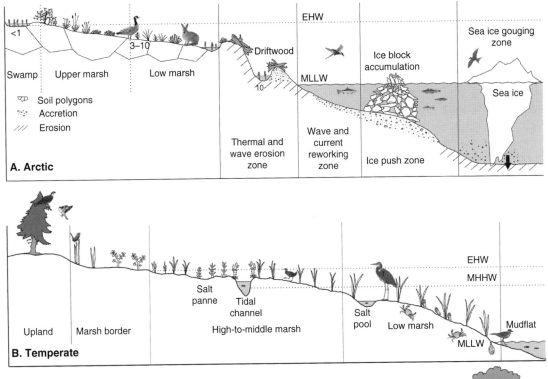

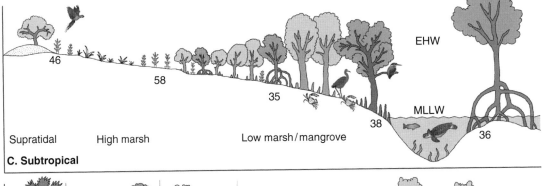

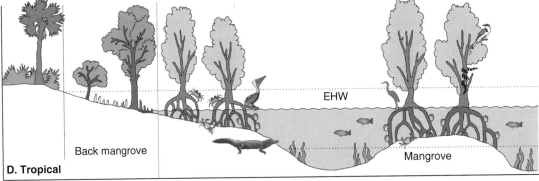

Figure 6.1 Conceptual diagrams of the major world coastal wetland ecosystems and critical habitats. A: polar coastal marshes, e.g. Mackenzie Delta, Canada; B: temperate salt marsh, e.g. Chezzetcook, Nova Scotia; C: subtropical mangrove–marsh ecosystem, e.g. Magdalena Bay, southwest Mexico; D: mangrove–seagrass ecosystem, e.g. Florida. Numbers indicate salinity (psu).

biosphere, and mitigation of climate-driven sea level rise and storminess, which can result in erosion and devastating coastal floods (Millennium Ecosystem Assessment, 2005). Barbier *et al.* (2011) provide further details of the quality of estuarine and coastal ecosystem services and they estimate the financial losses to be expected from elimination of these services.

Different types of tidal wetlands (see Chapter 2) form on low-lying coasts wherever there is sufficient sediment supply and wave protection. All coastal wetland plants must tolerate many physical extremes (Chapter 3), and are primarily azonal vegetation systems dependent more on soil conditions than climate (Peinado *et al.*, 2009); however, each latitudinal area has its own suite of characteristic conditions and biological diversity (zonobiomes), which generally increase in diversity from the poles to the Equator (Table 6.1). Each zone includes other climate-associated physical stressors, such as ice scour or glacier surge in polar biomes, and storm floods and wave erosion in the tropics.

As summarized in Table 6.1, polar (Arctic and subarctic) wetland ecosystems (Figure 6.1A) are characterized by a very low diversity (~5–10 genera) of low-growing grasses, sedges, rushes and herbs adapted to the harsh ice-dominated climate and short growing season (Martini *et al.*, 2009; Sergienko, 2013). Most of these plants have thick tap roots that anchor them in the heaving polygon soils, and they propagate by shallow underground rhizomes or surface stolons (Figure 6.2A–C). Arctic salt marsh plants also have ecophysiological adaptations that protect them from intense radiation and UV exposure (Markovskaya *et al.*, 2012); many species have either protective orange pigment, turning the landscape reddish in August (Figure 6.3B), or a waxy surface layer that gives them a grey appearance. Although these marshes are frozen and ice- or snow-covered most of the year, in summer, they become marshy bogs infested with many types of biting insects. Surprisingly, delicate azure dragonflies are also common in subarctic marshes (Figure 6.2D). Arctic coastal ponds also support brackish water submergent aquatics, including the 'mare's tail' genera *Hippuris* and *Myriophyllum*. The seagrass *Zostera marina* grows in subtidal areas south of the Arctic Circle (Figure 6.2E). Russia has by far the largest of the world's wetland tundra areas (Figure 6.3), with extensive mudflats and salt marshes in the macrotidal area of the western White Sea, and variable amounts of salt marshes on the deltas of Russia's large Arctic rivers – e.g., the Lena (largest, about four times bigger than the Mackenzie Delta), the Ob, the Yenesei and the Kolyma. There are no intertidal deposits or salt marshes north of ~73° N along the Siberian coast because the ice-free period is too short for mudflat formation and growth of stabilizing salt marshes. There are few English-language papers on the geomorphology of Russian–Siberian salt marshes, but some details on given in books by Eisma (1997) and by Kassens *et al.* (1999).

All polar coastal wetlands are vitally important for migratory waterfowl. From Europe to Russia, the birds sequentially occupy the marshes from south to north, following a 'green wave' of vegetation growth and temporary dieback from overgrazing (Figure 6.4). In Russia, the Taimyr wetlands are important feeding grounds for wild reindeer (*Rangifera tandarus*), which are the main source of food for the indigenous people. The reindeer populations are now declining because of industrial developments and early melting of river ice, both of which obstruct the deer migration and force them into smaller feeding areas that are easily damaged by overgrazing and trampling of the Arctic marshes (Malygina *et al.*, 2013). At the opposite pole in the Southern Hemisphere, subpolar marshes are dominated by

Table 6.1 Latitudes and conditions of coastal wetlands, with selected key references. The listed stressors exclude direct human alterations.

Latitude	Location	Climate conditions	Biological diversity	Physico-chemical stressors
1. Arctic and subpolar salt marshes				
70° N–66° N	North of Arctic Circle	Marshes frozen 9 months per year, extensive sea ice; arid, July <10 °C; effective sunlight less than 6 months per year	Low floral diversity; hardy, wiry perennial grasses and rushes; no annuals or seagrass; pondweeds in brackish water; migratory animals; anadromous fish; one small mollusc (*Macoma*), locally abundant amphipods and worms	Freeze–thaw soil heave; ice-scour; permafrost melting and thermokarst lakes/sinkholes; warming ocean, more storm erosion and use by sea ice mammals; heavy bird grazing and nitrogen loads; Hg releases; CHC pollution sink; large driftwood log pile-ups
66° N–50° N	Subarctic (e.g. NW Alaska, Hudson and James Bay, Canada; Scandinavia; mainland Russia)	Longer 'summer' growing season of 3–4 months, effective sunlight only 6 months per year; water salinity mostly low but high in dry summers	More diverse marsh flora with some succulent annual halophytes; dense seagrass (*Zostera*) in subtidal; similar animals but more molluscs and more diverse mudflat faunas	As for Arctic Circle; also emergent shores and low sediment supply in Eastern Canada, Europe; glacier discharge or surges in Alaska; soil creep; deposits of boulders from ice rafting (also see Scott and Martini, 1982)
50° S – 56° S	Subpolar South America (e.g. Tierra del Fuego, San Sebastian Bay), New Zealand (South Island)	Freeze–thaw patterns: freezing in winter, summer ice rafting; high UV-B from Antarctic ozone hole	Medium diversity of hardy marsh plants, often with succulent shrubby *Sarcocornia*; seagrass absent except *Ruppia* in ponds; abundant migratory birds feed on bivalves, polychaetes, crustaceans	High wind/wave energy; often mobile, gravelly soil; high winds 200 days/ yr; wind-borne salt; emergent coast; deflation ponds; oil slicks from drilling offshore; growth suppression by UV-B (Isla 2013; Isla *et al.*, 2005)
>66° S	Antarctic Circle: salt marshes absent	Icesheet edge, with floating with ice shelves and icebergs	No marsh plants; storm beaches with shell, seals	Persistent landfast sea ice; raised beaches with storm transported boulders (Hall and Denton, 1999)
2. Temperate (Boreal) marshes				
50° N–40° N	Between subtropical and polar regions of all continents except Antarctica	Relatively warm summers (July mean >10–20 °C); variable precipitation amount and seasonality	Mixture of southern and northern latitude plants and animals, with more diversity and larger animals toward lower latitudes; resident animals;	Impacts of winter ice, river floods or storms vary with latitude, tidal range and wave exposure; summer droughts and/or El Niño in

		(summer/winter/year-round)	higher microbial abundance and diversity than polar	Mediterranean climate regions; very high salinity in desert regions; highly impacted by human settlement except in South America
3. Subtropical marshes and mangroves				
~40° N/S – 23.5° N/S	Areas immediately bordering the Tropic of Cancer (Northern Hemisphere) and Tropic of Capricorn (Southern Hemisphere). Includes southern United States, mid-South America, areas around the Mediterranean and Red Seas, Southern Australia, SE Africa and SE China; Hawaii, northern New Zealand	Warm (above 10 °C monthly average), moist or with dry summer (Mediterranean climate), high sunlight and hot summer months; very rare frost; monsoons and tropical storms common	Very high diversity, with year-round growth of plants, a mixture of salt marsh grasses and relatively low diversity of mangrove trees, often taking the form of shrubs (woody plants less than 3 m high) near the northern/southern limits; estuaries and lagoons have large reptiles and seasonally may include both tropical and northern marine mammals	Susceptible to prolonged summer droughts during shifts of Pacific and Azores high pressure systems; trees and taller shrubs defoliated by hurricanes; lagoon barriers and tidal channels breached by storm waves and associated flash floods ('chubascos' in Mexico); lagoons with shifting barrier bars, prone to closure during long droughts
4. Tropical marshes and mangroves				
23.5° N – 23.5° S	Equatorial areas of Central America, northern South America, central Africa, Indonesia, Malaysia, southern India, northern Australia	Warm to hot (mean monthly temperature over 18 °C), and moist, with little seasonal variation in temperature; most of the annual rainfall occurs in 2–3 months; monsoons common	Very high diversity of mangrove trees, shrubs and palms; ferns grow on the trees; diverse seagrasses; salt marsh mostly in a narrow zone around EHW; diverse large reptiles and tree-dwelling animals; many large birds; abundant intertidal crustaceans and fish	Mangrove peat susceptible to shrinkage during droughts, with consequent increase of seawater flooding; prone to erosion by typhoon storm waves; rising temperatures and associated die-back of protecting coral reef barriers; increased storminess, and rising sea level threatens wide-scale destruction

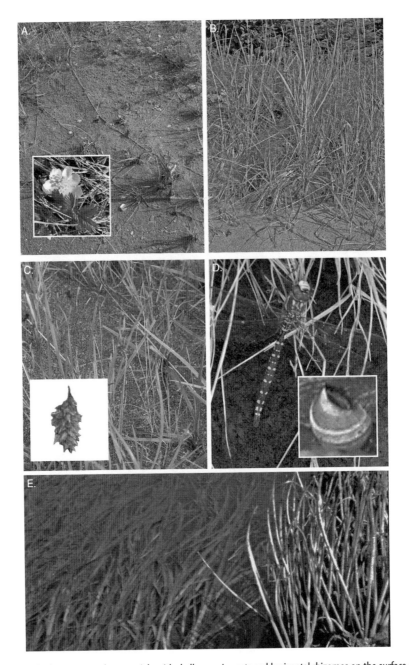

Figure 6.2   Arctic salt marsh plants are mostly perennials with shallow main roots and horizontal rhizomes on the surface or near-subsurface that allow them to spread over the unstable soil. If broken by frost heave or ice, the plants can regenerate from rhizome fragments, as well as seeds. A–D: low marsh: sedge (*Carex*) with V-shaped leaves, grass (*Poa phryganoides*) and arrow-grass with grey leaves, and cinquefoil (*Potentilla*, with inset yellow flower) near Anchorage, Alaska; B: high marsh, with taller lyme grass (*Elymus*); C: sedge and plantain with characteristically orange-pigmented leaves for protection from excessive UV and PAR; D: blue darner dragonfly feeding on lyme grass, above dense mats of blue-green algae grazed by gastropods (inset); E: *Zostera marina* and arrowroot (*Trichlochin maritima*) in a subtidal channel (photos by P. J. Mudie). See also colour plates section.

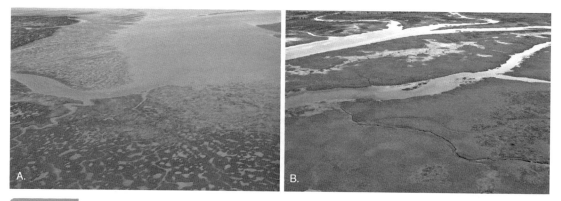

Figure 6.3 A: salt marsh vegetation of the Russian Arctic tundra estuaries at Baydaratskaya Bay (from Morozova and Ektova, 2013); B: *Suaeda* and *Spergularia* in Alaskan marshes have reddish leaves; also note the large driftwood logs (photos by P. J. Mudie). See also colour plates section.

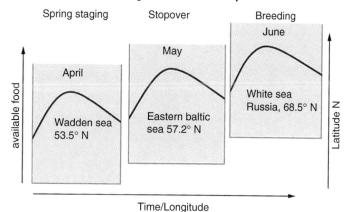

Barnacle goose: *Branta leucopsis*

Figure 6.4 Barnacle geese surf green waves of shifts in peaks of plant food availability on their spring migration from Europe to Russia (data from van der Graaf *et al.*, 2006).

freeze–thaw processes, with summer months of ice-rafting, very strong winds, and in recent years, high exposure to UV-B from the Antarctic ozone layer (Bianciotto *et al.*, 2003).

The most widespread wetland types are the temperate salt marshes dominated by halophytic grasses and succulent herbs. These marshes generally occur between latitudes 50° and 35° south or north of the subpolar regions. Both plant and animal life is more diverse than the polar salt marshes (Figure 6.1B). Subregional differences are largely related to the soil salinity regimes which vary with net moisture (i.e. precipitation minus evaporation). All the marshes, however, have large proportions of grasses (graminoids) with sturdy soil-binding rhizomes and leaves with salt glands that eliminate excess salt, while succulents store salt in stems or leaves during summer and shed this salt as dead tissue during winter dieback. Other plants in temperate marshes avoid salinity, either by growing rapidly during the rainy season (e.g. annuals with shallow roots) or by growing on the inland edge of the supratidal marsh where deep root systems can reach low-salinity subterranean water sources. Subtidal areas support several genera of the seagrasses *Zostera* and *Phyllodoce*; intertidal ponds contain *Ruppia* or *Hippuris*.

The floras and faunas of subtropical coastal wetlands (Figure 6.1C) overlap those of the adjoining temperate and tropical regions. The floral and faunal diversity therefore is often higher in these warm, seasonally wet coastal environments. However, salt marshes may be completely replaced by mangrove shrubs on their upland borders, while low marsh grasses may be shaded out by tall mangrove trees. The mangrove trees and shrubs have adaptations of root and salt excretion systems that parallel those of the salt marsh plants shown in Figures 3.4 and 3.5, and the main types are detailed in Chapters 8 and 9. In addition, mangrove and palm trees have reproductive systems adapted to survival and flotation in seawater (see Figure 3.6).

The boundary for the tropical coastal wetlands marsh zone varies latitudinally and corresponds to the northern or southern limit of mangrove vegetation, which is typically 25° to 0° north or south of the Equator. The biodiversity of tropical coastal wetlands (Figure 6.1D) varies more than that of the temperate regions. For example, mangrove diversity is greatest in Southeast Asia, where 268 species are found (Giesen *et al.*, 2007), whereas the American continents have only about 20 species. As in the subtropics, the tropical salt marshes consist of a relatively restricted intertidal flora with many of the temperate region salt marsh genera being confined to a narrow zone around the extreme high tide or storm tide level (the upper few decimeters of the tidal range) if they are present at all. Primary productivity, mostly from the mangrove trees, is very high because the high insolation and year-round warmth and humidity conditions potentially allow for active plant production during all seasons.

6.2 Biogeographic variation

The Ramsar Wetlands of International Importance (Ramsar Sites) are grouped into global administrative regions which correspond closely to seven of the eight widely recognized biogeographic realms – Australasian, Antarctic, Afrotropical, Indo-Malayan, Nearctic, Neotropical, Oceania and Palearctic (Figure 6.5). No marshes are known for Antarctica, although this continent has raised storm beaches with marine shells, seal skin and elephant seal remains dated as about 6400 ^{14}C years BP (Hall and Denton, 1999). Biogeographic realms are the fundamental units for comparative global ecological studies and they are an essential tool in conservation planning (Holt *et al.*, 2013). Recently, 11 global zoogeographic realms have been recognized by slightly different criteria that delineate regions of highest evolutionary turnover in terrestrial animals – but these are not easily related to wetland systems, so are not used here. The seven 'Ramsar realms' broadly reflect the long-term geological evolution (plate tectonics) of continents and oceans from the Cretaceous ('dinosaur' time) ~60 million years ago to the closing of the Mediterranean Tethys Seaway between the Indo-Pacific and Atlantic oceans (Figure 6.6), which started about 35 million

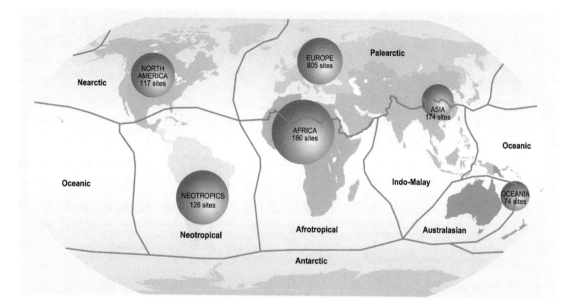

Figure 6.5 Distribution and biogeographic realms of wetlands listed as Internationally Important Ramsar Sites (from MEA, 2005). The amounts shown include all Ramsar wetland types; circle sizes range from 8–37 million ha, from smallest to largest (estimated in 2005). Only about 1–10% of the Ramsar temperate region wetlands are tidal salt marshes and about 25% of the tropical-subtropical sites are mangrove wetlands. See also colour plates section.

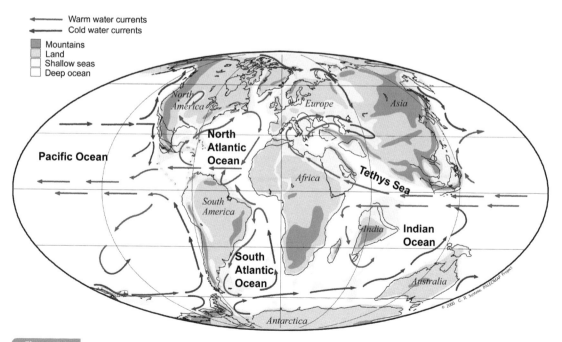

Figure 6.6 Late Cretaceous (66 Ma) ocean circulation allowed interocean migration of water-borne mangrove propagules, but it separated the northern and southern continents, which therefore developed different species diversity and regional adaptations (adapted from C. Scotese, PALEOMAP project). See also colour plates section.

Table 6.2 Occurrences of coastal wetland geobotanical units as products of Pleistocene–Recent tectonic settings and associated landforms, drainage, sediments and coastal geomorphology (modified from Lara *et al.*, 2009).

Driving process	Convergent/collision margin	Passive/trailing margin
Tectonics	Active, earthquakes, volcanoes	Stable, earthquakes rare
Landforms	Mountains, narrow continental shelf, deep valleys/canyons	Coastal aggradation plains, wide continental shelf, wide winding valleys
Weathering	Erosion, mass sediment movement	Chemical, fluvial
Drainage	Mostly short steep streams	Long meandering rivers
Sediment volume	Low (except near glaciers)	High
Sediment texture	Fine to coarse	Fine or diamictic (mixed with pebbles and boulders in glaciated regions)
Coastal morphology	Rocky, few large sandy beaches	Large delta and sandy barrier beaches
Wave attenuation	Low	Moderate to high
Geobotanical Unit		
Salt marsh	Restricted occurrence and mostly small size (rarely large, on wave- protected deltas)	Common, often extensive
Mangrove + marsh	Scarce, in isolated estuaries or barrier reefs	Widespread and extensive, especially on large deltas

years ago (Chapman, 1977). These biogeographical realms are much larger global-scale units than the geobotanical units discussed by Lara *et al.* (2009), who relate the local structure of coastal salt marshes and mangroves to their present-day tectonic settings (see Table 6.2).

The present-day biogeographic realms were established after the final closing of the Tethys Sea (Figure 6.6) and the Paratethys Seaway (from the Mediterranean to North Atlantic Ocean) about 5 million years ago. There is ongoing debate over Chapman's (1977) idea that the mangroves evolved in the Indo-West Pacific region (Indo-Malay+Oceania realms), where their diversity is highest, and that they migrated westwards, so fewer species evolved in the Atlantic Afrotropical and Neotropical regions. The alternative idea of Ellison *et al.* (1999) is based on studies of fossil mangrove pollen, leaves and flowers and suggests that: (1) mangroves and their associated gastropod faunas were actually spread throughout the entire shoreline region of the Tethys Sea and (2) the modern species evolved separately in the different regions after the closure of the Tethys and Paratethys Seas. A few million years later, the Pleistocene glaciations began when the permanent Greenland Icesheet formed, followed by fluctuating continent-wide icesheets over Siberia and Canada. The icesheets stopped the migration of temperate region biota between the Palearctic and Nearctic realms. For example, until 4–5 million years ago, redwood (*Sequoia*) forests grew across the whole Northern Hemisphere; however, these forests died in the Palearctic realm during the earliest glaciations, and today survive only in the unglaciated humid boreal Nearctic region of Northern California.

All present-day global coastal wetlands have developed during the past 20 000 years or less since the end of the last Pleistocene ice age. The apparent similarities of Afrotropical, Indo-Malay and Oceania (West Pacific) mangrove trees, which actually belong to different genera and species, are good examples of the convergent evolution demonstrated by azonal vegetation ecosystems adapted to the same range of extreme soil conditions. In contrast to this pan-global evolutionary divergence in mangrove tree species, Scott *et al.* (1990) show that the foraminifera of Nearctic North America are almost identical to those of subarctic and temperate Neotropical South America faunas, suggesting that evolutionary divergence has been slower in these marine microorganisms than in their associated wetland vegetation habitats.

Classical biogeographical studies of coastal wetlands use a complicated phytosociological method to group marshes into *seres* and *communities* (see Beeftink, 1968 in Chapman, 1977). *Seres* are successions from pioneer plants on new mudflats to mature ('climax') salt marsh or mangrove vegetation. *Communities* are based on the Zurich–Montpellier ecological naming system (Braun-Blanquet, 1928), which sorts vegetation into classes, orders and alliances according to: (1) life form 'class', e.g. subtidal aquatic; mudflat colonizer; (2) dominant species (orders) and (3) groups of species within the same ecological niche (alliances and associations). This complex system is still commonly used by European ecologists, so an example is shown here (Box 6.1 Phytosociological groups of Tasmania; from Tannheiser and Haacks, 2004). However, this method can obscure the linkages between coastal wetlands and climate zones because it focusses on succession rather than geography. The phytosociological method is therefore being replaced by statistical methods, e.g. average linkage clustering, which better reveals links to macroclimates and biogeographic provinces (Peinado *et al.*, 2009).

Data from the North American continent will be used to demonstrate the value of phytogeographic studies for in-depth understanding of linkages between coastal wetland distributions and latitude, which can vary from one side of a continent to the other. The Ramsar Nearctic realm (North America) is chosen for demonstration of phytogeographic studies because most of its coastal wetlands have been studied in detail and data are easily accessible; however, the method can be used for other continents or terrestrial biogeographical realms worldwide. Particular attention is given to the Pacific Coast, which extends over >8000 km from the tropics to the Arctic (Figure 6.7).

Studies of the latitudinal distribution of salt marsh molluscs from ~50 to 28° N (Macdonald, 1969) show that there is a strong correlation between the boundaries of molluscan provinces and the Californian and Oregonian phytogeographic provinces. In contrast, high marsh foraminiferan faunas (Scott *et al.*, 1996) do not vary notably, except for a switch from brackish to more saline marshes at ~40° N, near the Californian–Oregonian border. Macdonald and Barbour (1974) reported a gradual change in salt marsh species with latitude from 70 to 23° N, with least difference apparent among the low marsh floras, but larger differences in the middle–high marsh floral provinces of Alaskan, Oregonian, Californian and tropical floristic zones. Updated studies of the salt marshes (Peinado *et al.*, 2007, 2009; Gonzales-Zamorano *et al.*, 2011; Mudie *et al.*, 2014) reveal a closer correspondence

Phytosociological classification of coastal wetland according to Thannheiser and Haachs (2004)

Tasmania is in the Australasian biogeographical zone of the southwestern Pacific Ocean, far removed from North America. The coastal wetlands of Tasmania still have almost 80% of their native flora, compared to New Zealand, where over 50% of the marsh species are alien introductions. The profile for the temperate climate Blackmans Bay marsh (42.99° S, 147.32° E) shows a succession of vegetation communities, as classified by the phytosociological method, from (1) subtidal seagrass growth form, through 'meadows' of low-growing marsh perennial succulents (2 and 3), to high marsh with shrubs (4), sedge (6) and rush communities (7). A similar low-middle marsh succession is seen in Southern California of the Nearctic realm. However, in Tasmania, the high marsh includes a low-growing succulent called 'bonking grass' (*Selliera radicans*), which belongs to an endemic genus found only in Australia, within the almost endemic family Goodeniaceae. All the main *genera* below MHHW and at EHW also occur in North America, but the most Tasmanian marsh *species* are uniquely Australasian (*Zostera muelleri, Sarcocornia quinqueflora, Suaeda australis*), while *Juncus kraussii* also occurs in South Africa and South America. In the high marsh, however, both genera and species are Australian endemics, including the succulent shrub *Sclerostegia arbuscula* (Chenopodiaceae) and the grass *Gahnia filum*.

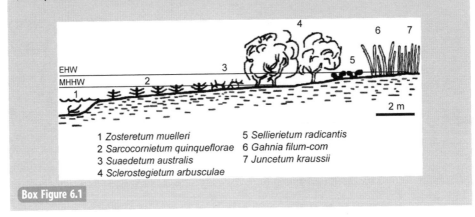

1 *Zosteretum muelleri*   5 *Sellierietum radicantis*
2 *Sarcocornietum quinqueflorae* 6 *Gahnia filum-com*
3 *Suaedetum australis*   7 *Juncetum kraussii*
4 *Sclerostegietum arbusculae*

Box Figure 6.1

between the wetland floras and regional paleofloristic boundaries (Figure 6.7), despite the recent origin of these wetlands and the limited gene exchange because of marsh separation by steep mountain ridges. Even during times of lower sea level, when the continental shelf may have been a migration route, the presence of deep canyons would have limited marsh expansion (Callaway and Zedler, 2009).

6.3 Subregional salinity variations: effects on plant assemblages

Graphs of latitudinal changes in temperature, net moisture and salt marsh soil salinity for the west and east coasts of North America (Figure 6.8) show the large differences in zonal climatic regimes and the contrasts in high marsh soil salinity regimes of wetlands on the

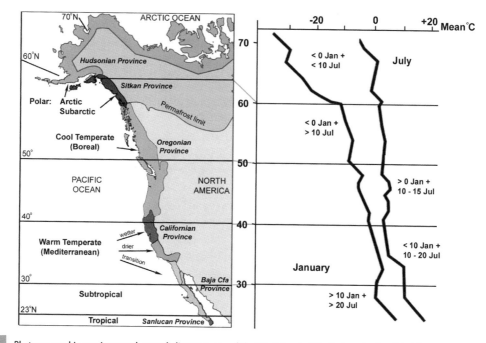

Figure 6.7 Phytogeographic provinces and coastal climate zones of the West Coast of North America (modified from Peinado *et al.*, 2009, with permission).

Pacific and Atlantic shores. Overall, although climatic zones dictate the major types of coastal wetlands found globally, the plant composition in any particular region varies according to daily and annual salinity regimes. The range of salinity tolerance of individual salt marsh species largely influences their zonation (high to low marsh) along the wetland tidal gradient. For example, in North America (Table 6.3), cordgrasses in the genus *Spartina* can grow in a wide range of salinities, but individual species such as *S. cynosuroides*, *S. patens* and *S. alterniflora* and/or *S. foliacea* have much narrower salinity tolerances (see Table 7.4). Succulent coastal *Salicornia* and *Suaeda* spp. cannot survive where the average water salinity is lower than about 15 psu, but can otherwise persist in various soil salinities up to periodic levels of >80 psu.

In the high marsh, there is a larger diversity of plants in both brackish and high-salinity areas. Commonly there are a few species of Asteraceae (daisies) like *Solidago* (golden rod), several grasses like *Puccinellia* spp. (alkali grass) and *Distichlis*, and graminoids such as rushes (*Juncus* spp.), sedges (*Carex* spp.), plantains (*Plantago* spp.) and arrow grass (*Triglochin*) that generally cannot survive exposure to higher than normal ocean salinities (i.e. 35 psu). However, high marsh plants in more arid areas can cope with a wider range of salinities from hyper- to hypo-salinity, depending on the season (e.g. *Salicornia* spp., *Batis maritima* and some mangroves). Clearly, salt marsh zonation within any region is not governed simply by climate or tidal regime but is the net result of a combination of multiple factors, as demonstrated by Silvestri *et al.* (2005) for the salt marshes of Venice Lagoon (Italy). The list of plants in Table 6.3 therefore is a highly simplified guide to complex inter- and subregional variations.

Table 6.3 Main native salt marsh and mangrove plant species in North America, their zonation and salinity (psu)

Region	Plant name	Habitat	Salinity (psu)
Polar region 50–70° N	*Puccinellia, Stellaria, Plantago, Carex, Juncus, Festuca, Triglochin*	Little vertical zonation, low plant diversity	5–32
Nova Scotia to North Carolina 45–35° N	*Spartina cynosuroides, Sarcocornia/ Salicornia, Spartina patens, Spartina alterniflora*	High marsh Middle marsh Low marsh	0–20 20–35 30–33
Southern California to Central Baja California 35–28° N	*Sarcocornia pacifica, Batis maritima, Frankenia, Spartina foliosa*	High marsh Middle marsh Low marsh	30–60 30–35 35–36
Florida and southwards, Southern Baja California 25–23° N	*Rhizophora mangle, Avicennia germinans, Laguncularia racemosa, Spartina cynosuroides* (east) or *S. foliosa* (west) *Sarcocornia, Batis*	Low–mid marsh Low–high marsh High marsh Brackish high marsh	30–35 30–45 20–35 0–20

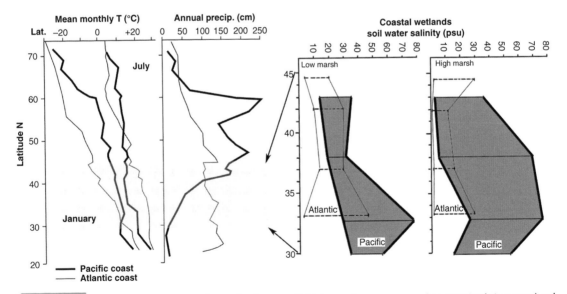

  Figure 6.8 Comparison of mean winter (January) and summer (July) temperatures, mean annual precipitation (rain + snow) and soil salinity ranges in low and high marsh zones of the Pacific and Atlantic coasts of North America.

6.4 What in the world to expect next?

In Chapters 7–11, we present examples of coastal wetlands from different regions of the world: the west and east coasts of North America and South America; five African marsh areas on all coasts, including the Nile River Delta; remnant marshes of Europe and the Iraqi Tigris–Euphrates Delta; the Indian coast, the Ganges, Mekong and Yangtze River Deltas; and the continental region of Australasia (Australia and New Zealand). These examples will be used to demonstrate: (1) the environment–plant–animal interactions characterizing the different global zonobiomes; (2) the biological diversification evident in the different paleogeographic regions of the world and (3) types and degrees of impacts resulting from human alteration of coastal wetlands worldwide.

7

Examples of North American salt marshes and coastal wetlands

Key points

Arctic wetlands are especially susceptible to damage from climate warming impacts, including permafrost melting, erosion and storm surges; subarctic wetlands are shaped by freeze–thaw cycles and overgrazing by waterfowl; West Coast earthquakes, tsunamis and glacier surges are added stressors in subarctic marshes; temperate marshes of the mountainous West Coast are in isolated valleys, sometimes disconnected from tidal exchange; temperate East Coast marshes are extensive, but altered by farming, frequent hurricanes and 'nor'easter' storms; Bay of Fundy megatidal marshes sequester much carbon despite erosion by massive winter ice blocks; subtropical wetlands of Florida and Mexico have shrubby mangroves fringed by brackish water swamp cedars or desert palms and cactus scrub; microtidal Mississippi Delta marshes include floating vegetation islands; tropical mangroves have a high diversity of tall trees and shrubs, particularly in Central America.

7.1 Arctic Coast: Mackenzie Delta region of the Beaufort Sea

The Arctic is warming faster than any other region on Earth, bringing dramatic reductions in sea ice extent, altered weather, and thawing permafrost. Implications of these changes include rapid coastal erosion... and unpredictable impacts on subsistence activities and critical social needs. (Clement *et al.*, 2013, in a report to the President.)

Arctic salt marshes are literally few and far between (Figure 1.1), occupying scattered segments of the 25 000 km-long Russian coastline (Lantuit *et al.*, 2012; Sergienko, 2013), but covering only about 60 km^2 of the Canadian Arctic Islands and small parts of the Arctic shores of Canada and Alaska (Mendelssohn and McKee, 2000; Martini *et al.*, 2009; Jorgenson, 2011). However, there is a growing interest in their ecology and geological history because of their increased contributions to atmospheric greenhouse gases as the permafrost melts (see Section 5.4), and because they occupy deltas and estuaries where gas and oil exploration and shipping terminals are rapidly changing the landscape. Types of possible impacts are outlined in Chapter 5, but here we present selected examples to demonstrate the magnitude of the unfolding events. Jorgenson (2011) illustrates a wide range of impacts to the Arctic Alaskan marsh landscape – from melting of ice by dust accumulations (which absorb more heat than clean ice) to soil upheaval associated with pipeline construction. Offshore, over 150 exploratory hydrocarbon wells have been drilled since early 1970, including 19 in the Kendall Island Bird Sanctuary of the Mackenzie Delta, which is the second largest delta in North America (Johnstone and Kokelj, 2008). Drilling in the Beaufort Sea includes many hazards not found outside the Arctic, such as the unpredictable and abrupt emergence of pingo-like features (PLFs) which are 12−40 m-high cones of ice-bonded sediments propelled upwards by methane gas released when seawater becomes warmer (Box 7.1 Vanishing Arctic villages). The danger of oil spills further increases as oil

Box 7.1   **Vanishing Arctic villages and sea ice; pop-up pingos**

Box Figure 7.1A and B   A. Courtesy of Department of Municipal and Community Affairs, Government of Northwest Territories.

Tuktoyaktuk (69.45° N, 133.03° W) is the northernmost community in mainland Canada. The hamlet is located around a microtidal thermokarst lagoon ('thaw lake') on the shore of the Beaufort Sea (A above, from Manson *et al.*, 2005) and it was inhabited for thousands of years by native caribou and whale hunters before becoming a port for the Hudson Bay Trading Company in 1937, and then a supply station for the offshore oil and gas industry. Since the 1970s, however, this community has been threatened by shoreline erosion because of permafrost melting and large waves in the summer open water which erode the shore and flood the salt marshes despite the armour (B above) of sand bags, rip-rap and concrete slabs. On the Mackenzie Delta, permafrost melt and waves undercut cliffs causing massive slumping and terracing (C below); offshore dangers include gas blow-outs and more than 1350 PLFs, each abruptly rising above the seabed when icy mud is squeezed up by gas from melting subsurface hydrates. Each PLF is surrounded by a moat (M) containing marine sediment and has a vent (VG) for methane gas release (D below, from Paull *et al.*, 2007). PLFs are a potential hazard to drill platforms or small boats and add to greenhouse gases that drive global warming.

Box Figure 7.1C and D

and gas exploration and drilling expands into deeper water, following the >50% reduction in summer sea ice since 1978, which now also allows summer travel of oil tankers through the hazardous Northwest Passage.

Under changing climate conditions, Arctic coasts will likely become one of the most impacted environments on Earth, despite their low human population density. In Canada, the western Arctic shorelines are among the highest risk areas for negative impacts of future climate warming and sea level rise (Solomon *et al.*, 2005; IPCC, 2007; Vermaire *et al.*, 2013). The Beaufort Sea coast is the most rapidly changing shoreline in Canada. Here many salt marshes are flanked by extensive barrier lagoons (Forbes and Hansom, 2011), but coastal erosion can remove up to 20 m of sediment in a single storm. Storm waves are eroding the Inuvialiut hamlet of Tuktoyaktuk (Box 7.1 Vanishing Arctic villages), encroaching on houses and infrastructure, and extensively flooding low-lying wetland, including the Anderson River wildlife refuge and Kendall Island preserve (Borstad *et al.*, 2008; Pisaric *et al.*, 2011). Pilkey and Young (2009) describe similar problems impacting the Inupiat villages of Kivalina and Sishmaref on the barrier islands of the Kotzebue Peninsula in the Chukchi Sea, Alaska.

The importance of the Mackenzie Delta region for waterfowl (Chapter 4; Box 4.2 Animals of polar and subarctic marshes), involves critical bird habitat both onshore and offshore. Polynya (permanently ice-free areas of water) and leads (openings in sea ice floes) off the Mackenzie Delta are very important habitats for sea ducks during spring migration. In summer, the shoreline water provides key habitat for >100 000 pre-moulting and moulting ducks. Oil spills could seriously reduce or eliminate some of these congregating birds and have a lasting impact on the entire population of migratory birds in the region (Hines and Wiebe Robertson, 2006).

Detailed studies were made of the history and climatic resilience of salt marshes in the Mackenzie Delta (Figure 7.1) because exploration drilling in this megadelta is damaging the slow-growing plants, and because the rapid shoreline erosion and flooding by wave surges is associated with the increase of ice-free summer water. The salt marsh study was made on Richards Island near 70° N, about 300 km north of the Arctic Circle. The tidal range in the delta is less than 50 cm and storm tides normally reach to only about 2.5 m above MSL (Figure 7.2), but the height can increase to ~8 m during northwesterly wind storms. Because the coastal waters are dominated by runoff from the Mackenzie River (the second largest river in North America), the water covering the salt marshes normally ranges from almost fresh to brackish (5–12 psu). Peak river flooding happens in late May when the vegetation and ponds are still ice-covered, so the 0.5 m-high flood peak just sweeps over the icy surface and is funnelled seawards in the river channels. In contrast, summer open-water storm tides wash saltwater over the marsh; winter freeze-in of salts deposited in the intertidal ponds results in bottom water salinities up to 31 psu (Solomon *et al.*, 2000).

Because of the low salinity of the coastal water, the Low Arctic salt marsh vegetation in outer Mackenzie Delta is more diverse than other parts of the northern Canadian coastline (Glooschenko *et al.*, 1988). In ~1977, these salt marshes were dominated by the turf-forming salt grass *Puccinellia phryganodes* and by sedges – *Carex subspathacea* (brant grass) and its hybrid, *Carex ramenskii* – with two small perennial herbs (*Stellaria humifusa*, *Cochlearia officinalis*) on the mudflats of sandy open coasts (Jefferies, 1977). Six other grasses and sedge species grew in more sheltered inlets, along with mare's tail pondweed

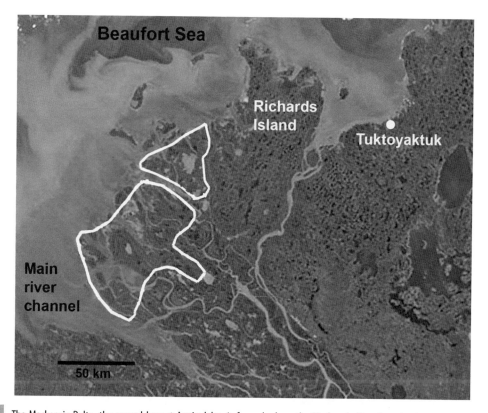

Figure 7.1 The Mackenzie Delta, the second largest Arctic delta, is formed where the Mackenzie River branches into tributaries and flows into the Beaufort Sea as a plume of muddy water that mixes with coastal water, giving it a low salinity in summer. Richards Island, the largest delta island, has salt marsh along the coast; inland is tundra marsh chequered by hundreds of thermokarst ponds. The thick white lines delimit marsh damaged by the 1999 storm surge; the northern polygon includes the Kendall Island Bird Sanctuary (photo by S. M. Solomon, GSCA).

(*Hippuris vulgaris* and *H. tetraphylla*) in ponds behind low beach ridges. In 1991, however, salt marsh transects made on Richards Island (Figure 7.2) showed a greater diversity of high marsh species, including beach grass (*Elymus arenaria*), rush (*Juncus* sp.) and the herbs *Potentilla* and *Cerastium*. Some arctic buttercup (*Ranunculus pallasii*) was also found with the grass in the low marsh. It is not known if these species have spread into the delta because of the warming climate or the increased presence of people from southern regions, but the recent increase in biodiversity deserves further monitoring. Drill sites on the delta are marked by sump ponds which are dug when drilling starts, then gradually filled with the drilling residue and capped with sediment when drilling ends. Domestic grass seed and nitrogen fertilizer are then applied to the bare cap soil to stimulate vegetation growth, and it is assumed that the sumps will be stabilized by regrowth of the permafrost. In fact, 30 years later, many sump caps either remain bare because of surface salt accumulation, or when the permafrost does not form, they collapse, becoming saline ponds with bare mud margins (Johnstone and Kokelj, 2008). The materials used for cap seeding possibly explain the apparent increase in high marsh plant diversity on Richards Island since 1977.

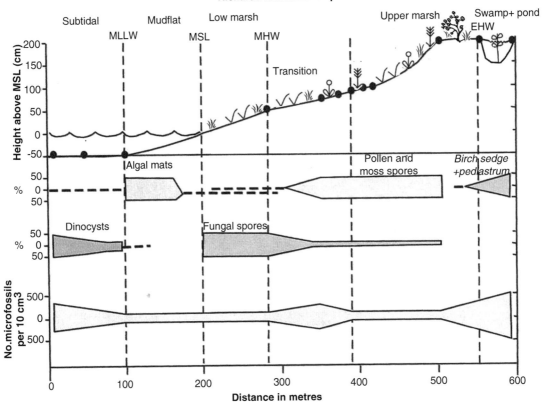

Figure 7.2 The Mackenzie Delta salt marsh and tidal zone, showing the distribution of salt marsh vegetation, pollen, algae, fungal spores and foraminiferal markers that distinguish different zones and can be traced in the sediment records.

The history of the Richards Island salt marshes was investigated by seismic surveys (Figure 7.3) and coring of several thermokarst lakes that mark areas of buried ice which thawed out after the last glaciation (Solomon *et al.*, 2000). Using pollen, algal spores and microfossils from surface samples along a salt marsh transect (Figure 7.2), fossils in the sediment cores were analysed and interpreted in terms of changes in salinity with time. Sediment ages were determined by radiocarbon dating of peat samples and by measurement of the atomic-bomb radionuclide ^{137}Cs (Section 12.3). The study shows that the outer Richards Island ponds contain 10–20 m-thick sequences of organic-rich mud deposited slowly over thousands of years in shallow seawater. During the past 50 years, however, the amount of sediment deposited from bigger river floods and/or more frequent coastal storms has greatly increased. Recently, Pisaric *et al.* (2011) made a higher resolution study of storm surges over the past 80–1000 years as recorded by tree rings in alder shrubs on the flood plain and by brackish water diatoms in sediments of ponds near Kendall Island. The tree rings show that an exceptionally large storm surge in 1999 immediately killed over 50% of the alders, and another ~40% of the shrubs died within the next five years (Vermaire *et al.*, 2013). After 13 years, much of the delta soil at and above EHW still has high sea salt

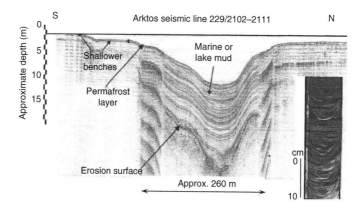

Figure 7.3 Acoustic profile of ice-bonded permafrost layered sediments filling a now submerged delta pond with a 15 m-thick sequence of brackish to marine sediments deposited over the eroded Kittigazuit bedrock sand during the past ~6000 years (Solomon *et al.*, 2000). The inset is a close-up of a section from a ~80 m-long drillhole (Unipkat 1) showing details of permafrost sediments frozen in time since the last glacial stage more than 10 000 years ago.

concentrations. The diatom records show that this ecological disaster is the largest in the past millennium, and it is predicted that the Mackenzie Delta ecosystems are on a new trajectory towards salt marsh expansion that is being driven by the warming climate, longer ice-free season and increased frequency of megastorm surges. Jenner and Hill (1998) also show that despite the high river sediment load of the Mackenzie River, shoreline erosion at the mouth of the main channel exceeds delta growth. They point out that levees of Arctic delta tributaries are low and flooding by seawater during summer storms is an important source of sediment supply to the wetlands.

7.2 Subarctic salt marshes: West versus East Coast

Geographically, the Arctic Circle at 66.3° N is the latitude that separates the polar Arctic region from the subarctic (subpolar) region. However, as shown in Chapter 6, salt marsh ecosystems tend to follow the temperature gradients established by the coastal currents of the major oceans rather than strict latitudinal delineations. In North America, the West Alaskan Current moves relatively warm water northwards in the Bering Sea, along the west coast of Alaska. Conversely, the Baffin and Labrador Currents transport very cold Arctic Ocean overflow water southwards on the East Coast of North America. As a result, the subarctic marshes on the West Coast extend almost to the Arctic Circle, whereas on the East Coast they lie much further south, ranging from about 60° N to the boreal region around 50° N. Martini *et al.* (2009) group together the salt marshes of all the arctic and subarctic regions, referring to them as polar coastal wetlands. In this book, we chose to keep the categories separate because of the difficulty of comparing the conditions encountered in High Arctic salt marshes of the cold desert tundra on Ellesmere Island (<10 °C in the warmest month, <250 mm/yr precipitation) with the forest-bordered subarctic salt marshes in southern Alaska and James Bay (July mean 12 °C and ~1360 mm/yr at Juneau).

The west and east coasts of North America also differ in their tectonic settings. Many subarctic marshes of southern Alaska are part of the relatively young (60 million years or less) tectonically active 'Pacific Ring of Fire' (see Chapter 12), where earthquakes, tsunamis and volcanic eruptions are common catastrophic events. Also, high mountains border most of this coastline, with glaciers extending to the coastal plains that deposit vast amounts of fine silt (glacial rock-flour) on the shores and periodically surge over salt marshes. In contrast, the East Coast is in a tectonically stable region of ancient Precambrian Shield Rock, including the oldest rocks on Earth (>3 billion years old). Here large earthquakes are extremely rare and earth movement is mostly a slow rise or fall in response to crustal adjustments (= isostatic rebound) following the final melting of the Laurentide Icesheet about 8000 years ago. There are no glaciers near salt marshes in the eastern region.

7.2.1 West Coast: Cook Inlet on the 'Pacific Ring of Fire'

In western North America, subarctic salt marshes have a discontinuous distribution in large fiords, such as Kotzbue and Norton Sound, and they cover the Yukon-Kuskokwim Delta which is the largest delta in western North America and one of the most important waterbird areas on the continent. There is a series of tidal and brackish water marshes in Cook Inlet west of Anchorage, Alaska (Figure 7.4). The Cook Inlet is another region near the Arctic Circle where exploration for gas and petroleum is occurring, not far from the infamous Exxon Valdez oil spill disaster in 1989, from which the biology of the region has not yet fully recovered (Section 5.7). Additionally, this is an area of recurrent earthquakes, including one of the largest of the twentieth century (see Section 12.1). The Cook Inlet has semi-diurnal megatides with a range of up to 11 m, summer water temperatures averaging ~13 °C, and salinity in the summer meltwater season of 5–15 psu, rising to 17−22 psu during the winter (Vince and Snow, 1984). Today the salt marshes of the inner Cook Inlet include eight vegetation zones, ranging from outer mudflats with a lawn-like cover of the short salt grasses *Puccinellia phryganoides* and *P. nutkaensis*, arrow grass (*Triglochin maritima*) and the salt marsh cinquefoil, *Potentilla egedii* (=*Argentina egedii*), to high marsh borders dominated by the sedge, *Carex lyngbyaei* in waterlogged areas or in drier areas, by sedge spp., arrow grass, cinquefoil and the daisy *Chrysanthemum arcticum*. The storm tide zone includes the tall beach grass, and other shorter grasses together with the invasive European introduction *Ligusticum scoticum* and a few common beach weeds. Another sedge *Carex ramenskii* dominates the middle marsh, along with salt marsh plantain (*Plantago maritima*) and beach milkweed, *Glaux maritima*. This beach species and the salt marsh halophytes *Salicornia europaea* and *Suaeda depressa* reach their northernmost distributions at this latitude.

Knowledge of this West Coast subarctic shoreline plant zonation and its associated salinity provide the information needed to interpret the geological impacts of catastrophic earthquake events in this region, as explained in Section 12.1. Recently, detailed studies have also begun to monitor changes in the salt marsh vegetation of the Katmai National Park

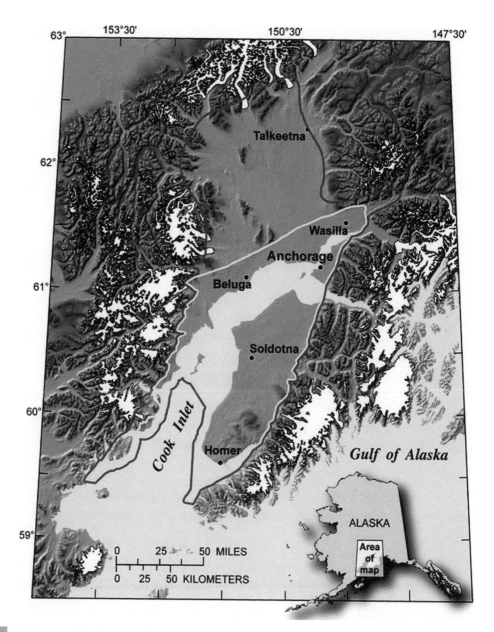

Figure 7.4   Northeastern Alaska showing Cook Inlet southwest of Anchorage. Girdwood marsh on Turnabout Arm (the fiord southeast of Anchorage). This area is targeted for oil and gas exploration – white line encircles the area of conventional gas and oil potential; dark grey line outlines areas of possible coalbed gas (Stanley *et al.*, 2011, USGS).

and Preserve on the volcanically active Alaskan Peninsula in Southwest Alaska (Jorgenson *et al.*, 2009), using a combination of marsh transects and remote sensing imagery. Here Ramenski's sedge dominates high marshes and is an important source of food for omnivorous brown bears that graze on the marshes, in addition to eating crabs and sea urchins.

7.2.2 West Coast: impacts of avalanches and glacial surges in Southern Alaska

The relatively warm Alaskan Current flowing northwards from British Columbia keeps the coast of Southeastern Alaska (a.k.a. the Alaskan Panhandle) warmer than northern Alaska. This subregion has many large rivers with extensive deltas fronted by tidal mudflats, such as the Copper River and Alsek River deltas. This coast also has many fiords with retreating glaciers that have deposited large volumes of fine-grained glacial sediment along the shorelines (Figure 7.5). These glacial sediments form extensive mudflats at the heads of the ice-scoured fiords where the glaciers have now moved far back from the shoreline, e.g. at Skagway and southern Glacier Bay and Taylor Bay (Figure 7.6). Where the salinity is above ~21 psu, these mudflats are colonized by the succulent perennial halophyte locally called beach asparagus (*Sarcocornia perennis*), and they grade upwards into high marshes with grasses and sedges. In the deltas of the large rivers, salinity is much lower (~5 psu); here sedges and rushes are the dominant colonizers and true (obligate) halophytes are not present. Mapping from 2004−2010 indicates that brackish and saltwater marshes make up 10% of the total Alaskan shoreline from the southern Alaskan Peninsula to British Columbia, but are more common (up to 46%) in the southern sector. Two kinds of seagrasses (*Zostera* and *Posidonia*) also grow along 20% of the southeastern coast (TNC, 2011).

Although this segment of the Alaskan coast is not directly part of the volcanic 'Pacific Ring of Fire', here the Pacific Plate is thrusting up against North America and sliding northwards in a plate collision zone. This movement can easily trigger mudslides or snow avalanches at the base of retreating glaciers. Some mudslides are so large that they

Figure 7.5   Alaskan marshes north of Glacier Bay where many Alaskan coastal wetlands are at the entrances of inlets with glaciers calving into the headwaters. Left: Calving glaciers and debris-laden ice at Lake Alsek near the head of Dry Bay. Right: Aerial view of the entrance to Lituya Bay (50 km south of Dry Bay), with extensive sedge marsh inside the beach barrier and a large plume of silt marking the outflow of glacial sediment to the Pacific Ocean (photos by P. J. Mudie). See also colour plates section.

Figure 7.6 Some Alaskan marshes are heavily grazed by migratory geese and brown bears, as evident in large areas of sparse, close-cropped vegetation. A: at Taylor Bay, sparsely vegetated marsh colonizes grey mudflats in front of the advancing Brady Glacier (centre white band); tree-covered boulders between marsh and the ice margin mark moraine deposits left when the glacier retreated more than 450 years ago; B: the marsh has grazed arrow grass (*Triglochin*, by arrow head) and bare ground between cropped tufts of *Puccinellia nutkaensis* alkali grass, where large goose droppings (black) and white feathers provide nutrients to the inorganic glacier rock-flour sediment (photos by P. J. Mudie).

create tsunami waves up to 50 m high (see Chapter 12). Surging glaciers form temporary ice dams across tidal fiords, thereby drowning salt marshes and impounding freshwater lakes behind the ice barrier. At Yakutat, surging of Hubbard Glacier in 1986 created a freshwater lake about 28 m deep from June until October. The ice dam then burst, releasing a torrent of freshwater that churned up the tidal fiord and bay sediments, tossing deep-water shrimp onto the land (Connor and O'Haire, 1988). A second surge and outburst flood event occurred in June to mid-August, 2002 (Willems *et al.*, 2011). By 2006, sparse salt marshes had recolonized the bouldery sediments at Yakutat, but the vegetation is low in diversity compared to those in undisrupted areas (Figure 7.7). Grazing by white-tail deer, however, may also contribute to the low diversity of the Yakutat marsh.

In this region of complex mountain topography, glaciers are both retreating and advancing over time and space (Capps *et al.*, 2011), and it is very difficult to unravel the long-term history of the coastal wetlands from sediment core archives. On a worldwide basis, the melting of Alaskan glaciers provides the largest measured glacial contribution to recent global sea level rise. Locally, however, the shorelines from Glacier Bay to Juneau are rising in some areas and subsiding in others. Overall, the coast of Southeastern Alaska from Glacier Bay northwards is presently the world's fastest region of glacio-isostatic uplift (Larsen *et al.*, 2005), and there are many raised tidal flats and beach ridges around Yakutat (P. J. Mudie, personal observation, 2006). In Glacier Bay, salt marshes have colonized large areas of mudflats and moraine sediments from the fiord entrance at Icy Strait to Berg Bay (mid-fiord) since the retreat of a 1.5 km-high glacier that filled the fiord at

Figure 7.7   Contrasting marsh vegetation downstream of a surging glacier and a retreating glacier. A: Disenchantment
Bay salt marsh near Yakutat, four years after the last surge of Hubbard Glacier, grazed by white-tail deer;
B: gravelly sediment and sparse marsh vegetation (salt grass, plantain and spurrey) at Disenchantment Bay; C: Eagle
Beach marsh seaward of the retreating Eagle–Mendenhall Glacier system, near Juneau; D: dense vegetation
dominated by beach asparagus and beach milkweed in a 'floating' salt marsh on quick-sand at Eagle Beach (photos by
P. J. Mudie).

the end of the Little Ice Age (LIA) about 250 years ago (Connor *et al.*, 2009). In contrast, at
Eagle Creek marsh near Juneau (Figure 7.7B), the mudflat and salt marsh were about 7 m
above present MLLW at the end of the LIA (Connor and O'Haire, 1988), but today, this high
marsh is about 2 m lower.

The dynamic nature of the Southeastern Alaskan coastline and limited accessibility of
this sparsely inhabited region means that relatively little is known about the zonation and
composition of the Southeast Alaskan and British Columbian salt marshes, and important
discoveries about the subarctic salt marsh plant distributions are still being made (Mudie
et al., 2014). The often-cited work of Macdonald and Barbour (1974) gives a generalized
list of salt marsh species for just two sites in each of Southeastern Alaska and British
Columbia. In contrast, Stone (1993) describes four different types of salt marshes within
the Juneau area of Alaska alone, and she shows how the vegetation and marsh zonations
vary with shoreline physiography, including sediment texture, sediment source, wave
exposure and water salinity. Thilenius (1990) describes change in the Copper River Delta

Figure 7.8 From Glacier Bay to Juneau, Alaska, shorelines and salt marshes are rapidly either uplifting or subsiding, and salt marshes are forming on exposed mudflats. A: at outer Glacier Bay, the shoreline is rising at a rate of 3.2–7.5 cm/yr after a mid-Holocene period of submergence that killed the spruce forest, seen here as large stumps in the high marsh of the rocky intertidal area, below the driftline beach grass zone; B: at Dyea near Skagway, rows of rotting wharf pilings in the high marsh are all that remains of a mile-long pier built above the mudflats in 1898, at the start of the Chilkoot Trail goldrush; here the Alaskan coast is rising at a rate of about 2.5 cm/yr (photos by P. J. Mudie).

wetlands decades after the 1964 Alaskan earthquake raised the shore by ~2–3.5 m. Despite the initially high salinity of the raised intertidal sediments, the marshes of the new bar-built estuarine lagoon are covered by brackish water sedges, and halophytes are absent, probably because the high rainfall (~1 m per year) and large river runoff volume. Recent surveys made by Boggs *et al.* (2007) and unpublished reports by P. Mudie and J. H. Dickson (2002–2005) show that sedges and grasses dominate the coastal wetlands in estuaries and bays north of Glacier Bay (Figure 7.8), while *Sarcocornia*-dominated marsh with other succulent halophytes (*Salicornia, Suaeda, Atriplex*) grows in and around Glacier Bay, and dominates many estuaries and bays southwards to the Fraser Delta, near the BC/USA border.

7.2.3 East Coast: Hudson Plains and James Bay

The Hudson Plains encompass all the coastal areas of James Bay and southern Hudson Bay and its wetlands cover, ~4% of Canada's total area (Government of Canada, 1996). The coastal wetlands are intensively used by migratory birds (Jefferies 1977; Jefferies *et al.*, 1979; 1992; Box 4.2 Animals of polar and subarctic marshes). The James Bay region of northern Quebec (Figure 1.1 and Figure 7.9A) is a remote wilderness of taiga/boreal forest (the Hudson Plains Ecozone) bordering southern Hudson Bay. This ecozone includes the Polar Bear Provincial Park, which is receiving increasing use by the marine bears because of the shrinking sea ice. In this region, the glacial ice of the Laurentide Icesheet was over 1-km thick during the last glaciation, depressing the Hudson Bay area by 100–300 m (Stewart and

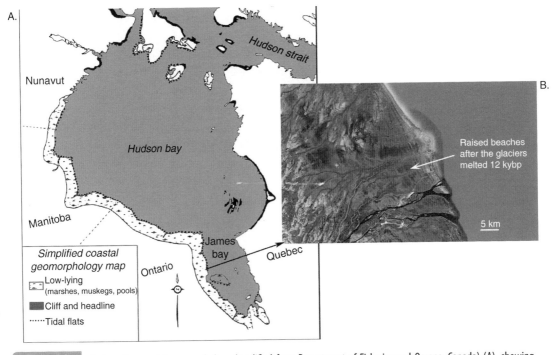

Figure 7.9 Hudson Bay coastal geomorphology (modified from Department of Fisheries and Oceans, Canada) (A), showing location of linear rows of raised beaches around James Bay (B) that support older Holocene salt marshes.

Lockhart, 2005). However, after the icesheet thinned 12 000–8 000 years ago, the land rebounded when the ice load was removed, leading to raised beaches surrounding James Bay, where land is emerging at up to 1 m/yr (Figure 7.9B). Commonly, a new beach ridge forms every 45 years, and some marshes are prograding at a rate of ~300 m over two decades (see Martini *et al.*, 1980). The former ice-age salt marshes are now well above EHW, but mollusc shells and foraminifera in the peat suggest they probably had much the same vegetation as the modern salt marshes.

Today, the climate of this eastern subarctic region is dominated by the conditions over Hudson Bay, which produce cold, moisture-laden, low-pressure systems alternating with polar high-pressure air masses. The cold winters have mean January temperatures near −19 °C, while the cool, short summers have a mean July temperature range of 12–16 °C; total precipitation is 400–700 mm/yr. The tidal range in these salt marshes is ~1.5 to 3 m, depending on the location. The salinity of incoming tides ranges from less than 25 in James Bay to ~28 psu in western Hudson Bay; moreover, within the salt marsh, salinity varies depending on proximity to large rivers.

The subarctic salt marsh vegetation (Table 7.1) is described in detail by Glooschenko *et al.* (1988), who show that it ranges in species diversity from that of the Low Arctic Mackenzie Delta to a richer flora in sheltered areas of James Bay, where there are boreal-type salt marshes. Both arctic and subarctic low marshes are dominated by the salt grass that can reproduce by runner stems (= stolons), even when the tidal movement of sea ice scours

Table 7.1 Zones of principal coastal wetland habitat used by birds in the James Bay region of the Hudson Bay ecosystem (from Martini *et al.*, 1980)

rocky areas	Brackish marsh	Salt marsh		Sand flat		
		Upper	Lower	High tidal flat	Sand flat	Zostera marina zone
Shorebirds						
Semipalmated plover				▬▬▬		
Black-bellied plover		▪▪▪▪▪▪				▬▬▬
Ruddy turnstone			▪▪▪			
Common snipe	▬▬▬▬▬			▪▪▪▪▪▪		
Greater yellowlegs		▬▬▬▬▬		▬▬▬▬		
Lesser yellowlegs		▬▬▬▬▬		▬▬▬▬		
Red knot						▬▬▬
Pectoral sandpiper		▬▬▬▬▬				
White-rumped sandpiper				▪▪▪▪▪		▬▬▬
Dunlin				▪▪▪▪▪▪		▬▬▬
Semipalmated sandpiper				▬▬▬▬	▪▪▪▪▪	▬▬▬
Hudsonian godwit				▪▪▪▪▪▪		▬▬▬
Sanderling					▬▬▬	
Geese						
Canada goose		▪▪▪▪▪				▬▬▬
Brant						
Lesser Snow goose		▪▪▪▪	▪▪▪▪	▬		
Ducks						
Black duck				▪▪▪▪▪▪	▪▪▪▪▪▬▬	
Pintail				▪▪▪▪▪▪▪	▬▬▬▬	
Green-winged teal		▪▪▪▪▪▪▪▪▪				

the marsh surface, cutting the runners into fragments. Ice melt-out creates shallow ponds that are habitats for submerged brackish water plants like ditch grass (*Ruppia maritima*) and horned pondweed (*Zanichellia palustris*), in addition to mare's tail pondweed. In eastern James Bay, eelgrass also grows in the subtidal area. Around 0.5 m above MHW, the salt grass is replaced by a mixture of other grasses (*Puccinellia lucida*, *Festuca rubra* and *Hordeum jubatum*) and perennial herbs, including plantain, arrowgrass and *Potentilla egedii*, with the annual glasswort *Salicornia rubra* occurring in open salt pans. Landward edges of the salt marshes are marked by dense growth of Baltic rush, sedges, the daisy *Senecio congestus* and by hardy dwarf willow shrubs.

The Hudson and James Bay salt marshes are important feeding grounds for millions of migratory waterfowl that breed and nest in the Hudson Plains. Important waterfowl include snow geese (Figure 7.10), Canada geese, king eider ducks and swans; other salt marsh birds are the gyrfalcon and peregrine falcon. Table 7.1 shows how different birds depend on foods from the various zones within the salt marsh. The area is a particularly important breeding area for the lesser snow goose (*Chen caerulescens caerulescens*). At La Pérouse Bay near Churchill in western Hudson Bay, Jefferies and Rockwell (2002) discovered an important co-dependence between the geese and salt marsh plant populations. Normally, the low marsh grasses benefit from nitrogen- and phosphate-rich feces left by the birds foraging

Figure 7.10 There are two colour variants (called morphs) of the lesser snow goose. Left: typical white morph in Chester, Nova Scotia, Canada (photo by Dori Chandler); Right: less common blue morph from Gloucestershire, England (photo by Adrian Pingstone).

on the plants. Recently, however, the bird populations exploded by ~7% per year because the geese are feeding on southern farm crops in winter and along their migration routes. The long-term effect of the goose population increase in the La Pérouse Bay marshes from 1985–1999 is a 98% loss of intertidal salt marsh vegetation because of vegetation grubbing, leaving extensive areas of bare, salty soil with little resistance to erosion by ice and tidal inundation. Martini *et al.* (2009) give further details on the vital importance of these subarctic wetlands as staging areas for waterfowl migrations, and the complex interactions between grazing and survival of the salt marsh plants. Similar interactions occur in the Canadian High Arctic salt marshes of Devon and Somerset Islands (Glooschenko *et al.*, 1988), and new research in Svalbard (Norway) shows that the seabird colonies play a significant role in emissions of the greenhouse gases methane (CH_4) and nitrous oxide (N_2O). Such methane 'hotspots' will likely be enhanced by continued climate warming, and this additional source of greenhouse gases must be factored into global climate change models (Zhu *et al.*, 2012).

The Hudson Plains represent the southernmost extent of the yearly migration of polar bears, which ends on the northwestern coast of James Bay. Normally, polar bears migrate onto shorefast ice off the coast of Hudson Bay to feed on marine mammals in autumn. Recently, however, there has been little or no sea ice and the bears tend to stay near villages where they can forage on domestic animals and human garbage in addition to berries and small arctic mammals. Other large mammals, including moose, caribou and black bears, also live in the relatively mild subarctic environment, together with smaller carnivores: muskrats, weasels, arctic foxes and martens. The adjacent marine ecozone supports walrus and seals (bearded, harbour and ringed). Plant growth patterns largely determine the ability of different species to survive under increased grazing pressure, with perennials reproducing by surface runners and short-lived annual species being at a disadvantage compared to deep-rooted sedges. Although the vegetation has been resilient to large climatic changes since the last glaciation, the order-of-magnitude faster rate of change since ~1980 is causing unprecedented large 'boom–bust' cycles of herbivore populations. A similar conclusion is reached by Mudie *et al.* (2005) based on a 6000-year record of proxy-phytoplankton and archeological data from the Baffin Bay region. With continued rapid global warming, the delicate balance between plant growth and grazing by macrofaunal vertebrates and insects and by meiofaunal soil nematodes will not be just a simple replacement by northward migration of boreal forest vegetation, as often depicted, but will involve variable adjustment responses for each marsh species, depending on their individual growth and reproductive characteristics (Jefferies *et al.*, 1992).

Conversely, in Eastern North America there are very few subarctic salt marshes south of Ungava Bay because the predominantly rocky shoreline, heavy sea-ice cover and cold polar waters do not provide suitable coastal habitats. Forbes (2011) gives a detailed classification of these glaciated coastline environments and some histories of the changes in their salt marshes during the Holocene marine transgression.

7.3 Temperate marshes: West versus East Coast

Both the West Coast (from Alaska to California) and the East Coast (from Newfoundland to South Carolina) of North America have numerous cool to warm temperate marshes (see Glooschenko *et al.*, 1988; Callaway and Zedler, 2009). In the western mountainous region, located on the leading edge of a tectonically active coast, however, the marshes are confined in river valleys separated by high mountain ridges that produce variable precipitation in the discontinuous marshlands. In contrast, on the trailing continental margin of eastern North America, many salt marshes are largely continuous because they occupy large, fairly flat coastal plains that allow expansion of marsh vegetation. In the following sections, we first cover the characteristics of the West Coast temperate wetlands, which have many globally unique species; subsequently we describe selected examples of the temperate East Coast salt marshes, which are quite similar to the well-known salt marshes of Europe and have been described in most other books on coastal wetlands.

7.3.1 Temperate wetlands of Western North America: Alaska to Mexico

In addition to the large between-marsh diversity on the rocky Pacific West Coast of North America, most of the temperate wetlands have highly altered hydrological regimes because of anthropogenic changes (Jacobs *et al.*, 2010). These changes began mainly during the Missionary Period (~1769 to 1823 CE) when about 90% of the native hunter-gathering people were displaced by European settlers who introduced horses, cattle and farming. The land-use changes on the grassy hills profoundly altered the native vegetation and the sediment dynamics of the coastal marshes, as shown in Section 12.2. Today, ~37 million people live within 50 km of the coastline in the state of California and 85% of this coast is now actively eroding from natural and anthropogenic changes (Griggs, 1994). Throughout the region, decadal-scale periodic floods occur during La Niña cool intervals, which alternate with drought conditions during the warm El Niño intervals; in addition, there are massive megafloods every ~200 years. The combination of soil erosion in the watersheds and restriction of coastal lagoon entrances by construction of roads, railroads and urban development led to the infill and periodic closure of many smaller estuaries, particularly in Southern California, as illustrated by the history of Los Peñasquitos Lagoon (Section 7.3.1.5). Other major changes are associated with dredging of the estuaries for use as ports, enclosure of tidal flats for salt production, and industrial or urban infill (San Francisco and San Diego Bays, Section 7.3.1.4). Willapa Bay and Grays Harbor are bar-built bays in Washington State, extensively used for oyster farming (Section 7.3.1.1). Netarts Bay (Section 7.3.1.2) is an example of a small bar-built estuary in the central Oregon coast, north of the largest Oregon marshland located in the drowned estuary at Coos Bay. The estuary-lagoon system at Humboldt Bay, Northern California, is the second largest Californian estuary (Section 7.3.1.3) after San Francisco Bay. The characteristics of these Northwest Pacific tidal wetlands and their vegetation are documented by Seliskar and Gallagher (1983), while those of smaller estuaries and barrier lagoons in the summer dry regions south of San Francisco Bay are outlined by Jacobs *et al.* (2010).

As explained in Chapter 6, the temperate climate region of the Pacific Coast varies from cool moist conditions in the north to semi-desert conditions in the south, with the major transition occurring between Oregon and San Francisco Bay where the boreal province gives way to summer-dry Mediterranean conditions. We therefore present selected case histories in sequence from the north, where salt marshes are very similar to those in British Columbia and subarctic Alaska, to the south, at the Mexican border at the transition from temperate to subtropical coastal wetlands.

7.3.1.1 Willapa Bay, Washington State

Willapa Bay (46.67° N, 124.00° W) is a shallow drowned megaestuary formed when the Long Beach Peninsula sand spit extended northwards from the Columbia River, partly enclosing the estuaries of several smaller rivers (Figure 7.11). Cores from the salt marsh revealed that several major earthquakes have occurred in the last few hundred years (Atwater, 1987; Atwater and Hemphill-Haley, 1997; Atwater *et al.*, 2004). Most of the bay shoreline remains relatively unaltered by development. However, in the past 150 years, much of the bay has been used for oyster farming and is the base of a seafood processing industry that together with Grays Harbor 50 km further north, produces ~9% of the oysters in the USA (Figure 7.12). To manage this industry, the pesticide carbaryl (Sevin) is sprayed over hundreds of acres of oysters to kill native ghost shrimp and mudflat worms that stir up mud and can coat the oysters with silt; this pesticide also kills other marsh crustaceans, amphipods and worms (University of Washington, 2013). Elsewhere, the bay wetlands are managed by governmental Wildlife Services to increase the carrying capacity for Pacific brant geese and to sustain overwintering waterfowl, particularly Canada geese, wigeons and canvasbacks. The marshlands are also important for survival and production of other birds, especially bald eagles, and marsh and wading birds.

The salt marsh at Willapa Bay is beyond the northern growth limit of the West Coast low marsh native California cordgrass, *Spartina foliosa*. Before 1894, the low marsh here

 Figure 7.11 Willapa Bay, Washington state is the second largest US Pacific Coast estuary. Left: oblique aerial view of the bay with the coastal Olympic Mountains just behind the inlet (photo by Sam Beebe); Right: details of the tidal wetlands and brackish marsh at the river mouth, eastern bay (photo by USFWS and Washington State Department of Ecology).

Figure 7.12 Oyster farming in Pacific salt marshes. Left: lines of oyster shell ('seed') are used to provide a firm surface for the seeded oyster larvae ('spat') in Drake's Estero, California, 1986 (photo by P. J. Mudie); Right: in unusually warm El Niño years, closure of oyster farms in Washington State is needed because of outbreaks of toxic red tide algae; in 1995, the Grays Harbor oyster beds were also closed for harvest 10 times from pollution by a sewage treatment plant and pulpmill effluent (photo by Kai Schumann). See also colour plates section.

and at Grays Harbor to the north was colonized by the low-growing succulents *Sarcocornia* and *Salicornia*. The narrow high marsh zone is dominated variably by hairgrass (*Deschampsia cespitosa*), saltgrass (*Distichlis spicata*), silverweed (*Potentilla anserina*), jaumea (*Jaumea carnosa*), gumplant (*Grindelia stricta* subsp. *stricta*) and perennial pickleweed (*Sarcocornia pacifica*, commonly but incorrectly called *Salicornia virginica*), with many herbs and rushes present at low abundance. The eastern and southern shores of Willapa Bay have many estuaries with silt loads that are deposited and then are colonized by diverse brackish marsh plants (*Aster* sp., *Carex lyngbyei*, *Achillea*, *Angelica* spp.) downstream from the discharge points of the rivers. The vegetation composition of these low-salinity native marshes is like that of the Alaskan—BC marshes, but with the addition of the saltgrass, jaumea and gumplant. Similar salt marsh vegetation extends southward in Oregon and along California's north coast to the Suisun Marsh in northern San Francisco Bay.

Since 1894, however, the low marshes of northwestern USA have been invaded by the East Coast introduction *Spartina alterniflora* (smooth cordgrass), which present a major environmental problem to managers of the wildlife refuge salt marshes and mud flats. The non-native smooth cordgrass rapidly invaded the Willapa Bay mudflats, destroying migratory bird, anadromous fish and shellfish habitats, as well as marine organism and salt marsh communities. Efforts are being made to control the invasive *Spartina* by mechanical methods and herbicides. Tracked vehicles used to crush the cordgrass disturb almost flightless marsh birds like the Virginia rail, and soil compression and decay of dead plant leaves result in low oxygen levels in marsh soils. Herbicides like imazapyr and glyphosate have low immediate toxicity to wildlife, but they can move long distances in groundwater and, like DDT, they have a long active persistence in plant and soil food chains.

7.3.1.2 Netarts Bay, Oregon

The coast of Oregon is marked by high coastal headlands interspersed with rolling forested hills, broad floodplains, and sandy beaches and spits sheltering some large lagoons and salt marshes. Erosion-resistant coastal lava rock headlands separate large salt marshes at Coos Bay,

Figure 7.13   Aerial photo of the bar-built lagoon at Netarts Bay, Oregon; a salt marsh core recorded an earthquake sequence dated ~1670 yr BP (see Section 12.1) (photo by Don Best, Best Impressions Photography).

Tillamook Bay and Netarts Bay, about 100 km south of the Columbia River estuary. The Netarts Bay marshes have developed behind a large sand spit enclosing a lagoon that is heavily used for recreation. The mixed mesotidal regime in this region is complicated: it includes both a semi-diurnal tide twice a day and a diurnal component once every 25 hours. The tidal range is ~3.3 m and the average RSL increase for the last ~80 years is 3.1 mm/yr, but higher tides are associated with El Niño events and associated upwelling of coastal water (Bromirski *et al.*, 2012). However, the salt marsh vegetation is similar in composition to that in Willapa Bay, and marsh peat and microfossils in sediment cores here provide important geological archives of earthquake activity and tsunamis on the Cascadia Margin (see Section 12.1).

7.3.1.3 Humboldt Bay and Eel River, Northern California

Located in one of the more tectonically active areas on the Californian coast, there are extensive marshes behind the barrier beaches at Humboldt Bay in Northern California, near Eureka (Figure 7.14). Here the native California cordgrass has its northernmost distribution. It is the primary colonist on the mudflats, while pickleweed and saltgrass dominate the high marsh in place of the more northern beach asparagus and alkali grass (Macdonald, 1977). Additional high marsh plants in the Humboldt Bay marshes are widespread orache/beach spinach *Atriplex patula*, small mouse-ear weed *Myosurus minimus* and salt marsh quillwort *Triglochin maritimum*, and several southwestern endemics – salt marsh bird's beak *Cordylanthus maritimus*, California sea lavender *Limonium californicum* and the parasitic salt marsh dodder *Cuscuta salina*.

Eel River, located a few kilometres below the south end of Humboldt Bay, is the third largest river located entirely in California, but its delta is now extensively farmed and little

Figure 7.14 The marsh at Humboldt Bay–Eel River, just south of the inlet, records a 6.9 earthquake that did extensive damage, but no loss of life, in 2010. Core studies in 2008 (see Section 12.1) showed a seismic event predating the 2010 earthquake (photo by Robert Campbell, US Army Corps of Engineers).

salt marsh remains. This seismically active area experiences almost one 5–6 magnitude earthquake each year (see Section 12.1), and the Eel River marsh near Ferndale contains records of prehistoric and historic earthquakes.

7.3.1.4 San Francisco Bay: transition from cool to warm temperate regions

San Francisco Bay is the largest bay on the West Coast of North America and the second largest in the USA. The bay is a tectonic estuarine system, probably formed by down-warping of the Earth's crust between two crustal faults: the San Andreas and Hayward Faults. During the LGM, the bay was a valley with two large rivers flowing into the Pacific Ocean in a canyon cut through the Golden Gate entrance to San Francisco Bay. As the glaciers melted, the sea level rose 91 m over 4000 years, and the coastal valley was flooded by seawater, forming the present-day bay, within which scattered islands represent the tops of drowned hills of the former valley.

San Francisco Bay is the southernmost of the large, cool temperate region salt marshes on the West Coast. The northern bay wetlands occupy the estuaries of the eastern Sacramento and San Joaquin rivers, which drain ~40% of California, and flow into an inner bay (Suisun Bay) and then are joined by the northern Napa River in San Pablo Bay. This estuarine complex supports large semi-natural saline and brackish water wetlands. Formerly, there were also large salt marshes in southern San Francisco Bay but these are now mostly reclaimed for salt production ponds which are fringed with halophytic vegetation (Figure 7.15). Historically, the bay was an important hunting and shellfish gathering area for Native Americans from ~4000–1500 yr BP, and huge shellmounds testify to the richness of the productive beds of oysters (*Ostrya lurida*) and bent-nose clams (*Macoma nausta*) (Culleton *et al.*, 2009). Today, however, the bay shoreline is occupied by several large cities including San Francisco, Oakland and San Jose. The bay still contains about half of the salt marshes in California (~2350 km^2, including mudflats), but only about 5% of the original marsh remains after being dredged, diked or filled in for commercial or urban development (MacDonald, 1977; Callaway and Zedler, 2009).

San Francisco Bay has experienced some large earthquakes, such as the famous 1906 earthquake of ~8.0 magnitude. However, the San Andreas Fault events are generally not vertical upheavals, but strike-slip faults with north–south lateral movement that does not generate large tsunami waves (see Section 12.1). A 4000-year record of benthic foraminifera, their oxygen and carbon isotopes and their shell chemistry (ratios of Mg:Ca and Sr:Ca) provides a detailed record of oscillating intervals of warmer, drier and colder climates in south San Francisco Bay, and shows that the recent global warming has been accompanied by a strong decline in oxygen availability and increase of pollution indicators, such as the invasive Japanese species *Trochammina hadai* (McGann, 2008). A similar history of salinity fluctuations in the Sacramento Delta brackish marshes is documented by Byrne *et al.* (2001), using pollen, peat, diatoms and carbon isotopes as proxies of salinity change. The Sacramento Delta data suggest that the recent increase in marsh salinity primarily reflects upstream water storage and diversion within the Sacramento–San Joaquin watershed. New studies by Watson and Byrne (2009) show that the relationship between salinity and halophyte growth in these marshes is non-linear, involving complex species interaction,

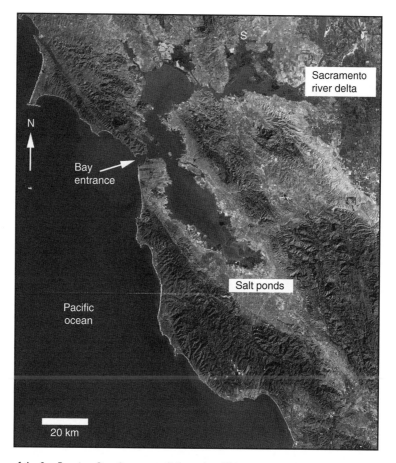

Figure 7.15 Aerial view of the San Francisco Bay–Sacramento Delta region. White area at south end is where salt ponds have largely replaced marshland; other white shoreline areas are urban developments. S indicates the location of Suisun Bay, fringed by diked marshland. Darkest areas around the bays are remnants of the original salt marshes (Image modified from USGS).

making it difficult to predict future wetland changes expected from continued global warming and sea level rise.

Within the San Francisco Estuary in the northern bay, the semi-diurnal mesotidal ranges and tidal datums vary significantly with distance away from the open bay and with proximity to the river channels. The marsh vegetation distribution varies with salinity, tidal drainage patterns, competition for light and other factors. Therefore, the tidal elevations in this marshland do not correlate precisely with vertical vegetation zones, although a broad division into low, middle and high tidal marsh can be made (Baye, 2007; Table 7.2). All the San Francisco Bay marshes are important for both migratory and resident waterfowl. Daiber (1977) explains how the salt marsh song sparrow (*Melospiza melodia maxillaris*) is adapted to the San Francisco marsh zonation and tidal regimes by nesting several weeks earlier than its terrestrial counterparts to take advantage of the lower tides during March. The

Zone	psu	Distinguishing flora (Macdonald, 1977)
Table 7.2 Marsh zonations of San Francisco Bay		
Tidal Marshes	20–34	Halophytic succulents and grasses
Low Marsh (MSL): submerged and emerged daily		
Middle Marsh (MHW–MHHW): submerged only by higher high tides		
High Marsh (>MHHW): flooded only during spring high tides:		
Located along levees of old tidal channels, dikes and railway embankments		Invasive weeds from adjacent terrestrial zones; high marsh daisies, jaumea, gumweed, quill grass, salt marsh heath (*Frankenia salina =F. grandifolia*); most at northern limits
Brackish Marsh	<18	Tules/bulrushes (*Scirpus, Schoenoplectus, Bolboschoenus*) to rushes in high marsh
Freshwater Marsh	<5	Cattail reeds (*Typha* spp.) and freshwater marsh plants

other birds in this marshland are more vulnerable to predation by gulls, hawks and owls because they congregate on exposed levees or drifting debris during extreme high tides.

The warm temperate coastal region of California south of San Francisco Bay is largely covered by urban development. An estimated 75–91% of the coastal wetlands has been lost (Greer and Stow, 2003) and the focus now is on preserving the remaining 19 wetland fragments and/or restoring highly altered areas. Jacobs *et al.* (2010) outline the main characteristics of these now small, isolated wetlands and the geomorphological characteristics that shaped their origin. These Southern Californian Mediterranean-type wetlands fall into eight groups, based on their natural closure patterns, which are controlled primarily by volume of stream flow and secondarily by the height of barriers to tidal inflow. Many of the smallest lagoons naturally stay closed for long periods (multiyear to multidecadal scale). Warme (1969) provides details on the geomorphological development, sediments, plants, and macro- and microfaunas of one such small wetland in the tide-dominated Mugu Lagoon near Santa Barbara. This lagoon presently is kept open by dredging, but formerly was closed for several months per year.

In the following sections (7.3.1.5 and 7.3.1.6), we outline important features for the two best-documented Southern California coastal wetlands. We compare the small Los Peñasquitos Lagoon with tidal circulation heavily impacted by railroad and road construction with the large Tijuana Estuary, which has minimal anthropogenic disturbance of the tidal channel entrance. These two examples represent endpoints of the range of warm temperate salt marshes found from Mugu southwards to Guerro Negro (Scammons) lagoon in Baja California, Mexico. These wetlands provide good examples of Mediterranean climate salt marshes, which in the 1960s–1970s were in danger of extinction by urban and industrial growth, but are now protected as wildlife preserves and are being restored as fish nurseries and waterfowl refuges. Other publications (Mudie, 1970; Mudie *et al.*, 1974;

1976) describe the ecology, microfossils and the recent geological history of other San Diego County lagoons (Scott *et al.*, 1976, 2001, 2011).

7.3.1.5 Los Peñasquitos Lagoon case history, Southern California

Los Peñasquitos Lagoon is a barrier beach lagoon in San Diego city, on the north side of Torrey Pines State Park where the 257 ha of wetlands have been preserved despite the encroaching development (Figure 7.16). This small lagoon is a vital part of a chain of wetlands needed to sustain migratory seabirds along the largely urbanized Southern California coastline. The lagoon marsh complex was formed after the last glaciation when the rising sea flooded the Los Peñasquitos River estuary, forming a deep bay. About 8000 years ago, however, river sediment began to fill the embayment, forming extensive mudflats dissected by tidal channels (see Section 12.2 for details). By the start of Spanish Missionary time in 1769, the lagoon was largely filled by wetland vegetation and there was more freshwater in the river, which today is often almost dry in summer. In winter, the Los Peñasquitos River transports freshwater to the lagoon from a watershed of ~6 000 000 ha and treated sewage effluent is also discharged into the inner lagoon.

Old maps show that the nineteenth-century lagoon entrance was originally in the southwest corner of the wetlands, but it meandered northward through the beach dunes, so that when a railroad was built across the wetland in 1888, the entrance was at the northwest edge of the valley. In 1909, a road was built that crossed the main channel on a bridge, but up to 1925, the natural lagoon drainage was not much changed. Construction of the Santa Fe Railroad, however, divided the lagoon in half and significantly altered the tidal current pattern, reducing the effective tidal flow by about 25%, and by 1928, the entrance channel was filled with sand. In the 1930s, the beach barrier height was increased for a wider coastal highway. As a result, the lagoon entrance shifted southward about 70 m relative to its present location (Figure 7.16A).

Long before these recent landscape alterations, microfossils and pollen in cores from the lagoon indicate that higher rainfall, larger channels and a wider entrance were enough to keep the main channel open most of the time. Continuous tidal connection is needed for survival of a normal lagoon flora and fauna, and shellfish and fish-bones in nearby shell middens indicate that the lagoon mouth was permanently open thousands of years ago. In recent times, the small lagoons have experienced only short periods of ocean connection alternating with longer periods of closure and stagnation, despite a 25-year-long programme of regular sand removal started in 1968 (Greer and Stow, 2003). After 1992, however, the lagoon entrance stayed open 85% of the time.

During periods of no tidal exchange, evaporation increases lagoon salinities to high values (Figure 3.7), rising from the normal 34 of seawater to 63.9 psu over the eight-month warm dry season. At this high salinity, only the hardiest marine animals survive: California killifish, bay topsmelt and the California mudsucker fish. The absence of tidal water exchange and the death of less hardy fish and shellfish produce anoxic water and sulfide-rich black sediments. In contrast, when the freshwater inflow volume exceeds the amount of water lost by evaporation, the channel water is diluted and the marine biota becomes stressed. When sewage effluent was added in 1962, the surface water salinity dropped to ~13 psu within four years. The low salt content of the surface water then prevented survival of many marine species, but the surface

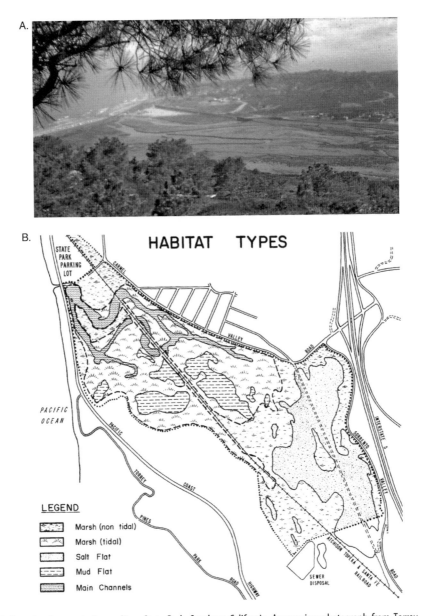

Figure 7.16 Los Peñasquitos Lagoon in Torrey Pines State Park, Southern California. A: overview photograph from Torrey Pines State Park (photo by P. J. Mudie), looking north towards the city of Del Mar, and showing the marsh, road and rail features, and the excavated beach parking lot; B: sketch map of Los Peñasquitos Marsh Natural Preserve using data from Mudie *et al.* (1974) to show the several lagoonal environments, including strandline beach, relict sand dunes, mudflats, marsh areas (stippled), salt pans and man-made objects (road, railway and parking lot).

Figure 7.17 Excessive plant growth associated with effluent discharge and low salinity in Los Peñasquitos lagoon when the channel entrance is closed. Left: surface scum of *Enteromorpha* algae beginning to decay in the main channel at onset of summer; Right: strands of ditchgrass mixed with masses of blue-green algae grazed by *Melampus* snails; white bubbles are hydrogen sulfide from black masses of decaying plants (photos by P. J. Mudie). See also colour plates section.

brackish water was ideal for growth of the pondweed, ditchgrass (*Ruppia*), which covered the entire water surface in 1966 (Figure 7.17). When a high spring rainfall breached the sandbar in 1967, however, renewed tidal exchange increased the channel salinity and the ditchgrass gradually disappeared. Much of the normal estuarine biota also recolonized the channels, although low marsh plants like cordgrass and eelgrass, and some animals, including California hornshell (*Cerithidea californica*) and smooth chione (*Chione fluctifraga*), have not yet returned.

In Los Peñasquitos Lagoon, the dissolved oxygen level is generally high during the day, but low at night when photosynthesis and the oxygen output of plants stops during the absence of daylight. The naturally low oxygen levels in bottom waters reflect high rates of organic decomposition associated with the wetland's high productivity rates. In 1972, nitrate and phosphate levels were also much higher in this lagoon compared with coastal water, presumably because of the input of sewage effluent. Nutrients trigger blooms of the green macroalga *Enteromorpha*, which dies from excessive heat and salinity when the lagoon is closed, adding to the overenrichment of sediments (Figure 7.17). The barrier lagoon environment is thus a highly variable habitat compared with the open ocean, and to survive this unstable environment, the plants and animals have specialized physiology and growth/ behaviour patterns, as outlined in Chapters 3 and 4.

In 1974, there were a total of 30 species of salt marsh plants in Los Peñasquitos Lagoon (Mudie *et al.*, 1976). Closest to the lagoon channels, pickleweed dominated and there were patches of alkali heath (Figure 3.4c). At higher marsh elevations covered by saltwater only during the high spring tides, there were several other plants in open spaces among the pickleweed and alkali heath, including salt grass, sea lavender and California seablite. The uppermost salt marsh, wetted only by extreme high spring tides and storm waves, was marked by circular islands of the shrubby glasswort, *Arthrocnemon subterminale* (formerly called *Salicornia subterminalis*), encircled by annual glassworts (*Salicornia europea*)

Figure 7.18 Saltflat plants at Los Peñasquitos in 1974. Circular colonies (left) of *Arthrocnemon* (alkali bush), ringed by the yellow annual daisy *Lasthenia* (centre) in early spring, which sets seed before salt build-up and is followed by the more salt-resistant annual little ice-plant (right), with both succulent leaves and bubble-like salt glands (photos by P. J. Mudie).

(Figure 7.18). On large saltpans, two other spring annuals appeared in colourful succession. First, the yellow salt marsh daisy (*Lasthenia glabrata*) germinates as soon as winter rains wash off the previous summer's salt crust; secondly, when the salt resurfaces in late spring, the succulent little ice plant (*Mesembryanthemum crystallinum*) appears, with white flowers and purple leaves. Another saltflat species was the salt marsh bird's beak (*Cordylanthus maritimus* subsp. *maritimus*), which is an endangered plant species with a scattered distribution from Morro Bay to San Quentin (northwest Baja California). This rare toadflax is nearly eliminated throughout its restricted range and has not been recorded in Los Peñasquitos since 1942. A recent aerial photograph survey of the lagoon (Greer and Stow, 2003) shows that from 1974 to 1999, there was a 20% increase in brackish and riparian marsh vegetation, and a corresponding decline in saltflat vegetation. This change is linked to lowering of soil salinity as a result of increased dry-season runoff of freshwater from urban development of the watershed.

Table 7.3 lists a selection of animals from Los Peñasquitos Lagoon (Mudie *et al*., 1974). Twenty-three species of invertebrates were recorded, and in addition to those listed in the table, there are clams, scallop, cockle and several shrimp species (both swimming and burrowing species), and several species of gastropod snails. The fish in shallow pools and channels are often hardy species, or juveniles of commercially important species. Lagoon mammals mostly appear to live in the grassland, brush and dry bank areas around the marshland and forage in the marsh during low tides. Additionally, common visitors in the upland marsh areas are the Audubon cottontail and brush rabbits, ground squirrels and pocket gophers.

In 1967−1970, 68 species of water-related birds were recorded for Los Peñasquitos lagoon, with peak populations of around 1500 (as listed in Mudie *et al*., 1974). Four endangered birds live and/or nest in the lagoon: the light-footed clapper rail, the California least tern, Belding's savannah sparrow and the long-billed curlew. In addition to migratory waterfowl and shorebirds, diving birds, such as cormorants, grebes and

Table 7.3 Selection of common animals from Los Peñasquitos Lagoon, California (data from Mudie *et al.*, 1974)

Animal	Scientific Name	Notes
Invertebrates (23 species)		
Native oyster	*Ostrea lurida*	
Bay mussel	*Mytilus edulis*	Tolerates subaerial exposure at low tide
Common littleneck clam	*Leukoma staminea*	Both clams survive in the black, oxygen-poor, sulfide-rich mud
Jackknife clam	*Tagelus subteres*	
Shore crabs	*Pachygrapsus crassipes* & two *Hemigrapsus* spp.	
Fiddler crab	*Uca crenulata*	Mud burrower
Crayfish	*Procambarus*	
Burrowing anemone	*Harenactis attenuata*	
Fish (15+ species in shallow pools and channels)		
California killifish	*Fundulus parvipinnis*	Tolerant of summer salinities to 80 psu
California mudsucker	*Gillichthys mirabilis*	Capable of being out of water for up to a week if in damp seaweed; hibernates in bottom sediment in winter; valuable bait species
Bay topsmelt	*Atherinops affinis littoralis*	Tolerant of wide salinity ranges
Blenny	*Hypsoblennius* sp.	
Pipefish	*Syngnathus* sp.	
California halibut	*Paralichthys californicus*	Juveniles of commercially important species.
Diamond turbot	*Hypsopsetta guttulata*	
Spotfin croaker	*Roncador stearnsii*	Commercial importance
Sandbass (2)	*Paralabrax* spp.	Commercial importance
Mammals		
Mule deer	*Odocoileus hemionus*	Saltflat grazers south of the lagoon
Coyote	*Canis latrans*	
Bobcat	*Lynx rufus*	Tracks common; foragers in marsh
Racoon	*Procyon lotor*	
Ornate shrew	*Sorex ornatus*	
Western harvest mouse	*Reithrodontomys megalotis*	All rodents forage for seeds, shoots and insects in the high marsh
White-bellied deer mouse	*Peromyscus maniculatus*	
Pocket mouse (2)	*Perognathus* spp.	
House mouse	*Mus musculus*	

pelicans feed in the deepest tidal channels, while ducks congregate in the brackish water area of the inner estuary. Resident birds include great blue herons, black-necked stilts and killdeer (which nest on the saltflats), savannah sparrows, western meadowlarks, black-crowned night heron and red-winged blackbirds. Predators include the northern harrier and the black-shouldered kite (once in danger of extinction), and hawks, California brown pelicans and ospreys are occasional visitors.

The 'State Natural Preserve' status of Los Peñasquitos Lagoon ensures the future protection of the lagoon flora and fauna from direct public misuse. However, the boundaries of the lagoon ecosystem do not coincide with the fences around the preserve, but extend upstream to the headwaters of creeks and the tops of the mesas surrounding the marshland. Therefore, the lagoon is still vulnerable to man-made changes, including excessive silting of the channels and accelerated filling-in of the marshland (Figure 7.19). Ongoing watershed development and periodic floods continue to build up the inner saltflats over and above the 600% increase in sedimentation rate associated with European colonization (see Section 12.3). In particular, from 1968 to 1985, the elevation of the saltflats in the

Figure 7.19   Urban and industrial developments on the hillsides and canyons draining to Los Peñasquitos Lagoon have accelerated soil erosion and infill of the lagoon channels and saltflats. Top: hillside erosion (~1 m) associated with housing development, ~1968; Bottom left: mudslides from arroyos in Carmel Valley engulfed cars and spilled onto the saltflat; Bottom right: bulldozer clearing a mudslide near the lagoon entrance, which filled some of the channel and middle marsh (photos by P. J. Mudie).

northeastern part of the lagoon increased by ~3 m, enabling invasion of non-native plants that will probably replace the unique saltflat flora if the soil erosion and summer drainage issues are not corrected.

7.3.1.6 Tijuana Estuary: California–Mexico boundary salt marsh reserve

The geological history, environmental and historical factors that have shaped the landscape and ecology of the Tijuana Estuary are described in several documents by Zedler and Nordby (1986), Falk *et al.* (2006) and Callaway and Zedler (2009). The Tijuana Estuary is located at the mouth of the Tijuana River, which has its headwaters in the mountain ranges of Northern Mexico and therefore has most of its watercourse beyond the jurisdiction and protection of the United States (Figure 7.20). The tidal range at the entrance to the estuary is 1.63 m, with MHW at 1.22 m, and the channel water salinity ranges from a normal 35 psu up to 180 psu after prolonged closure of the channel entrance. Tijuana Estuary has a much wider entrance channel than Los Peñasquitos Lagoon and this is not obstructed by bridges or railroad embankments, allowing the channel to migrate along the sandy beach according to wind and wave forces. The tidal wetlands and surrounding saltflats also cover a much greater area (>1000 ha) than at Los Peñasquitos Lagoon and, since 1982, the wetlands have been protected by the NOAA as a Wildlife Wetlands Sanctuary. For many years, the Tijuana Estuary has been the focus of numerous important pioneer studies in the ecology of temperate region salt marshes and the base of experiments in restoration ecology done by San Diego State University (Figure 7.21; Section 14.2).

Tijuana Estuary is a national preserve for the endangered salt marsh birdsbeak flower and it is near the coastal limits of several other subtropical salt marsh plant species such as iodine bush (*Allenrolfea occidentalis*) and Palmer's seaheath (*Frankenia palmeri*). The estuary also protects nine animal species of special concern, including (in 1995), light-footed clapper rails, snowy plovers and Belding's savannah sparrows. Other species with special

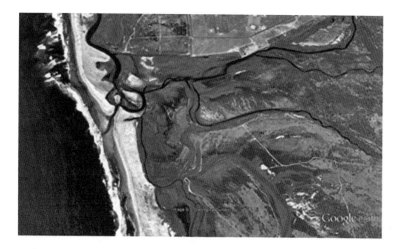

Figure 7.20 Satellite view of the Tijuana River National Estuarine Research Reserve and the mouth of the Tijuana River breaking over a sandbar and entering the Pacific Ocean (Google Earth image).

Figure 7.21   A: low marsh cordgrass community at Tijuana Estuary supports various snails, crabs and other crustaceans that are food for the California least tern and (see B) the light-footed clapper rail; B: the secretive light-brown rail with black chicks uses cordgrass for cover, food and nesting material, making a floating nest with a protective arch woven from cordgrass leaves. Bottom images show representative animals and rare plants. C: left to right – horn snail (*Cerithidea californica*), male sea spider with eggs (*Ammothella bioguiculata* var. *californica*), white bubble snail (*Haminoea vesicula*), fire worm (*Nephtys caecoides*), spionid polychete worm (*Polydora* sp.), hermit crab (*Pagurus hirsutiusculus*), broken back shrimp (*Spirontocaris palpator*), amphipod (*Jassa falcata*); D: high marsh with California ground squirrel behind the succulent *Allenrolfea subterminalis* shrub; the endangered small bird's beak flower is lower left, below the inset San Diego horned lizard (from Zedler and Nordby, 1986, USFWS).

protection status that feed in the estuary are the California brown pelican, California least tern, American peregrine falcon, reddish egret, long-billed curlew and the large-billed savannah sparrow. Sampling in 1986–1989 collected 21 fish species from 15 families, the main ones being the arrow goby, topsmelt and California killifish. Various studies since 1969 document changes in the benthic invertebrate community structure after catastrophic events. In normal years, however, e.g. September 1986–June 1988, sampling of the northern tidal channels collected 58 taxa of polychaetes (predominantly Capitellids and Spionids), bivalves, and the California ghost shrimp. The highest bivalve densities were at the deepest channel site, also farthest from sewage inflow. An insect survey from autumn 1988 to spring 1989 lists 16 different orders for the beach dunes, pickleweed marsh, marsh transition and adjacent upland. Beetles, springtails, flies, leafhoppers and plant hoppers, bees and wasps, butterflies and moths were most abundant; the salt marsh wandering skipper butterfly was present, also globose dune beetle was found on the dunes. Other wildlife reported for 1988−1989 include four species of frogs (including the introduced African clawed frog and the bullfrog), three snake species and six species of lizards, including the endangered San Diego coastal horned lizard.

Tijuana Estuary is one of very few San Diego County inlets that remain open to the ocean year-round, even during the dry summer. Foraminifera in surface samples here include some species not described before 1973. These microfossils are not present in geological records for the past 3000 years (Scott et al., 2011), apparently because migration of the main channels affects peat production and associated microfossils over time. The average sedimentation rates over the 2500 years prior to the mid eighteenth century were 9.0–12.8 cm/century. In contrast, vertical accretion rates of 0.7–1.2 cm/yr were measured from 1963 to 1998 by ^{137}Cs dating of the low and middle marsh sediments. Measurements made using feldspar marker layers yielded much higher rates of 2–8 cm/yr during the 1993 winter flooding of the low marsh, and 9.5 cm/yr (for each storm season) in the north tidal channel from 2000 to 2004.

7.3.2 Temperate wetlands of Eastern North America: Nova Scotia to Chesapeake Bay

7.3.2.1 Chezzetcook Marsh, Southeastern Canada

The Chezzetcook wetland (Figure 7.22 and Box Figure 14.1) is an intensively studied small marsh in the cold temperate region of Eastern Canada and provides a sharp contrast to the small warm temperate West Coast salt marsh described for Los Peñasquitos Lagoon in Section 7.3.1.5. The average temperature at Chezzetcook ranges from –6 °C in the winter (January) to about 19 °C in summer (August), though there are extremes close to –30 °C in winter and a few hot days of 35 °C in the summer. Chezzetcook Inlet (44.70° N, 63.25° W) is located approximately 45 km east of Halifax, Nova Scotia, and is about 7 km long and 2 km wide at the mouth of the inlet. The wetland is a partially drowned drumlin field at the mouth of the Chezzetcook River, and it is common to find glacier-transported boulders throughout the upper areas of the marsh. The geomorphological evolution of the marsh is described by Forbes (2011). This marsh is thousands of years old at the head of the inlet and is currently growing seaward. Located along the Atlantic coast of Nova Scotia, it is a mesotidal marsh without the dramatic macrotidal influences seen on the Bay of Fundy coast (see Section 7.3.2.2). Semi-diurnal tides range from 0.4 m to 2 m, with averages between 0.6 and 1.8 m. As a result, the marsh zones occur between 90 cm at the highest marsh to 0–10 cm

Figure 7.22   Chezzetcook Marsh, Nova Scotia, Canada. A: high and middle marsh with deep tidal channel; B: low marsh (photos by J. Frail-Gauthier); C: outer harbour (photo by D. B. Scott).

above MSL for the low marsh and mudflat zones. Salinity in Chezzetcook inlet changes with season, precipitation, freshwater runoff, and elevation within the marsh (see Scott *et al.*, 2001). Salinity values are highest in the summer, when evaporation is high and also in winter when the marsh is ice-covered; the values are lower in the spring during the snowmelt and in the fall, when rainfall is higher (Scott and Medioli, 1980). Total precipitation averages over 1400 mm of rain plus snow per year. Salinity is usually over 25 psu in the lower marsh, and 0–5 psu at the head of the marsh, which, in conjunction with the length of tidal submergence, influences the vertical zonation of flora and fauna of the marsh.

The boreal region Chezzetcook salt marsh has higher plant diversity than the Eastern Canadian subarctic marshes of James Bay described in Section 7.2.3. At Chezzetcook, the low marsh is almost exclusively dominated by cordgrass (*Spartina alterniflora*), with annual glassworts, sea spurrey and seablite on open areas of the mudflats. Species of sedge, spearleaf, goldenrod (*Solidago*) and another cordgrass (*Spartina cynosuroides*) dominate in the upper marsh, while salt hay (*Spartina patens*), arrowgrass, cinquefoil and plantain are the dominants in the middle marsh. Within the middle and high marshes, there are open areas formed by either stranded muddy channel ice or kelp deposited during storms; these open areas are rapidly colonized by herbs such as spearscale (*Atriplex hastata*) or *Salicornia*.

This boreal marsh has been studied for over 35 years and much of the sea level detection method and development of Atlantic Coast sea level curves (Scott *et al.*, 2001) came from this inlet, showing how benthic foraminifera can reconstruct accurate paleo-sea level curves (±10 cm, Scott and Medioli, 1978; 1980). Most recently, Gehrels *et al.* (2005) have derived quantitative transfer functions (see Box 12.2 Paleoecological transfer functions) from the Chezzetcook transect data and they have applied these equations to a high-resolution microfossil record for the past ~1200 years. This new work establishes that the ongoing rapid sea level rise in the western North Atlantic Ocean is closely linked with the recent global warming: the pre-1900s rate of 1.6 mm/yr doubled to 3.2 mm/yr from 1900 to 1920. Sedimentological and pollen data were used as proxies of geological and vegetation changes in a study of the salt marsh development from a 2000-year record for the outer estuary (Jennings *et al.*, 1993). These paleoecological records show that the changes at the estuary mouth reflect an over-riding role of both coastal and estuarine sedimentation processes in shaping regressive marsh–sediment complexes, despite the overall RSL rise of up to 3.8 mm/yr during the late Holocene. In contrast, at the inner estuary, the salt marsh developed rapidly over the last 200 years because of sediment in-wash from European land use following about 5000 years of stable tidal-flat conditions. These paleoecological studies highlight the differences in prevailing sediment sources and distribution mechanisms operating in the outer and inner sections of the estuary.

Other sediment cores collected from the salt marsh in the West Head of Chezzetcook Inlet reveal the recent changes in geochemistry resulting from both natural and anthropogenic influences. Ultra-high-resolution ^{137}Cs dating was used to determine the accretion rates (see Box Figure 12.1). The results show that sedimentation rates have been slightly higher in the high marsh than in the middle marsh for the last 30 years, probably because of increased terrestrial sediment influx from the adjacent roads and farms. Element distribution in the marsh sediments is mostly related to differences in sedimentary mineral matter abundance and variety (e.g. mica versus halite clays). However, elevated concentrations of Fe, Ni, Co, Mo and As near the surface of the core at the landward edge of the marsh reflect anthropogenic inputs associated with recent road construction. At and below about 1 m core depth, however, the element enrichment is attributed to chemical alterations (diagenetic processes), which cause precipitation as sulfide minerals at the boundary between the upper oxidized and the lower reduced sediment layers (Chagué-Goff *et al.*, 2001).

7.3.2.2 Bay of Fundy, Southeastern Canada

In contrast to the small Chezzetcook estuarine salt marsh with a mesotidal range of ~2 m, the Bay of Fundy has the world's largest tidal range: 100 billion tonnes of water move in and out of Canada's Bay of Fundy twice a day. The Bay of Fundy tide is five times higher than the Atlantic coast average: its tides range from 3.5 m at the southern end to 16 m in the north; it has the world's greatest mean spring range with 14.5 m, and an extreme range of 16.3 m. This natural phenomenon provides a fleeting opportunity to explore the marshes and abundant mudflat biota at low tide, but there is significant danger of getting too far offshore before the high tide sweeps back as a tidal bore that is as fast and high as an incoming ocean surf wave.

The Bay of Fundy (Figure 7.23) lies in an ancient rift valley called the Fundy Basin; as the rift began to separate from mainland North America about 190 million years ago, tectonic activity formed volcanoes and flood basalts. These flood basalts poured out over much of southern Nova Scotia and formed the basaltic North Mountain range along the south shore of the bay. Much of the basin floor is made of tholeiitic basalts, which gives it a reddish-brown colour. The rift valley eventually stopped moving as the Mid-Atlantic Ridge seafloor spreading-centre pushed away North America from Europe and Africa, and the Fundy Bay valley became flooded with seawater over time. The shoreline has been shaped by the tides based on the resistance of the rocks to erosion. Winter freezing and thawing accelerates this process, which also contributes much of the sediment to the tidal flats in the Minas Basin and in Chignecto Bay. Ice formation in the marsh tidal channels can result in collapse of enormous blocks of sediment from the channel slopes (Figure 7.24), which add more silt and clay to the extensive, shifting mudwaves of the intertidal flats. The sediment-laden 'ice-cakes' form by repeated freezing and thawing of muddy water and sediment, and can range in size from sheets about 100 m long and 15 cm thick to cottage-sized rectangular blocks of 100 m^3 volume and 100 tonnes weight (Sanders, 2011). This type of ice formation is extremely rare in temperate regions, but also known for the subarctic marshes in Cook Inlet, Alaska.

A good example of the rapid sedimentation within the Bay of Fundy can be seen at the Windsor causeway in the upper reaches of the Bay of Fundy, at the Minas Basin. The size of the Windsor salt marsh increased greatly over about 50 years after a causeway was built across the Avon River in 1970. Because of the high sediment load in the upper Bay of Fundy, the barricaded river resulted in a 15 cm/month initial deposition of sediment downstream of the causeway. About 10 years later, cordgrass appeared in an isolated patch, and by 1992, there were over 30 patches. Between 1995 and 2001, the *Spartina* covered from 41 000 m^2 to over 390 000 m^2 of the mud surface (Daborn *et al.*, 2003). Based on satellite images in 2013, the latest estimate of *Spartina* cover approaches 1 km^2, not including cover along the outer banks.

The nutrient-rich Bay of Fundy supports abundant marine life, including the finback whale, the world's second largest animal, which comes to the Bay of Fundy to feed along with the minke, humpback and endangered right whales. Parts of the inner Bay of Fundy (Shepody Bay and Minas Basin) form one of six Canadian sites in the Western Hemisphere Shorebird Reserve Network. The reserve is jointly administered by the provinces of New

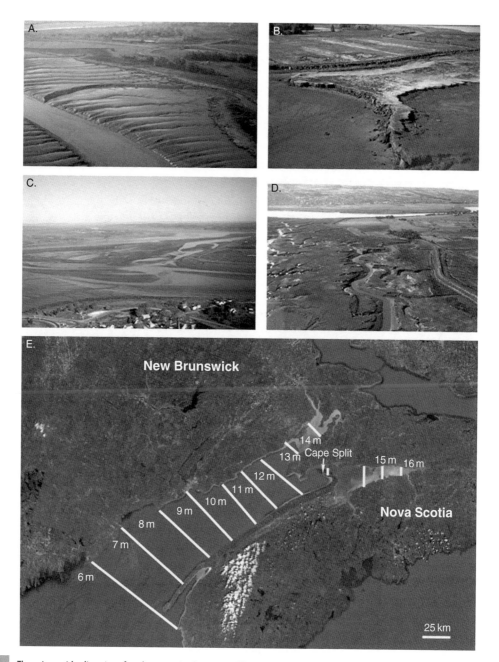

Figure 7.23 There is a wide diversity of anthropogenic changes visible in the Bay of Fundy. A: dissected mudflats extend in front of dikes protecting farmland; B: old Acadian dikes used in the seventeenth and eighteenth centuries now also a shoreline armour, but this provides little protection; C: seaward view of the mudflats from Windsor, Nova Scotia; D: meandering marsh channels characterize tidally dominated environments (Photos by D. B. Scott); E: map showing how the Fundy tidal range increases landwards as the bay narrows and the tidal bore is constrained by Cape Split. The Bay of Fundy has amplified tidal ranges because its basin has the same period as the tidal surge of 16–17 m during at spring tidal levels (modified from Google Earth).

Figure 7.24 Sediment-laden ice erodes the marsh channel walls and adds sediment to the aggrading mudflats. Top: channel 'ice-cake' in the Shubenacadie River at the head of Minas Bay; Bottom: huge (2+ m) chunks of sediment-laden channel ice can scour the mudflats until it melts out in early summer. Note the scale given by the metre-high camera tripod (photos by Richard Sanders).

Brunswick and Nova Scotia, and the Canadian Wildlife Service, and is managed in conjunction with Ducks Unlimited Canada and the Nature Conservancy of Canada.

The highest water level ever recorded in the Bay of Fundy system occurred at the head of the Minas Basin on the night of 4–5 October 1869, during a tropical cyclone named the 'Saxby Gale'. The tidal surge of 21.6 m resulted from the combination of high winds, abnormally low atmospheric pressure and a spring tide. Oceanographers explain the normal extremely high tide levels by tidal resonance resulting from a coincidence of timing in a funnel-shaped bay. The time it takes a large wave to travel from the mouth of the bay to the inner shore and back is almost the same as the time from one high tide to the next, so the waters are always piled up (Figure 7.23E). Additionally, tidal bores are a common feature here because the narrow channel funnels the incoming flood tide (Figure 7.25). Because water current velocity is always in excess of 3 m/s, the Bay of Fundy is a strong candidate for tidal power (see Box 7.2 Tidal power), a non-fossil-fuel renewable resource.

Figure 7.25   Tidal bore coming up the Peticodiac River, New Brunswick. Before the construction of a causeway in 1968, this area had one of the world's largest tidal bores (1–2 m at 5–13 km/h) (photo from Tripsister.com with permission).

Box 7.2   **Tidal power and estuarine wetlands**

There have been several proposals to build tidal harnesses for electrical power generation in the Bay of Fundy, with a production estimate of 7000 MW of electricity. These proposals have involved building barrages, which cut off part of the bay and extract power from water flowing through sluice gates, usually only in one direction (ebb or flood tides). Experimental turbines were anchored off Cape Split (Figure 7.23E) but the tidal bore currents removed blades within two tidal cycles. A facility currently operating at Annapolis Royal, Nova Scotia, consists of a dam and 18 MW power house on the Annapolis River, but construction of larger systems have been hampered by several problems, including environmental concerns of erosion and siltation. The current engineering designs for harnessing power in the Bay of Fundy include a combination of in-stream turbines along the seabed, which could generate 90% of Nova Scotia's current electricity needs (Fundy Ocean Research Center for Energy, 2013). Environmental concerns such as sedimentation, noise and electromagnetic emissions are still being examined prior to implementation.

As described in Chapter 5, land reclamation of salt marshes for farming has been carried out in parts of the Bay of Fundy since its settlement in 1604. Marshes at the head of the bay are classical sites of Atlantic Canada RSL studies, using fossil trees, peat, pollen and foraminifera (Shaw and Ceman, 1999). New studies by Graf and Chmura (2006), however, show that adjustments to the RSL curves may be needed to accommodate anthropogenic influences. Connor *et al.* (2001) examine the possibility that reversion of the 'reclaimed' farmland to tidal marshes would add to the value of the extensive Fundy marshes as giant carbon storage sinks. Comparison of marsh carbon accumulation rates in different parts of the bay shows that carbon accumulation is higher where sedimentation is greater, being more than twice as high in the inner bay marshes than in the outer bay. The carbon storage capacities of reclaimed marshes and tidal marshes are similar, but the tidal wetlands can continue growing and sequestering carbon with rising sea level, whereas diked marsh soils have a fixed value and may shrink over time if the water table drops. The potential release of methane and carbon dioxide from the Fundy marshes seems to be variable, but some natural tidal marshes periodically release large amounts of these gases on disturbance (Pitcher *et al.*, 2005). Overall, however, the Fundy wetlands act as an important sink for the sequestration of the greenhouse gases carbon dioxide and methane.

7.3.2.3 New England–Chesapeake Bay, Northeastern United States

The Chesapeake Bay tidal wetlands at ~35–38° N in Maryland and Virginia are the largest of a group of Atlantic East Coast marshes, covering a total area of approximately 0.5 million ha in 2000, down from 2.5 million in 1780 (Mitsch and Gosselink, 2007). At the north end of this region, there is a subgroup of colder winter New England salt marshes, including the marshes of Long Island, New York, in addition to those extending from the northern states of Maine to Connecticut. This northern group of coastal salt marshes is described in detail by John and Mildred Teal (1969) in their book, *The Life and Death of the Salt Marsh*, and they will not be covered here.

The southern, warm temperate section of the East Coast region is best represented by the well-known wetlands of Chesapeake Bay, between Maryland and Virginia, located about 270 km south of New York. Chesapeake Bay (~35.2–35.5° N) is the largest estuary in the USA and is also a densely populated commercial waterway that includes the location of the USA capital city of Washington, DC. The bay is a *ria*, or drowned valley, of the Susquehanna River which was formed by river erosion when the sea level was lower after the last glaciation and also includes an ancient bolide impact crater formed ~35.5 million years ago. Much of the bay is shallow: the average depth is 7 m and about 25% of the estuary is less than 2 m deep. The bay area has been intensively settled for several centuries and was the focus of several large naval battles, including the Battle of Chesapeake in 1781. Today, the bay water is used to cool a nuclear power plant reactor. To try to protect the remaining coastal wetlands of Maryland, three reserves have been established that encompass the diversity of estuarine systems, providing essential habitat for commercial and recreationally important fish and crab resources, and to maintain the tidal marshes that reduce shoreline erosion and flooding.

Chesapeake Bay is about 300 km long and 4.5 km wide, on average. The climate of the bay area is warm and temperate, with hot, humid summers and cold to mild winters, with several months of frost covering the wetlands. The mouth of the Susquehanna River sometimes freezes in winter, but this has rarely happened since the end of the LIA about a century ago; the last freeze-up was winter 1976–1977. The shallowness of the estuary makes it vertically stratified, meaning that the upstream section is entirely freshwater, the middle section is brackish and the entrance area has saltwater. Within the brackish water segment, there is a lateral salt gradient from low-salinity oligohaline water (0.5–10 psu), where freshwater species can live, to a mesohaline zone with salinity 10.7−18 psu, and from the

Rappahannock River mouth to the ocean, there is a polyhaline zone with salinities of 18.7−36 psu. Over 300 fish and shellfish species live in the bay or annually migrate into the estuary, including Atlantic menhaden, striped bass (rockfish), American eel, eastern oysters (*Crassostrea virginica*), and blue crabs and clams. The bay supports a lucrative oyster farming industry, and locally the oysters are called 'Chesapeake white gold' because of the large profits to be made. Pollution as a result of runoff from farms, urban and industrial developments in the watershed has frequently triggered algal blooms in the bay because phytoplankton blooms are stimulated by either excess phosphorus and nitrogen (diatoms) or by humic compounds from soil erosion (heterotrophic dinoflagellates). Some of the algae are toxic to zooplankton and/or fish, shellfish and birds, and massive kills (Figure 7.26) lead to rapid accumulation of organic matter that deoxygenates the water, creating widespread 'dead zones' without any oxygen (= hypoxic). In the 1970s, the hypoxia in summer caused the death of tonnes of clams and worms, and crabs would mass onto the shore to get air. In the watershed, soil erosion results in high suspended sediment loads that hinder growth of the filter-feeding oysters and contaminate seagrass beds. Beds of eelgrass in the southern bay have shrunk by more than half since the early 1970s, and the loss of aquatic vegetation in turn has depleted the habitat used by young fish and crab larvae.

The parasitic dinoflagellate *Pfiesteria piscicida* contains harmful toxins that can infect fish and humans. *Pfiesteria* came to the world's attention in the 1990s when a series of large

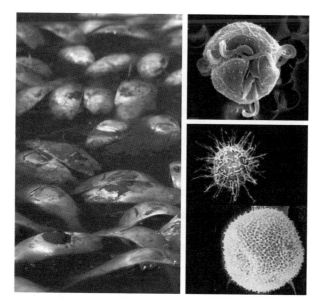

Figure 7.26 Chesapeake Bay is one of the largest bird sanctuaries in North America and an important fish nursery that experiences periodic fish kills related to toxic algae (left) (courtesy of Chesapeake Bay Foundation/CBF Staff/cbf.org). Right: *Pfiesteria piscicida* plankton stage cell (top) and resting stage (bottom); this dinoflagellate is dubbed the 'cell from hell' because its resting stage cells seemed to take up to ten different shapes; DNA studies now show that more than one genus and species is involved – including *Pseudofiesteria shumwayea* (bottom right) (photos by J. Burkholder; Citizendium free images).

phytoplankton blooms started killing many fish and giving swimmers rashes. Nutrient runoff from pig and chicken farms was the cause of the plankton growth. A good account of the difficulty of tracing the source of the illness to the highly polymorphic (shape-shifting) dinoflagellate vector is given by Rodney Barker (1998) in a book called *And the Waters Turned to Blood*.

In 2006, the cost of restoring the health of the Chesapeake Bay ecosystem was estimated at $15 billion. By 2010, although the health of the bay had improved, the value of the Maryland oyster fishery was reduced by about 88% and the reduction in oyster growing area (~80%) changed the rate of water cleaning by particle filtration from less than a week in pre-colonial times to a year. Overharvesting and outbreaks of two shellfish diseases (MSX = *Haplosporidium nelsoni* and Dermo = *Perkinsus marinus*) caused by parasitic single-celled organisms have also contributed to the demise of the shellfish industry. Recently, efforts have been made to repopulate the estuary with hatchery-grown oysters. In 2004, six million oysters were seeded on 32 000 m^2 of the Trent Hall marine sanctuary, which yielded an improvement of ~80 million oysters by 2012, although this number is still much lower than the early industry days (NOAA, 2013).

The marsh zonations in Chesapeake Bay are described by Perry *et al.* (2001), who give details of nine plant zones and their relation to tidal inundation (Table 7.4) and to farming use (Figure 7.27). About 50% of the salt marshes throughout the warm temperate region are degraded by anthropogenic impacts, as outlined in several papers by Silliman *et al.* (2009). These scientists also emphasize that the complexity of this region, which is transitional from northern areas, where ice and kelp (wrack) deposits create open areas in tidal wetlands, to southern areas where heat stress creates open hypersaline salt pans and year-round consumer pressure (intensified by man-made changes) is a pervading factor. Associated with these climatic and vegetation shifts are switches from: (1) slow detritus decomposition and thick peaty soils in the north to fast decomposition and less peaty soils in the south and (2) from more palatable grass-dominated marsh plants to less palatable succulent and shrubby marsh vegetation (Pennings and Bertness, 2001).

Although subtropical hurricanes are more intense further south (see Section 7.4), the extremely high population density of Chesapeake Bay and New Jersey, Long Island, New York, and New England to the north make this area highly vulnerable to damage from hurricane landfall. This region is considered a 'global hotspot' for impacts of future sea level rise because sea level here has risen four times faster than the global average over the past 40 years (Fischetti, 2013). Reasons for this particular 'hot spot' include glacio-tectonic subsidence in the wake of the last ice age, combined with accelerated melting of Arctic glaciers in the North Atlantic region. In addition, these temperate climate regions are prone to damage from mid-latitude storms called 'northeasters/nor'easters', which have a longer duration and cover larger areas than hurricanes (Coch, 1994; Dolan and Davis, 1994). For example, Superstorm Sandy in October 2012, the second most damaging storm ever recorded in the USA, went ashore at the coastal town of Mantoloking on a barrier island along the New Jersey coastline (see Figure 2.4). Wave erosion opened a new lagoon entrance and swept away the road bridge and blocks of houses (Figure 7.28), as well as causing other destruction by fires from ruptured gas lines. This superstorm was a hybrid storm that combined a tropical hurricane with a 'nor'easter' system, making it spread over a much wider area than the typical 'eye' of a

Table 7.4 Chesapeake marsh vegetation (from Perry *et al.*, 2001)

Marsh community	Dominant species	Associated species	Zonal range	Region
I. Saltmarsh cordgrass	*Spartina alterniflora*	*Spartina patens* *Distichlis spicata* *Limonium carolinianum* *Borrichia frutescens* *Solidago/Myrica gale* *Juncus roemerianus/* *Juncus gerardii*	MSL to MHW	Canada–Florida
II. Saltmeadow	*Spartina patens* *Distichlis spicata*	*Iva frutescens* *Baccharis halimifolia/* *Solidago sempervirens*	MHW to EHW	Canada–Florida
III. Black needlerush	*Juncus roemerianus*	Pure stands	MHW to EHW	Maryland– Texas
IV. Saltbush	*Iva frutescens* *Baccharis halimifolia*	*Spartina patens* *Borrichia frutescens*	EHW to upland	Massachusetts– Texas
V. Big cordgrass	*Spartina cynosuroides*	Pure stands	MHW to EHW	Massachusetts– Texas
VI. Cattail	*Typha angustifolia*	*Typha latifolia*	EHW to upland	Canada–Florida
VII. Reed grass	*Phragmites australis*	Invasive, pure stands	MHW to upland	All areas
VIII. Saltwort	*Sarcocornia pacifica* *Salicornia depressa/* *Salicornia bigelovii*	*Spartina alterniflora* *Distichlis spicata*	MHW (pans)	Canada–Texas
IX. Mixed brackish	No dominant species	*Scirpus, Juncus* and various grasses and herbs	MHW to upland	Canada–Texas

hurricane. The fact that much of New York City is already below sea level and is expected to have one superstorm every two years by 2100 is forcing a re-evaluation of the traditional shoreline construction defence mechanisms. Scientists like Fischetti (2013) point to the need for a shift towards superstorm defenses that rely on unhindered recovery of coastal salt marsh barriers and the relocation of people to higher land.

7.4 North American subtropical marshes

7.4.1 Carolina and Georgia salt marshes: Cape Hatteras to Northern Florida

South of Chesapeake Bay are the coastal wetlands of Virginia, and North and South Carolina. The salt marshes of the southeastern USA coastal plain are extensive, accounting

Figure 7.27   Inner Chesapeake Bay salt marshes and tidal channels. Straight lines radiating from meandering channels are spoil-lined ditches dug for mosquito control, which reduces invertebrate and fish populations and allows invasions of shrubs like marsh elder *(Iva)* (image courtesy of Chesapeake Bay Program). See also colour plates section.

Figure 7.28   The coastal town of Mantoloking, New Jersey, before and after Superstorm Sandy. Left: view on 18 March 2007; Right: view after landfall of the storm here on 31 October 2012 (aerial photographs by NASA Remote Sensing Division).

for about 21% of the total coastal wetland area in the USA (Mendelssohn and Mckee, 2000). The >400 km-long coastline flanks a low Quaternary plain with many sandy coastal barrier islands intersected and backed by estuaries and bays, merging inland to freshwater marshes (Hayes, 2010). Many rivers drain into estuaries in the region from Cape Henry (Virginia) to Cape Hatteras (North Carolina) and this area supports extensive microtidal–mesotidal brackish water marshes dominated by the black spikerush, *Juncus roemerianus*, and sometimes including large areas of either meadow or smooth cordgrass. The climate of the region is humid, subtropical with mild winters (13 °C average) and with long, hot summers (daytime average ~31 °C, up to 41 °C). Rainfall is high year-round (191 mm/month), and there are frequent hurricanes and/or tropical cyclones in late summer and fall. The vulnerability of the coastline to storm damage varies subregionally with wind and wave exposure (Gornitz *et al.*, 1994), but about 50% of the shore length has a very high impact risk and storm damage is often extensive. For example, the long barrier beach enclosing Pamlico Sound shelters salt marshes behind the narrow dunes, but these are constantly changing in response to damage from hurricanes.

The warm temperate East Coast wetlands have different vegetation than marshes farther north, although the high marsh contains some of the same species. Details of the marsh flora, fauna and ecology are given by Wiegert and Freeman (1990), while Godfrey and Godfrey (1974) show how the marsh vegetation responds to the complex dynamics of alternating barrier-island storm overwash versus inlet erosion cycles. Overall, in the Carolinian marshes, big cordgrass (*Spartina cynosuroides*) occupies the lowest salinity areas at the highest tide level, while spikerushes dominate slightly lower in the marsh, interspersed with bare areas of high salinity that are colonized by sea ox-eye shrubs (*Borrichia* spp.), glassworts, saltwort (*Batis maritima*) and salt grasses (*Distichlis* spp.). The low marsh is almost exclusively smooth cordgrass. Some of the wetland plants here do not extend further north, e.g. *Batis*, *Borrichia* and *Distichlis* spp., and the brackish water bald/swamp cypress tree (*Taxodium distichum*) found in sheltered areas is also near its northern limit. Osgood and Silliman (2009) discuss the potential causes for extensive dieback of salt marshes along this and the Gulf Coasts. The diebacks can be triggered by various stresses, including climate warming and drought, metal toxicity, fungal pathogens, intense consumer grazing pressure (e.g. by periwinkle invasions) and/or loss of seagrass or kelp beds. The result is a cascade or domino effect in a chain of harmful events that is initiated primarily by climate change and human interference.

The history of sea level change in this warm temperate region is considered particularly important because it is south of the influence of ice-sheet loading during glacial stages (Kemp *et al.*, 2009). Consequently, various microfossils, including foraminifera, diatoms and pollen grains, have long been studied in surface samples to test their value for reconstructing past sea levels. In addition, possible use of speedier geochemistry methods using stable carbon isotope (δ^{13}C) and C:N ratios has been investigated because black spikerush and smooth cordgrass are C_3 and C_4 plants, respectively, with different δ^{13}C signatures (Kemp *et al.*, 2010). The results of Kemp's studies indicate that benthic foraminifera are the most reliable proxy indictors of past change, producing RSL records close to those of modern tide-gauge measurements. The limit of precision, however, may be constrained by the new finding that this segment of the North American coastline is not as tectonically stable as previously assumed, but contains a mantle convection-driven component that must be considered when estimating amounts of RSL change over the past ~3 million years (Rowley *et al.*, 2013).

Foraminifera have also been used to trace the history of hurricane strikes in South Carolina, using sediment cores from non-tidal ponds or beach ridge environments (see Scott *et al.*, 2001). The evolution of the beach ridges was studied at North Inlet, Price's Inlet (Figure 7.29, top) and near Seabrook Island south of Charleston, where Moslow (1984) shows how they migrated slowly (3–4 m/decade) seawards over a shelly transgressive

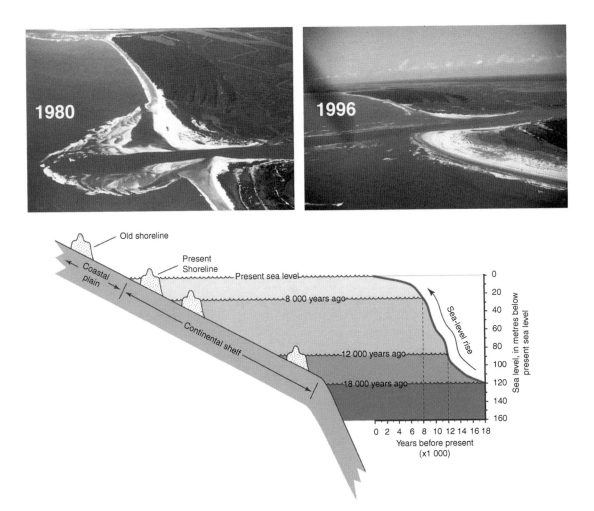

Figure 7.29 Price's Inlet, South Carolina. Top: view of Price's Inlet looking south and showing the progressive formation of beach ridges since sea level started to drop ~5000 yr BP; also note the offset of the southern beach from erosion eight years after Hurricane Hugo in 1989 (photos by D. B. Scott); Bottom: profile of a section across the wide, shallow continental shelf off the Carolinas, showing the succession of beach ridges formed at different times of slower sea level change during the postglacial marine transgression. There is an irregular rate of sea level rise from ~120 m below the present level at the end of the Ice Age; in the steep section from 12–8 kyr BP, the transgression was too fast for dunes to form. Old shoreline deposits on the coastal plain are relicts from at least 125 000 years ago when sea level was higher than it is today (Barnhardt, 2009, USGS).

shell-lag dated ~4.5 kyr BP as sea level dropped after ~5.0 kyr BP, after an earlier period of shoreward marine transgression (Figure 7.29, bottom). Goat, Kiawah and Seabrook islands are relicts of mid-late Holocene beach ridges that are now above the present sea level. Elsewhere on this coast, however, the barrier islands are presently transgressing inland at a rate of >10 m/yr (Hayes, 2010), underscoring the complexity of the regional

 Figure 7.30 Inlets along the coast of South Carolina illustrate differences in beach widths, amount of offset and submergence of marsh offshore at Murrel Inlet (A) and Santee River (B) (photos by D. B. Scott).

geomorphology. Air photo series show how rapidly this coast changes: in 1980 there was no offset in the beaches north and south of Price's Inlet, but after Hurricane Hugo in 1989, the southern beach was eroded back by ~50 m.

An unusual feature of the coastal plain is the presence of Carolina Bays, the origins of which are uncertain. The 'bays' are freshwater ponds rimmed by sand that get their name from the presence of bay trees. The 'bays' tend to be oval, lining up in a northwest to southeast orientation on flat terrain where the soils consist of recent sand, silt and clay sediments. These unique 'bays' support many threatened plant species (Wiegert and Freeman, 1990). Inland of the coastal plain is the Sandhills region, where fossil coastal dunes mark a time when the land was submergent or sea level was much higher. Pilkey and Young (2009) show that a Charleston lighthouse built 300 m behind the beach in 1876 is now >300 m offshore because of coastal retreat and erosion following building of protective structures. Two inlets (Figure 7.30) illustrate the various configurations of the South Carolina coastal lagoon entrances and geomorphology.

7.4.2 Florida mangroves

The Carolinian-type temperate salt marshes continue southwards into Florida, but at ~30° N, they become mixed with shrubby black mangroves (*Avicennia germinans*) and by 28° N at the Florida Space Center, there are extensive areas of black mangrove, together with white mangroves (*Laguncularia racemosa*) and red mangroves (*Rhizophora mangle*). Further south there are large mangrove forests in Florida and the Florida Keys (Figure 7.31) which are similar to those that extend south of the Tropic of Cancer (23° N) into the Caribbean, Cuba, Mexico and northern South America. The Florida mangroves are at their northern natural limit on the mainland of eastern North America, although some were transplanted onto the Mississippi Delta and some black mangroves grow on the Gulf Coast of Louisiana and Texas, at about the same latitude. At the northernmost location in Crystal Bay, western Florida, the red mangroves are very stunted (0.5 m tall), whereas around the southern tip of

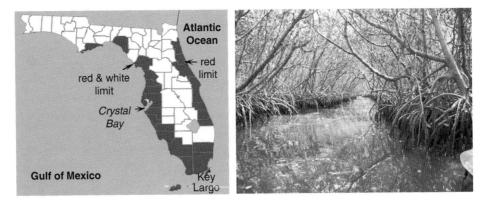

Figure 7.31 Left: map of mangrove distributions in Florida (Atlas of Florida Vascular Plants, 2013); arrows mark the northern limits of red mangrove on the Atlantic and Gulf coasts; Right: Florida red mangrove swamp, showing dense growth of 'stilt' roots' (photo by Epeterson1222, CC).

Florida, they are trees up to 12 m high. According to Chapman (1977), the vigorous growth of mangroves is limited to regions where the January temperature is greater than ~16 °C although stunted forms can survive to lower temperatures down to ~10 °C. Research on the west coast of Florida indicates that the red mangroves cannot survive more than 24 hours of continuous frost, although a few days of discontinuous light frost can be tolerated (Novitsky, 2010).

The Florida mangrove ecosystem includes the three common mangrove species and a mangrove associate, the buttonwood tree (*Conocarpus erectus*). These four mangrove species have differing adaptations to conditions along coasts, and are generally found in partially overlapping bands or zones. The red mangrove grows closest to open water (Figure 7.31) and has multiple stilt-like prop roots which help to stabilize and aerate the soil around its roots. Next is the black mangrove, which has aerial 'breathing' roots (pneumatophores) that grow above soil water level in addition to the underground anchor roots. White mangroves grow closest to the shore and have prop roots and/or pneumato-phores, depending on the substrate conditions. The buttonwood can grow in either shallow, brackish water or on dry land. The Florida Everglades are largely freshwater, with bald cypress trees at the upper margin, but the lower reaches are slightly brackish, with swamps dominated by tall sawgrass reeds (*Cladium jamacaiense*).

In 1981, mangrove communities covered an estimated 1700 to 2200 km^2 in Florida, with about half the mangal forests being managed by Federal, State and local governments, as well as private organizations. The largest areas are in Everglades National Park and there are mangroves throughout the Florida Keys, but coverage there and at Biscayne Bay is reduced by urban/tourist development. Elsewhere, mangals are not common, with the largest areas in the Indian River Lagoon on the east coast, and in a few estuaries and Tampa Bay on the west coast. Here there has been ~44% loss of mangals in the later twentieth century; in contrast, the total mangrove loss for Florida was only ~5% averaged over the entire century. During the twentieth century, three-quarters of the wetlands and mangals along the Indian River Lagoon were impounded (encircled by dikes and flooded)

Figure 7.32   A: Everglades swamp at the lower limit of non-marine plants; B: mangrove channel after Hurricane Andrew in 1992 (before the hurricane, the mangroves were choking the channel); C: germinated buds (propagules) waiting to fall and take root at low tide; D: mangrove with propagules over an oyster bed (photos by D. B. Scott).

for control of mosquitos that do not lay eggs in non-moving water, but in 2001, natural water flow was restored to some areas.

The Florida mangrove system is an important ecological habitat, acting as nursery grounds for over 75% of game and commercial fishes, as well as crustaceans and molluscs. Many fish feed in the mangrove forests, including snook (*Centropomus undecimalis*), snappers (grey, blue and mangrove snappers, *Lutjanus* spp.), tarpon, jack, sheepshead, red drum, hardhead silverside, juveniles of blue angelfish, porkfish and barracuda, the lined seahorse and scrawled cowfish (*Lactophrys quadricornis*). Commercially important invertebrates include shrimp and clams. Mangrove trees are habitats for many birds: brown pelican (*Oelicanus occidentalis*), roseate spoonbill (*Ajajia ajaia*), frigate bird (*Fregata magnificans*), double-crested cormorant (*Phalacrocorax carbo*), the brown noddy tern (*Anous stolidus*), four heron species, osprey, two egrets, the greater yellowlegs shorebird, American coot, peregrine falcon and bald eagle. Other mangrove swamp animals include the American crocodile (*Crocodylus acutus*), diamondback rattlesnake (*Crotalus adamanteus*), the mangrove and salt marsh snakes (*Nerodia clarkii* subsp. *compressicauda* and *taeniata*), the round-railed muskrat (*Neofiber allenii*) and various snails, crabs and spiders. Branches may be draped with bromeliads of the genus *Tillandsia*, including Spanish moss, which is a flowering plant and not a true moss like sphagnum peat moss. In the water, often encrusted on the mangrove roots, are sponges, anemones, corals, oysters, tunicates, mussels, sea stars, crabs, Florida spiny lobster (*Panulirus argus*) and seagrass.

7.4.3 The Deep South: Mississippi Delta and Louisiana wetlands

The most southerly delta in the United States is the Mississippi Delta system, including Balize Delta and the associated set of beach ridges that connect the mainland and the eastern delta along the Louisiana coastline. The Mississippi Delta is the largest in North America and is listed with the world's most vulnerable deltas with regard to impact of anthropogenic activity and future sea level rise (Blum and Roberts, 2009; Syvitski *et al.*, 2009). Here an area the size of a football-field (~70 000 m^2) sinks below sea level every day and tropical storms cause devastating floods in the city of New Orleans. The microtidal (30 cm) delta evolved as a shifting series of delta lobes over the last 6000 yrs (Figure 7.33A). Under natural conditions, about once every millennium, a new deltaic cycle builds another distributary lobe to provide a shorter route to the Gulf of Mexico down the steepening slope built up by sediments deposited on the aggrading delta. Abandonment of the older

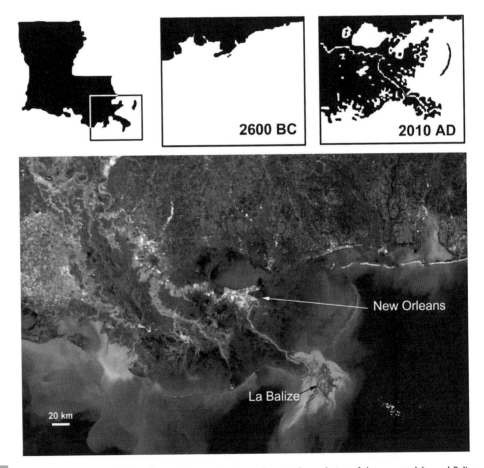

 Figure 7.33   Mississippi Delta region of the Gulf of Mexico. Top: Sketches showing the evolution of the arcuate delta and Balize 'bird's foot' distributary over the past 4.6 millennia. Bottom: Google Earth Image of the Mississippi Delta.

delta lobe then cuts off the main freshwater supply and the interdistributary wetlands dry out, undergoing compaction, subsidence and erosion. Seawater then transgresses into the abandoned delta lobe, forming bayous, bays and lakes along the coastline (see Figure 2.2C).

Major anthropogenic changes include accelerated subsidence from both channelization and sediment compaction. The Balize Delta lobe extension into the Gulf of Mexico began to develop about 700 years ago, but because high artificial levees were built to contain the river shipping to New Orleans, the river sediment was pushed seawards. The new delta therefore forms a long, narrow feature (a 'bird's-foot' delta lobe) protruding into the Gulf of Mexico in contrast to the normal broad-based, triangular shape of natural deltas. The high levees also prevent renewal of sediment over the delta floodplain and contribute to loss of wetlands.

The Mississippi River Delta is a biologically significant region which includes $12\,000\ \mathrm{km^2}$ of the coastal wetlands and 40% of the salt marshes along the total Gulf coastline, and it provides ~16% of the USA fishery harvest, including shrimp, crayfish and crabs. The wetlands of this subtropical delta are predominantly brackish or freshwater ecosystems in the lee of the tributary levees (see Figure 2.2), but the southwestern delta also includes salt marshes with salinities of 25−40 psu that support cordgrass, pickleweed, sea blite, saltwort (*Batis maritima*), salt grass and scattered black mangroves (Scott *et al.*, 1991). The brackish marshes (1–3 psu) are dominated by panic grass (*Panicum virgatum*) mixed with salt grass (90%) and small amounts of spikerushes. The freshwater marshes include floating mats (= 'flotantes'; Figure 7.34)

Figure 7.34 A: crevasse splays; B: one of many freshwater lakes on the delta; C: sediment core taken from 5 m beneath the raft of vegetation; D: floating marsh of tall grasses and herbs (photos by D. B. Scott).

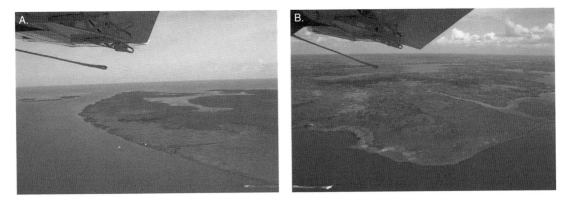

Figure 7.35 These two images show brown marsh in Louisiana marking the dieback of huge areas of cordgrass and meadow salt grass in subtropical salt marshes, which is associated with rising soil water salinity because of increased heat and lowered river water levels since 2000. Green areas are black mangroves and salt grass, which are more salt-tolerant species (USGS: Gabrielle Bodin images). See also colour plates section.

of vegetation ~45–60 cm thick, formed by the dense roots of panic grasses, sedges and a fern (*Thelypteris palustris*) that become detached from the shallow shores and float into deeper water, buoyed by the air in root and horizontal stem tissues. The 'flotantes' are interspersed with pondweeds such as *Sagittaria lancifolia* (see Sasser *et al.*, 1995 for details). The invasive reed *Phragmites communis* is introduced on the delta.

Today, massive floodgates confine the river flow and as a result, the sediment on which the delta complex depends for aggradation is shunted out to sea; this recent change is the leading driver of the region's significant land-loss rates. The vast network of shipping channels and thousands of kilometres of canals dug to accommodate oil and gas industry infrastructure and extraction allows saltwater to penetrate deep into the wetlands, increasing the salinity and killing the freshwater wetlands and/or draining the floating marshes. Death of the vegetation and rising sea level (Syvitski *et al.*, 2009) has resulted in increased erosion and soil subsidence, while the port of La Balize on the delta has been rebuilt every 50 years since 1699 in response to shifting sandbars and hurricane damage.

Flooding in the marshes of the Louisiana mainland appears to be a primary cause of extensive dieback of marsh vegetation, forming thousands of hectares of 'brown marsh' (Figure 7.35). In the nearshore, a 'dead zone' has been forming since 1985, and is increasing in frequency and size as a result of increased discharge of nitrogen and phosphorus from the Mississippi drainage basin (Osterman *et al.*, 2009). In Mobile Bay, Alabama, other studies of foraminifera in sediment cores dated by excess ^{210}Pb chronology show that expansion of the shipping channel and associated dredging have reduced estuarine mixing and changed sedimentation patterns, causing a complete faunal turnover within the bay during the past century. Since the 1950s, restricted tidal flushing and increased terrestrial organic matter have stimulated an increase in agglutinated foraminifera and also contributed to the nearshore dead zone (Osterman and Smith, 2012). Elsewhere in the Louisiana wetlands, dredging and harvesting of the swamp cypress caused surface scouring and, together with marsh drainage, has increased salt build-up and killed vegetation (Keddy *et al.*, 2009). In addition, overhunting of alligators that feed on marsh herbivores (the native muskrat and the intro-duced nutria) may have triggered runaway consumption of the marsh vegetation. As a result of the wetland vegetation die-off, there is now more extensive damage from storm winds and wave erosion, and a greater build-up of soil salinity, which contributes to the formation of brown marshes.

7.4.4 Subtropical coastal wetlands of Baja California peninsula, Mexico

The 1300 km-long peninsula of Baja California in northwest Mexico (Figure 7.36) is a sliver of the Mexican mainland that was isolated and pushed westwards 5–10 Myr ago by tectonic rifting during the formation of the Gulf of California (Sea of Cortés). The peninsula is an actively rising tectonic block containing a complex array of mountainous landscapes interspersed with sandy or saline deserts. The entire region is noted for its abundance of endemic plant species (Garcillán and Ezcurra, 2003; Peinado et al., 2008) and high marine biodiversity, including giant manta rays, gray whales, miniature harbour porpoises and several species of turtles. The subtropical coastal wetlands vary from treeless salt marshes in the Mediterranean and desert climate regions north of the Viscaino Peninsula (West

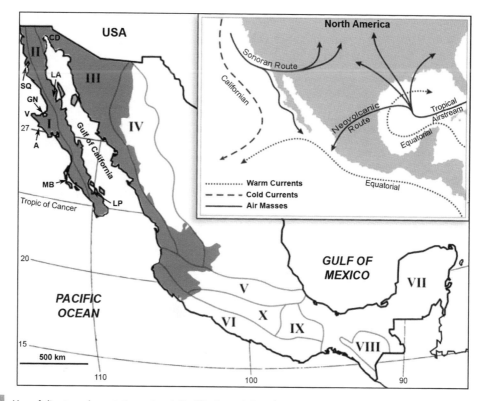

Figure 7.36 Map of climate and vegetation regions I–X of Mexico and Central America, showing how the vegetation zones are related to major ocean currents and weather systems (from Peinado et al., 2011, with permission from Springer Science) A = Laguna de Abreojos; CD = Colorado Delta; GN = Gurerro Negro; LA = Bahia de Los Angeles; LP = La Paz; SQ = San Quintin, V = Viscaino Peninsula.

Figure 7.37 Scrubby mangroves at their northern limits in 1973. Left: red mangrove ~1 m high at in the low marsh at Laguna del Abreojos on the Pacific Coast, surrounded by cordgrass, with pickleweed and *Sesuvium portulacastrum* in the high marsh; Right: sparsely vegetated tidal channel on the Sea of Cortez at Bahia del Los Angeles, where a meadow of cordgrass fringes wide mudflats and surrounds isolated clusters of stunted black mangroves in the low marsh (middle background) (photos by P. J. Mudie).

Coast) and Bahia de Los Angeles (East Coast) to mangrove scrub or forest in Baja California Sur (Figure 7.37). Most of the coastline region is fringed by rocky headlands, particularly along the Gulf of California and access to the shoreline is difficult. Hence, in contrast to the other regions of North America, little was known about the Baja California coastal wetlands until 1972–1974, when systematic surveys were made of 40 salt marshes, mangals and their associated soil salinities by Mudie *et al.* (partly summarized in Macdonald and Barbour, 1974). In 1973, the northern limit of mangroves was at Laguna de Abreojos (27.3° N) on the west coast and at Bahia de Los Angeles (28.9° N) on the Gulf coast. At that time, *Rhizophora mangle* was the northernmost species on the relatively cool, moist coast, but *Avicennia germinans* was furthest north on the warmer, drier Sea of Cortez.

From 1986 to 2001, however, significant reductions in the formerly pristine wetlands occurred as a result of tourist resort development, increased salt production (Figure 7.38) and the expansion of shellfish and fin fisheries. As a result, several large-scale surveys were made to document changes in the mangrove systems ecology and to find ways of curbing further damage and reducing pollution (e.g. Cartron *et al.*, 2005; CONABIO, 2009; Polidoro *et al.*, 2010; Pico *et al.*, 2011; Gonzales-Zamorano *et al.*, 2012). The surveys show ~32 100 ha of mangroves in isolated barrier lagoons and drowned estuaries, with ~98% of these being on the west coast, which has larger sandy barrier lagoons and a cooler, wetter climate than the Gulf Coast. The geomorphology of 51 wetlands with mangroves is now well-documented by Gonzales-Zamorano *et al.* (2011, 2012) and Dominguez-Cadena *et al.* (2011), who differentiate open drowned estuaries, barred drowned valleys, inner shelf barrier lagoons and cuspate lagoons (see Chapter 2). These teams found significant correlations between the mangrove communities and type of coastal geomorphology, and they consider that higher seasonal tides and temperatures on the steep-sided Gulf Coast impose more stress on the wetland vegetation than on the west coast. The lagoons along the gentle slopes of the sandy Pacific coastline also offer

Figure 7.38   Saltflats and salt ponds in the Viscaino Desert of northwestern Baja California. Left: salt harvesting industry at Guerro Negro: only the endemic Baja salt grass can grow on the low pond levees; Right: natural salt ponds behind the sand dunes at Laguna Mormona, north of Gurrero Negro; the white polygons are salt-encrusted mats of blue green algae forming microbialites similar to fossil stromatolites or thrombolites (photos by P. J. Mudie).

greater accommodation space and habitat continuity which favours the spread and colonization of new areas by mangrove propagules. On a microscale, white mangroves dominate the open coasts, red mangroves dominate channels and low-mid marsh zones with daily inundation, while black mangroves dominate the seasonally inundated high marsh.

On a regional scale, it appears that the northern limits of the Baja California mangroves are not simply regulated by winter temperatures, but by a combination of marine currents, continuity of suitable habitats for establishment of propagules, and species resistance to environmental stress. It is also notable that the recent surveys found *Conocarpus erecta* (buttonwood) only in the higher rainfall area of the eastern Cape Region, whereas in 1973, it was also growing at El Palmar near Magdalena Bay on the west coast of the Cape (see Table 7.4). In another recent survey, López-Medellín *et al.* (2011) compare aerial photos of the mangroves at Magdalena Bay and conclude that there was a significant increase in mangroves (mainly black mangrove seedlings) in the high marshes over the last 40 years. This spread is linked to both rising sea level and warm climate El Niño weather anomalies in the 1980s and 1990s. At the same time as the inland spread, however, the seaward mangrove fringe has been receding; therefore there is no net gain in mangrove vegetation in Magdalena Bay.

Other surveys of the salt marshes and mangals of Baja California (Peinado *et al.*, 1995; 2011) compare the salt marsh and mangrove plant species with those in California, Mexico, Florida and Central America (total >300 sites). Using the plant association (phytosociological) method described in Section 6.2, all the coastal wetlands north of 28° N were grouped together as a *Spartina foliosa–Salicornia bigelovii–Limonium–Frankenia salina* complex, and all the mangroves throughout North and Central America, including the Caribbean Islands were grouped in the *Rhizophora mangle–Laguncularia racemosa* order. In general, the subdivision of the Mexican salt marsh communities reflect the intertidal zones described for the Southern Californian marshes in Section 7.3.5, with the additional recognition of a high marsh *Monanthochloe littoralis–Arthrocnemon subterminale* association dominated

Salt marsh habitats of Northwest Mexico. Left: characteristic *Arthrocnemon–Allenrolfea–Monanthochloe* salt marsh at San Quintin, with the succulent subtropical-tropical species *Sesuvium portulacoides* in the foreground; Top right: meadow of purple sea lavender in *Sarcocornia* marsh between salt flats at San Quintin; Bottom right: dead tule reeds on the banks of a tidal channel at low tide, Colorado River Delta near El Golfo, Sonoran Mexico, 1971 (photos by P. J. Mudie). See also colour plates section.

by grass and succulent shrubs, including abundant *Allenrolfea occidentalis*, *Atriplex semi-baccata*, *A. watsonii*, *Cressa truxillensis*, *Distichlis spicata* and *Spergularia macrotheca* (Figure 7.39). In a statistical study of the Baja California coastal marshes, Peinado *et al.* (1995) makes some further distinctions for the Peninsula marshes based on sediment texture, as summarized in Figure 7.40, and notes the isolated occurrence of sweet mangrove (*Maytenus phyllanthoides*) in the southernmost mangrove communities, in addition to the rare buttonwood. Overall, however, it is interesting that in contrast to the extraordinary high occurrence of endemic plants in the upland floras of the Baja California Peninsula, there are only two recognized endemic coastal wetland species: the salt grasses *Distichlis bajaensis* and *Distichlis palmeri*, which occur in the most saline soils of the northwest coast (Bell, 2010) and Colorado River Delta, respectively (Figures 7.38 and 7.39).

The coastal wetlands of Baja California and northwestern mainland Mexico include several unusual features. First, highly saline ponds (up to ~150 psu) behind dune ridges on both west and east coasts (at Isla Angel de la Guarda) contain stromatolite-like growths of blue-green algae (cyanobacteria), where algal masses (microbialites) grow out from the pond margin to form salt-encrusted knobbly mats (Figure 7.38). These cyanobacterial colonies may be modern counterparts of fossil formations called thrombolites seen in

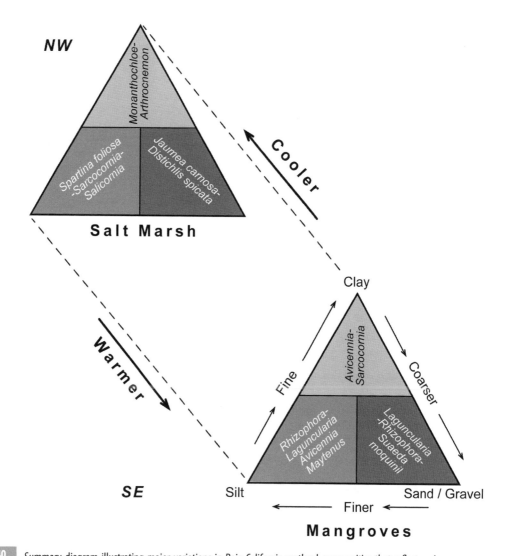

Figure 7.40   Summary diagram illustrating major variations in Baja California wetland communities that reflect major latitudinal and soil texture variations. Triangles are Shepard ternary diagrams for sediment grain size, where an apex represents 100% of either fine clay, medium silt, or coarse sand/gravel, and names indicate the dominant plants listed by Peinado *et al.* (1994) and Mudie (unpublished field notes).

some Pliocene rocks of the peninsula (Johnson *et al.*, 2012). The mats are formed from colonies of the single-celled alga *Chroococcidiopsis* interlayered with filamentous species e.g. *Oscillatoria* and *Spirulina*, and they may indicate the way in which very ancient stromatolites grew a billion years ago (see also Section 11.3). Second, in the Colorado River Delta, megatides of up to 7 m extend 56 km upstream at high tide and are deeply eroding the delta tributaries (Figure 7.39). The channel's banks support the endemic nipa saltgrass (*Distichlis palmeri*), which was harvested for grain by the Native American people (Luecke *et al.*, 1999). Before 1930, the Colorado River Delta was one of the world's largest desert estuaries (the size of Rhode Island), which provided feeding and nesting grounds for birds, spawning habitat for fish and crustaceans in the Sea of Cortez, and sheltered large populations of jaguars, beavers, deer and coyotes. However, construction of dams such as Hoover Dam in the 1930s and diversion of Colorado River water to farms and huge cities like Los Angeles cut off the sediment supply to the megatidal delta, resulting in its drying out and the loss of its brackish water tule reeds (*Scirpus* spp.). In 1993, the wetlands loss was so great that United Nations Educational, Scientific and Cultural Organisation (UNESCO) designated over 12 000 km^2 of the upper Gulf and the Colorado River Delta as a Biosphere Nature Reserve, including much of the lower delta, shoreline and adjacent marine area.

Baja California and adjacent Sonoran Mexico are the only regions of the world where mangrove vegetation is seen against a backdrop of desert cactus vegetation (Figure 7.41). Lagoon barriers across some estuaries in western Baja California Sur are formed by boulders from the La Purisima volcanic field, and these support only stunted white and red mangrove

Figure 7.41 Estuarine and lagoon landscapes of Baja California Sur. A: five episodes of volcanic eruption are the source of the rocky terrane at Rio de la Purisima, inland from San Ignacio Lagoon; creosote bush, short cholla cactuses and one tall saguaro cactus are in the foreground; citrus and date groves fringe the oasis; B: red mangroves fringe the channels at Magdalena Bay, with tall saguaro cacti on the raised beaches behind (photos by P. J. Mudie).

shrubs. In sandy Pacific bays like Bahia Magdalena, however, despite the extensive shrimp farms that now occupy the high marshes, there is a higher diversity (3−5 spp.) of tree-sized mangroves, dominated by *Rhizophora*, *Avicennia* or, rarely, *Laguncularia*. In Bahia del La Paz on the Gulf Coast, however, the black mangrove is the dominant or exclusive species in wetlands that are much altered by road construction for the tourist industry and by discharge of sewage effluent (Mendoza-Salgado *et al.*, 2011).

7.5 Tropical wetlands of North America and the Caribbean Islands

The tropical coastal wetlands are characterized by tall mangrove forests. The largest North American area of tropical coastal wetlands is in Mexico south of the Tropic of Cancer (23.5° N); however, the diversity of mangals is greater in Central America and in some Caribbean Islands, which are also included in this section. Central America is the tropical region from the southern border of Mexico in North America to the northern border of Columbia in South America, and it includes the countries of Guatemala, Belize, Honduras, El Salvador, Nicaragua, Costa Rica and Panama. The major Caribbean Islands are Cuba, Haiti–Dominion Republic, Puerto Rico, Jamaica, the Bahamas, and the Turks and Cacos Islands. The coastline of Mexico alone is about 12 000 km long and it includes a vast array of geomorphological features, including coral reefs and atolls, which cannot be covered adequately in this book. Therefore in the following section, only the salient characteristics of the southernmost regions of tropical North America are mentioned.

The geomorphology of the mainland region is described in several classical papers in the volume *Lagunas Costeras* (Phleger and Ayala-Castañares, 1969) and notes on the island coastlines are given by Spalding *et al.* (2010), who also describe the salient impacts of climate and anthropogenic changes. Twilley and Day (2013) provide the most up-to-date summary of the tropical mangrove ecosystems and adaptations, and give an eco-geomorphological classification of typical mangrove occurrences in deltaic, estuarine, lagoonal and insular (volcanic or coralline) settings. Peinado *et al.* (2011) show that subregional variations in the mainland mangrove communities are most strongly linked to climate, which ranges from subtropical desert to tropical rainforest (see Figure 7.36). In general, the Caribbean Islands have the same hot, humid temperature regime as the Central American mainland, but the island mangroves are smaller because of higher salinities associated with lower rainfall and absence of large rivers. As in Baja California, although the mangroves are valued as resource for fisheries and are the basis of a booming tourist industry, there has also been a significant loss of tropical coastal wetlands during the past 25 years from development of resorts and supporting infrastructure (Figure 7.42) – amounting to ~50% in Mexico alone (CONABIO, 2009). Other losses are because of agriculture and aquaculture – e.g. 95% loss to shrimp farming in the northern Gulf of Mexico (Glenn *et al.*, 2006) – and there are now substantial efforts to protect the unique tropical mangrove fauna, including American, Morolet and Cuban crocodiles, West Indian manatees, the estuarine dolphin, various *Hutia* species (a Caribbean rodent), Caribbean flamingos and the pygmy three-toed sloth (*Bradypus pygmaeus*), of which there are only about 100 still living on a

Figure 7.42 Tourism has been both a major attraction to the Neotropical mangroves of North America and the basis of their destruction. Left: permanently submerged stilt-roots of red mangrove in the Virgin Islands (photo by Caroline Rogers/ USGS Southeast Ecological Science Center); Right top: dead mangroves outside La Paz in 1973, following construction of a new highway; Right bottom: tidal inlet near St. Thomas, US Virgin Islands, with *Rhizophora* roots in silt-laden water (photos by P. J. Mudie). See also colour plates section.

tiny island off Panama. In the microtidal Gulf of Mexico and Caribbean Sea, the widespread red mangrove has adapted to growing with continuously submerged roots and these support an additional colourful fauna of invertebrates and small tropical fish (Figure 7.42).

Overall in the Central and South American tropics, tall mangrove forests (mangrove 'jungles') are confined to the warmest, wettest regions, from ~17° S in southern Mexico to about 4° S in northern Peru on the West Coast, and to about 10° S on the East Coast. The neotropical mangroves are everywhere dominated by the same species found in Florida – the red, black, white and buttonwood mangroves (*Rhizophora mangle*, *Avicennia geminans*, *Laguncularia racemosa* and *Conocarpus erectus*). However, two additional *Rhizophora* species – *R. harrisonii* and *R. racemosa* – are present from El Salvador (~15° N) southwards and their distributions overlap with *R. mangle* in Central America and the Caribbean Islands. Previously it was thought that *R. harrisonii* was a hybrid between *R. mangle* and the west Pacific species *Rhizophora stylosa* – an ancient relict of geological times when the Pacific continents were much closer together, about 60 million years ago. However, DNA

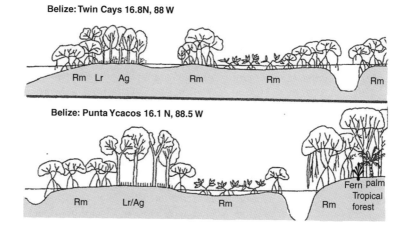

Figure 7.43 Almost all the tropical mangrove swamps in North and Central America include the four species *Rhizophora mangle* (Rm), *Avicennia germinans* (Ag), *Laguncularia racemosa* (Lr) and *Conocarpus erectus*, but height varies with exposure and substrate. Top: shrubs on the barrier reefs at Belize; Bottom: on Belize mainland and further south, mangroves are taller and communities are more diverse, including the mangrove fern and the introduced nypa palm (from Saintilan *et al.*, 2009).

studies now show that the species *R. harrisonii* evolved from introgression (i.e. repeated cross-breeding and back-crossing) between *R. mangle* and *R. stylosa*, probably within the New World tropics (Cerón-Souza *et al.*, 2010). Two other *Avicennia* species join the widespread black mangrove in the tropics: on the West Coast south of Mexico is *Avicennia bicolor* (including *A. tonduzii*, previously thought to be a separate Panamanian species) and *A. schauerana*; on the east coast, only *Avicennia schauerana* is found in addition to *A. germinans*. Both these mangrove species are IUCN Red-List Threatened Plant Species, along with a third exclusively New World genus *Pelliciera*, which has only one species, *P. rhizophoreae* (tea mangrove; manglar piñuelo). The tea mangrove is found mainly on the Pacific Coast from Costa Rica to Colombia and on the Galapagos Islands.

In general, the New World or Atlantic–East Pacific mangrove flora is much lower diversity than those of the Old World and Indo–West Pacific regions (see Box 9.1 Panmangal 3-D zonation). The height of the New World tropical mangrove 'forests' also varies over short latitudinal distances, depending on small differences in substrate and exposure, as shown here for mangals in Belize (Figure 7.43). Exposure to tropical storms can be a primary factor controlling the horizontal zonation, height and diversity of the mangroves, as shown by Piou *et al.* (2006) for Calabash Cay, a 3 km-long island in a Belizean coral atoll. Using an index of diversity based on the Simpson index, they found different patterns of recovery at four sites 41 years after the 1961 Hurricane Hattie hit the atoll, severely damaging the small island. These workers conclude that the post-hurricane recovery rate and subsequent succession of the four dominant mangrove species is actually most strongly linked to the number of surviving trees, being most rapid in the more sheltered locations where nutrient availability is also higher.

8

Examples of South American coastal wetlands

Key points

Most South American wetlands are on the low-lying eastern shores, with few on the tectonically active Andean coast; there are some extensive mangrove forests, but many are reduced by aquaculture and pollution; indigenous mangal people and endemic biota are now endangered; tropical mangroves thrive on deltas and beach barriers of huge rivers, including the Amazon; subtropical wetlands grow in lagoons ('gamboa') with small ocean entrances and tidal creeks – these are permanently closed in Uruguay; the southernmost red, white and black mangroves are at 28.9° S; temperate coastal lagoons of Argentina have cordgrass and pickleweed in the intertidal zone and halophytic herbs in high marshes; temperate wetlands along the desert coast of Chile support beds of sea anemones and mussels. Subarctic wetlands are sparse because strong winds, ozone-hole irradiation and oil spills are added stressors.

As in North America, the West and East Coasts of South America are very different from a geological viewpoint, which is reflected in the Neotropical coastal wetland distributions (Figure 8.1). Tides on both coasts are semi-diurnal or mixed and mostly of medium range (2–3 m) except for macrotidal areas in the northwest, southeast and at the mouth of the Amazon River (Eisma, 1997). The West Coast, part of the 'Pacific Ring of Fire', is tectonically active and bordered by the high Andean mountain ranges, which restrict the amount of lowland available for spread of intertidal wetlands. Here rivers and streams are small, wave erosion is high and earthquakes followed by tsunami waves can liquefy the marsh sediments, resulting in subsidence of up to 1.6 m over large areas (e.g. 200 km in 1979). In contrast, the East Coast passive margin is formed mainly from sedimentary basins and it has extensive low-lying plains crossed by long rivers. There are large deltas at the mouths of the Orinoco, Amazon and São Francisco rivers, which supply sediment to huge mudflats (Figure 8.2). The Amazon River is the source for vast amounts of sediment that coastal currents transport to the giant mudbanks ('slikke') off Surinam and French Guiana. Eisma (1997) summarizes the long-term history of these mudbanks and their alternation with cherniers (shelly sand dunes) during fluctuating climate and sea level changes over the past millennium. The present-day stability of the Guyana coast depends on longshore migration of the mudbanks and their stabilization by mangroves. This coast is now being transformed into fields and aquaculture ponds protected by coastal dikes, which dissipate less wave energy than mudbanks, and which hinder mangrove colonization by preventing transport of propagules from now-enclosed mangals to new mud banks (Anthony and Gratiot, 2010). South of the Amazon, other large rivers, including the São Francisco and Paraíba do Sul, have cuspate

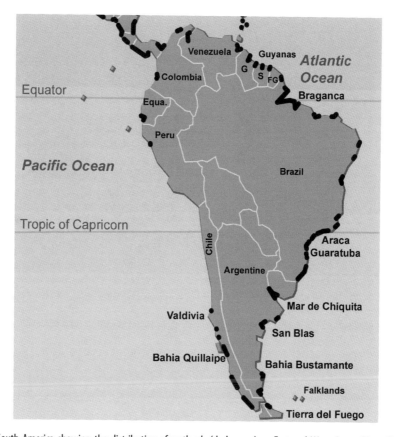

Figure 8.1 Map of South America showing the distribution of wetlands (dark areas) on East and West Coasts (From Ramsar, 2013) and location of places in text. Equa. = Equador; G = Guyana; S = Suriname; FG = French Guiana (modified from WRI/UNEP-WCMC).

deltas and barrier-island lagoonal systems (Figure 2.3) that were shaped by wave-transported sandy sediments during fluctuating late Quaternary sea level changes and are now prograding onto the continental shelf (Dominguez *et al.*, 1987).

West (1977) provides overviews of the tidal salt marsh and mangrove (mangal) forma-tions of South America, and Costa and Davy (1992) list the salt and brackish marsh plants for 31 sites, mostly on the East Coast. Costa *et al.* (2009) provide additional information on the South American salt marshes and their faunas, with a focus on the anthropogenic impacts of landfill, freshwater diversion and overgrazing, since European colonization in the fifteenth century. These authors also detail rapid changes since ~1945 when the South American population outgrew that of North America, industrialization advanced rapidly and pollution increased. In general, South American saltwater and brackish marshes prevail in the temperate-climate areas from 31° to 43° S (~200 000 ha) and mangroves fringed by subtropical-tropical salt marshes (>1.5 million ha) occur north of the Tropic of Capricorn. Tall mangrove forests, however, are only found in the tropics north of 10° S, which are

Figure 8.2 South American mangroves and marshes are concentrated on the east coasts of Colombia, the Guyanas and Brazil; in the Guyanas, mangroves colonize enormous mudbanks exposed at low tide off the mouths of large rivers north of the Amazon River (photo by IRD/C. Proisy).

warmed by the North Equatorial current. The diversity of the Neotropical South American mangroves is the lowest in the world (FAO, 2007).

8.1 South America's tropical coastal wetlands: composition, importance and changes

Mangroves and salt marshes can co-exist in suitable coastal habitats throughout the tropics wherever the mean sea surface temperature is at least 20 °C (Saintilan *et al.*, 2009). The diversity of Neotropical mangroves in Brazil is mostly the same as in Florida, with the addition of another black mangrove, *Avicennia schaueriana* (mangle preto). However, in Colombia, which has shorelines on both Pacific and Atlantic coasts, mangrove diversity is higher, including the eastern Pacific endemic species *Pellicieria rhizophorae* and *Mora oleifera* (Figure 8.3), and the rare shrub *Tabebuia palustris*. In the American tropics, salt marshes are generally rare on the Pacific coast, where tall mangroves grow in isolated sheltered coves separated by high cliffs. On the Atlantic coast, a southern cordgrass (sometimes called *Spartina brasiliensis*) colonizes aggrading mudflats in front of mangrove woodlands along the Atlantic coast of northern Brazil and the Guyanas (= Guyana, Surinam, French Guiana), wherever there is sufficient protection from waves (West, 1977; Costa *et al.*, 2009). Inland of the mangroves, there are salt marshes with cordgrass and tropical grasses like *Sporobolus virginicus*, *Paspalum vaginatum* (seashore paspalum) and salt grass, mixed with sedges and succulents (*Batis maritimus*, *Sesuvium portulacastrum*, *Blutaporon vermiculare* = *Philoxera vermicularis*) in the more saline areas, called *salina*,

Figure 8.3   Endemic East Pacific tropical mangroves. Left: tea mangrove *Pellicieria rhizophorae* with fluted buttress roots (courtesy Christy Holland); Right: *Mora oleifera*, a species in the bean family Cesalpiniaceae, has the largest seeds of all flowering plants (dicotyledons) (photo by Chad Husby, Montgomery Botanical Center).

salitrales, *salares*, which are only flooded at the spring tides. However, in more arid regions of the west coast, like the shallow Bay of Fonseca between El Salvador and Nicaragua, mudflats are bare beyond the mangrove- or marsh-fringed channels, probably because of the hypersaline soils with salinities up to 140 psu.

Near the Amazon Delta, the height of the mangrove trees varies with soil salinity in the range of ~30–80 psu, with the largest trees growing in the least saline areas (Lara and Cohen, 2006). Mixed salt marsh and mangroves grow where mangals have been destroyed by cutting. Without the shade of mangrove trees, mudflats are invaded by Brazilian cordgrass and salt marsh fern *Acrostichum aureum*. In French Guiana, when mangrove forests were cut off from tides by coastal developments, the soils dry out and mangroves die from soil compaction and increased acidity from the formations of hydrogen sulfide. The dead mangroves are then replaced by the seashore paspalum grass and marsh fern, which have shallow roots above the sulfide layer. The impact of these anthropogenic changes and efforts being made to restore the ecosystem are dramatically illustrated in a video presentation by Heliodora Sanchez (2009).

Brazil has the largest area of South American mangroves (8.5% of the world total), filling many estuaries and coastal lagoons along the northern coastline (Spalding *et al.*, 2010). Most of this forest is within the high autumn–winter rainfall region (120–280 cm/yr) around the Amazon River estuary. This river is tidal for several hundred kilometres inland, but the mangroves are restricted to the seaward margin because of the very low salinity in its freshwater-dominated estuary. Some of the earliest native South Americans, such as the 'Warao' boat people of the Orinoco Delta, have lived in the mangroves for over 7000 years and still use them as sources of giant crabs and other shellfish, timber, tannin and fuel. However, large areas of eastern Guyana mangroves and saltflats have been converted to agriculture, and land clearance for shrimp aquaculture is also widespread in Equador and Peru.

The faunal biodiversity of the Neotropical Guyanan mangrove forests is very rich (Table 8.1), being especially noted as a wintering area for migrating shorebirds from the Nearctic biogeographic region (Cleveland, 2008). In Suriname and Venezuela alone there are more than 118 species of birds, including 70 waterfowl species and millions of shorebirds. Mammals include opossums, bats and a variety of primates, carnivores and the world's largest rodent, the semi-aquatic, 0.6 m-high capybara. Reptiles and the only two amphibians tolerant to saltwater are also found here.

Using a 25-year time series of radar and satellite images, Cohen and Lara (2003) studied the mangrove forests along the North Brazilian coast and found significant changes from

Table 8.1 Animals of Neotropical Guyanan mangrove forests. E = endangered

Animal	Scientific name
Birds	
South American scarlet ibis	*Eudocimus ruber*
Black skimmer	*Rhynchops niger*
Gull-billed tern	*Sterna nilotica*
Short-billed dowitcher	*Limnodromus griseus*
Lesser and greater yellowlegs	*Tringa flavipes* and *T. melanoleuca*
Black-bellied whistling duck	*Dendrocygna autumnalis*
Tricoloured egret	*Egretta tricolor*
Mammals	
Capuchin monkey	*Cebus apella*
Squirrel monkey	*Saimira sciureus*
Howler monkey	*Alouatta seniculus*
Guyanan saki	*Pithecia pithecia*
Giant ant-eater	*Myrmecophaga triactyla*
Jaguar	*Panthera onca*
Puma	*Puma concolor*
Ocelot	*Leopardus pardalis*
Capybara	*Hydrochaeris hydrochaeris*
Reptiles	
Olive Ridley's turtle (E)	*Lepidochelys olivacea*
Green iguana	*Iguana iguana*
Spectacled caiman	*Caiman crocodilus*
Anaconda	*Eunectes murinus*
Amphibians	
Paradoxal frog	*Pseudis paradoxa*
Pipa frog	*Pipa pipa*

1972 to1997. On the Bragança Peninsula at the mouth of the Amazon River, the landward migration of sand covered the tidal mudflats and smothered the mangrove pneumatophores, thereby killing the trees and reducing the original forest area by ~42%. On the other hand, 19% of the mangrove forest area expanded over the mudflats, ~38% invaded the high marshes and ~39% of the mangals remained stable. Another study of recent ecosystem change focusses on the mangroves and marshes of the Caravelas estuary and the chernier plain of the Brazilian tropical coast ~17.75° S, in one of the oldest areas colonized by European settlers and impacted by anthropogenic changes for over 500 years (Sousa *et al.*, 2013). This study uses sediment grain size, 11 kinds of heavy metals (including toxins like arsenic and lead), carbon:nitrogen ratios, and foraminifera to investigate the environmental evolution of the Caravelas estuary over the past 70 years. The results suggest that the largest changes in the wetlands are the result of recent dredging rather than landscape changes and/ or urban pollution.

The main drivers of the tropical coastal wetland dynamics, however, seem to be longer-term trends in sea level rise, including large oscillations (2−3 m over 200−300 years) in a generally regressing sea level (Dominguez *et al.*, 1987). Proxy data from peat, mangrove and salt marsh pollen in radiocarbon-dated cores from Bragança Peninsula show that there has been a succession of advances and retreats in mangrove forests during the past 1000 years (Figure 8.4; Cohen *et al.*, 2005, 2009). The succession can be linked to both sea level changes and fluctuations in tropical rainfall associated with a temperature rise during the 'Mediaeval Warm Period' and a cooling during the 'Little Ice Age' (LIA) interval. In an

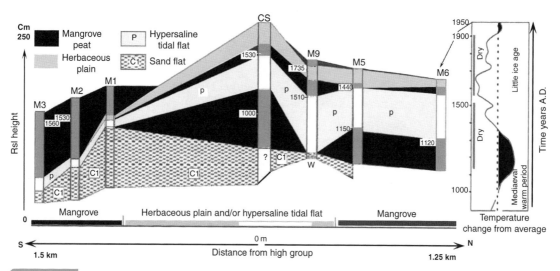

Figure 8.4 Sediments (columns) and pollen assemblages (legend) in series of cores from the Amazon Estuary show several shifts from mangrove forest vegetation (black) to high marsh hypersaline tidal flats (P) or upland herbs in over the past ~1000 years. Numbers indicate ages in years AD and show that mangrove growth was mainly during the Mediaeval Warm Period in low-lying terranes, but in higher areas to the north, this was interrupted by dry intervals when hypersaline soils developed. With the present climate warming, mangroves are spreading again (adapted from Cohen and Lara, 2003, with permission from Elsevier).

innovative follow-up study, Cohen's team also successfully tested the use of spectroscopic measurement of tannin concentrations in sediment cores as a quicker way of tracing mangrove histories than pollen or foraminiferal records. This method works because mangroves have very high tannin content (up to 20% leaf weight). Other mangrove pollen records are documented for Marajó Island, eastern Amazon estuary (França *et al.*, 2012), and for the Guyana coast, where baseline mangrove pollen studies were made by Muller in the Orinoco Delta (Blasco, 1984). The Marajó Island record also reveals that the recent disappearance of mangroves is linked to increased erosion from sea level rise and greater river discharge from Andean icefields, which are melting faster than any other tropical region. A pollen study of four longer Holocene cores from Marajó Island shows that mangrove vegetation was more widespread around 7200 and 2270 calendar years BP because of stronger marine and lower freshwater influences during those times (Smith *et al.*, 2011).

8.2 South American subtropical and temperate East Coast

Subtropical mangrove swamps and temperate salt marshes in the Southern Atlantic extend from 25 to 45° S, from Brazil to Argentina. The coastal wetlands of this summer-dry mesotidal region are mostly small coastal lagoons on large deltaic plains that are in various stages of infill (Isla, 2013). Many lagoons in Uruguay (between Brazil and Argentina) are permanently closed to tidal exchange. The northern wetlands (25−30° S) generally have more precipitation and much warmer climates, which allows the growth of mangrove forests. In this subregion, there is much concern over recent changes because new laws allow 10–35% of the associated saltflats to be converted to shrimp ponds (Rovai *et al.*, 2012). In the cooler latitudes to the south, there are fewer mangroves and more salt marshes with greater diversity than in the tropics. *Spartina braziliensis* and several other common marsh plant species (mostly *Salicornia* and its relatives) thrive in these more moderate temperate region climates. Isacch *et al.* (2006) summarize the main features of the habitats characterizing the Brazilian and Argentinian tidal salt marshes and they list the main species as smooth cordgrass and the native dense-flowered cordgrass (*Spartina densiflora*), pickle-weed, salt marsh sedges and bulrush, black spikerush, spiny rush (*Juncus acutus*), pampas-grass (*Cortadeira celloana*) and the invasive reed, *Phragmites australis*. The southernmost temperate wetlands are influenced by the cold Falkland Current and generally have less diversity because of more severe climate conditions that are similar to the subarctic salt marshes of North America. The microfossils (principally foraminifera) that live in these southernmost areas are not only less diverse, but have numbers less than 100 per 10 cm^2 of sediment (Scott *et al.*, 1990). These cool temperate marshes are frozen for much of the southern polar winter (July–September). The high and low marshes of this cooler region are heavily grazed by rabbits (*Oryctolagus* sp.) and various other wetland rodents (Costa *et al.*, 2009), which are prey for the thick-tailed possum (*Lustreolina crassicaudata*) and crab-eating fox (*Cerdocyon thous*). An abundance of the intertidal burrowing crab *Neohelice granata* (*Chasmognathis granulatus*) is also a characteristic feature of these marshes and may determine the vegetation succession (see Section 8.2.3).

8.2.1 Brazilian south coast

The estuaries and 70 km-long lagoons of the Brazilian Mar de Cananeia area ~25° S are the historically best-studied wetlands in this region. Detailed baseline studies were made in the 1950s and 1970s, after which river diversion by canals lowered the lagoon salinity and irreversibly changed the mangrove vegetation, although the salinity increased after the canal filled in (Por *et al.*, 1984). The three common mangroves in this lagoon (Figure 8.5) grade inland to hibiscus shrubs, spider lily (*Crinum*) and salt marsh fern, and the trees support abundant lichens, some orchids and bromeliads. Neritic (shallow-water) and open-ocean diatoms are abundant, as well as estuarine marine zooplankton (copepods and mysids). The mangrove roots host a wide range of benthic animals and various cushion-forming algae and small seaweeds that differ from most of those on the tropical Caribbean mangroves

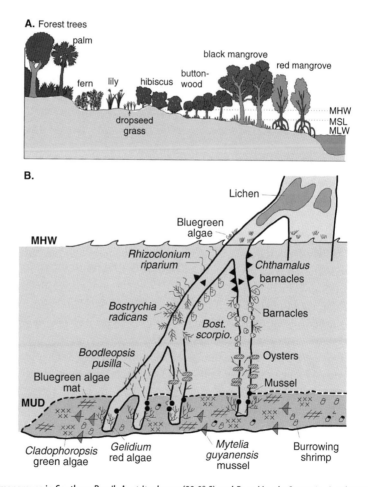

Figure 8.5 Subtropical mangroves in Southern Brazil, A: at Itanhaem (23.8° S) and B: at Mar de Cananeia. A: schematic profile of zonation for three low marsh mangrove trees R, A and L, and gradation to high marsh–EHW transition with *Hibiscus* shrubs, cordgrass, *Crinum* lily, *Acrostichum* fern and palms e.g. *Euterpe oleracea* at the border of the rainforest (Chapman 1984 data); B: zonation of algae, molluscs and crustaceans on *Rhizophora* roots (after Por et al., 1984). *Bost. scorpio.* = *Bostrychia scorpiodes*.

(Figure 7.43). The Cananeia mangrove-root vertical zonation of epiphytic plants and animals is illustrated in Figure 8.5B. In the semi-enclosed estuarine system of Paranaguá Bay, mangroves of the high energy, outer euhaline (25–30 psu) bay almost exclusively host the tree-dwelling sesarmid crab *Aratus pisonii*. Year-round, this herbivore consumes up to 6% of the leaf area of the mangroves, preferring white mangroves growing in more nutrient-rich soils where the trees are taller (Faraco and da Cunha Lana, 2004).

At Araçá Bay, ~23.75° S, a small relict mangrove stand supports a diverse and colourful marine ecosystem that is important for the local fishery and is being intensively monitored for impacts of sewage outfall and other anthropogenic changes relative to studies from 1919 to 1986 (Amaral *et al.*, 2010). There are now 34 new introduced species (mostly crustaceans and polychaete worms), while nine native species (mostly sponges) are near extinction. A study of the long-term history of mangroves at Cardoso Island, ~25° S, uses organic carbon and its isotopes, combined with pollen and diatom analyses, to show that the mangrove community in this area is probably relatively young, having replaced the rainforest vegetation between 5500 and 2200 yr BP (Pessenda *et al.*, 2012).

The most southerly Neotropical mangroves are found in the Guaratuba area of Brazil (Figure 8.6) at about 28.9° S. These wetlands were investigated, together with several salt marshes in Argentina (Scott *et al.*, 1990; Barbosa, 1995; Barbosa *et al.*, 2005) to determine the microfossil faunas of this region. Guaratuba has a robust shrub mangrove flora and diverse foraminiferal populations similar to those further north in South America, but less diverse than in the tropics. A mainly brackish wetland at Saquarema Lagoon near Rio de Janeiro had few mangroves at the time of the study.

To investigate the environmental response to anthropogenic influences, a comparison was made of the species dynamics in two mangals with different degrees of disturbance, degradation and anthropogenic influences. The Guaratuba wetland was relatively undisturbed compared to the Antonina mangle, which was altered and polluted by harbour usage. The structural and species composition of mangroves was made for 15 plots in each wetland. Roughly equal total counts (~500) of the red, white and mangal preto (*Avicennia shaueriana*) were tallied in both the Guaratuba and Antonina wetlands, but there were large differences in species dominance and distribution: (1) in the less-disturbed Guaratuba wetland, mangroves were evenly distributed, and mangal preto is most frequent and larger than at Antonina, (2) at the more disturbed Antonina wetland, there were few mangal preto and more red and white mangroves, with white mangrove dominating in the most disturbed area. These local differences highlight the requirements of each mangrove community and emphasize the need for specific ecosystem management plans.

8.2.2 Patos Lagoon: the world's largest 'choked' lagoon

Patos Lagoon is a large (10 000 km^2) coastal lagoon extending from 30 to 32° S and ending at the Brazilian–Argentinian border. This huge, warm-temperate lagoon complex is connected to the Atlantic Ocean only by the short, narrow Patos Lagoon Estuary, forming a 'choked' lagoon that is permanently open to limited amounts of tidal exchange (Costa *et al.*, 2003). Locally, over 80% of the marshes have been either lost or degraded by anthropogenic changes and increased erosion since ~1820, at which time *Salicornia ambigua* and *Sesuvium portulacastrum* dominated the low and high marshes, respectively. Today, the low marsh is dominated by cordgrass and the high marsh is sedge-dominated,

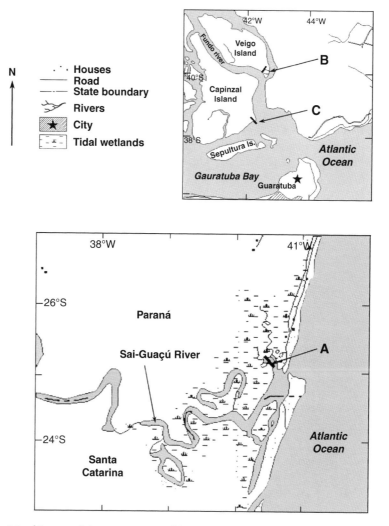

Figure 8.6   Map of the state of Parana and the mangrove area of Guaratuba Lagoon (after Scott *et al.*, 1990; Barbosa, 1995; Barbosa *et al.*, 2005, with permission of Cushman Foundation for Foraminiferal Research). A, B, C mark sites studied for microfossils.

grading landwards to a zone of the endemic shrub 'capororoca' (*Myrsine parviflora*) and giant leather-fern, *Acrostichum danaeifolium*. Seagrasses in the shallow lagoon (<1.5 m) are primarily ditchgrass, with horned pondweed (*Zannichellia palustris*) in oligohaline areas (Copertino and Seeliger, 2010). Studies of sediment porewater, crab feeding and transplants of cordgrass and sedges among four vegetation zones show that unlike fully tidal wetlands, there is no distinct gradient of physical stress with height above MLLW in these 'choked' salt marshes that are irregularly flooded. However, transplant experiments show that smooth cordgrass barely survives in the sedge marsh (*Scirpus maritimus*) zone, where it is preferentially eaten by the euryhaline Atlantic estuarine crab *Neohelice granulata* (*Chasmognathus granulatus*).

Parts of the Patos marshes are being eroded, apparently because of rising sea level (4.4 mm/yr) and increased frequency of northeasterly wind-driven waves (Marangoni and

Costa, 2009). Annual- to decadal-scale studies also show a correlation between expansion of southern cordgrass and frequency of flooding during El Niño weather events, with lateral growth being reduced by ~80% in the El Niño years. Growth of ditchgrass and macroalgae show quasi-decadal fluctuations that are related to river discharge, with the submerged aquatics disappearing for about five years after high discharge events.

8.2.3 Argentina to Tierra del Fuego and Mar Chiquita wetlands

Andrejo Bortolus and his coworkers (2009) outline the main features of a discontinuous string of cool-temperate to subarctic salt marshes extending from San Blas (40.5° S) in Southern Argentina to Rio Grande (~54° S) in Tierra del Fuego, on the Atlantic coast of South America. This region has smaller coastal plains than in either northern Argentina (Buenos Aires province) or in eastern North America. However, large tidal ranges (up to 14 m amplitude) often combine with low-slope shorelines to produce mudflats protected from wave action that support salt marshes. Other smaller areas have rocky/gravelly marshes. The marsh vegetation is dominated by cordgrasses to ~44° S, but in the cooler regions, the dominant is *Sarcocornia perennis*. This latitudinal shift is the same trend as found in northwestern North America (see Section 7.3.1). In South America, vegetation height and biomass is lower in the cooler, higher-latitude *Sarcocornia* marshes, but plant cover shows the opposite trend, with individuals growing closer together in the colder regions. Where there are cordgrass and pickleweed marshes at the same latitude, the cordgrass wetlands support more diverse macroinvertebrate faunas than the succulent pickleweed marshes. Overall, however, most of the wetlands along this coast are characterized by invertebrate assemblages of small shellfish, ostracods, barnacles, polychaete worms, mussels and other small bivalve molluscs.

Bortolus (2006) describes the history of the super-invasive southern/dense-flowered cordgrass, *Spartina densiflora*, that seems to have originated on the East Coast of South America, but has spread almost worldwide because it tolerates a broad spectrum of conditions and can reshape the structure of the communities its invades – not just mudflats, but also sandy, muddy and rocky shores, and even shifting cobble beaches. This invasive 'bioengineer' may have gained a large amount of genetic diversity as it invaded Chile, Spain and Morocco, so detailed local-scale research throughout its widespread distribution is needed to analyze the sources of this variation in the effort to develop effective conservation strategies.

Examples of the Argentinian temperate salt marshes are shown for Mar de Chiquita (Figure 8.7), San Blas (Figure 8.8) and Bahia Bustamante (Figure 8.9). The San Blas marsh area has an interesting feature of a raised mid-Holocene high stand, with razor clams well above the tide levels: these reefs were alive ~5 kyr BP. These sea level high stands extend from Tierra del Fuego to northern Brazil, with some coastal areas also containing anthropological mollusc shell heaps (middens) up to 10–15 m high. Farther south in Argentina, some salt marshes have a more diverse flora, including many of the species found in the Northern Hemisphere. For example, at Mar de Chiquita (Figure 8.7), there is mostly *Sarcocornia*, with some other succulents and grasses similar to those in the cool temperate parts of North America.

Box 8.1   **What's in a name?**

When researching information on the salt marsh vegetation of South America, one quickly runs into a confusion regarding the names given to the dominant salt marsh grass and succulents – the equivalents of cordgrass and pickleweed (*Salicornia*) in North America. 'A rose by any other name may smell as sweet', but all roses are not necessarily even remotely related botanically. It is necessary, therefore, to understand some boring but important details about South American salt marsh plant names. According to Isacch *et al.* (2006), there is no coherent systematic treatment of the flora of Latin America. Consequently, along the Atlantic coast, *Spartina alterniflora* and *Spartina densiflora* have been variously classified into several varieties, different species and hybrids, mainly based on differences of selected features of the flowering stem (= inflorescence). For practical field mapping, the dense-flowered *Spartina densiflora* Brong. can be treated as being the same (synonymous) as *Spartina montevidensis* Arech., *Spartina montevidensis* (Arech.) St. Yves, *Spartina patagonica* Speg. and *Spartina juncea* Willd. The other important low marsh cordgrass species *Spartina brasiliensis* Raddi. is considered synonymous with the smooth cordgrass *Spartina alterniflora* Loesel and with *Spartina maritima* var. *brasiliensis* (Raddi) St. Yves. Similar taxonomic uncertainties and ambiguities are found for *Sarcocornia perennis* (P. Mill.) A.J. Scott, which is treated as synonymous with *Sarcocornia fruticosa* (= *Salicornia fruticosa* L.), *Salicornia ambigua* Michx., *Salicornia gaudichaudiana* Mog. and *Salicornia virginica* L.

Figure 8.7   Mar del Chiquita coastal lagoon, Argentina, ~37.5° S; NASA image of the lagoon and surrounding farmland, July 2006.

At ~37.5° S, close to Mar del Plata, Argentina, there is an extensive coastal salt marsh at Mar Chiquita ('Little Sea' in Spanish), which is a periodically closed microtidal coastal lagoon now largely filled by sediments from four rivers (Isla, 2013). This temperate climate region has an annual mean temperature of 13.8 °C (monthly range 7.3–20 °C) and ~941 mm precipitation, mainly in summer (Latorre *et al.*, 2010). The salt marsh has several different plant communities: southern cordgrass and pickleweed dominate the marginal flats around the lagoon, surrounded by saltgrass meadow with spearscale (*Atriplex montevidensis*), smooth cordgrass, gumweed (*Grindelia discoidea*) and sea lavender (*Limonium brasiliense*)

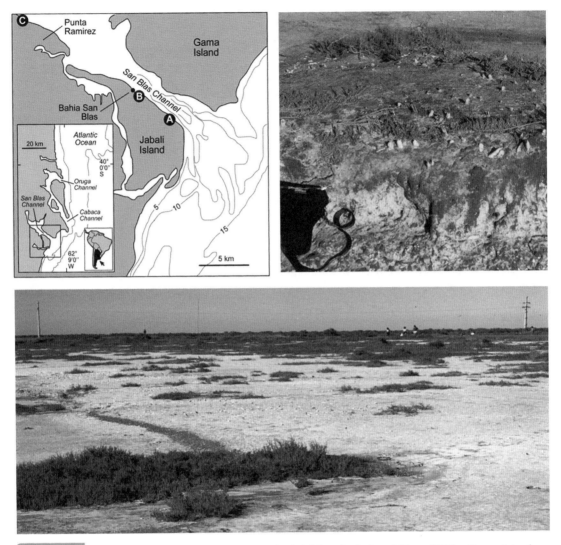

Figure 8.8 Top left: San Blas Channel and lagoon water depth (m) from Cuadrado and Gómez (2011), with permission from Elsevier; Top right: fossil razor clam shells on a raised saltflat mark the position of the higher mid-Holocene sea level. These molluscs would have lived in the lower intertidal zone 5000 yr BP. The salt marsh halophyte shrub *Sarcocornia* here is also now well above the high tide line; Bottom: high marsh, with shrubby halophytes on the saltflats (foreground) and in the middle marsh (background) (photos by D. B. Scott).

as the main species. In the highest salt marsh zone, spiny rush and the sedge *Scirpus cernuus* occur together with several southern herbs (*Ambrosia tenuifolia*, *Apium sellowianum*, *Samolus valerandi*) and paspalum grass.

Mar Chiquita coastal lagoon is located in the southeastern Pampas grasslands of Buenos Aires Province, Argentina, on plains built from late Pleistocene wind-eroded basins and paleodunes, representing Holocene beach ridge deposits formed during sea level fluctuations,

Figure 8.9 Bahia Bustamante *Sarcocornia* marsh with a penguin colony on the salt marsh (photo from Say Hueque Tours in Argentina, www.sayhueque.com). Bahia Bustamante is ~45.9° S.

6000 yr BP (Figure 8.7). Geological records of foraminifera and ostracods in a core from shallow water off the mouth of Rio del Plata just north of Mar Chiquita lagoon support other studies outlining a stepped model of late-Holocene sea level fall between 5511–5792 cal yr BP and 2854–3188 cal yr BP on the coast of Buenos Aires. Infilling of the lagoon was accelerated by introduction of an introduced polychaete *Ficopomatus enigmatus* that builds worm reefs and traps sediment (Laprida *et al.*, 2007).

Alberti *et al.* (2008) carried out an experiment in the Mar Chiquita high marsh (flooded 15 times/month) to examine the idea that invertebrate herbivory regulates the competitive and facilitative interactions driving vegetation succession in Argentinean cordgrass–pickleweed marshes. In this ecosystem, the cordgrass is preferred by the dominant herbivore, the crab *Chasmagnathus granulatus*. Experiments performed on crab feeding, plant structure and shading show that when crab herbivory is low, plant interactions prevail, but when herbivory is highest, survival of young cordgrass plants initially depends on association with pickleweed forbs, which reduce herbivory. However, as the grass patches expand, the larger cordgrasses can tolerate herbivory better and they out-compete the slower-growing, shorter pickleweed. In another important study, Daleo *et al.* (2007) show that good growth of the southern cordgrass also depends on presence of root fungi (arbuscular mycorrhiza), which in turn require the crab ecosystem engineers to supply sufficient soil oxygen by burrow formation.

8.2.4 San Blas, Argentina

There is an important salt marsh study area at San Blas, 40° S in Anegada Bay, Argentina, on the coast of southern Buenos Aires Province. This bay is shallow, with depths usually less than 5 m, and it is separated from the Atlantic Ocean by several islands, e.g. Gama Island (Figure 8.8). However, unlike Mar Chiquita and Patas lagoons, the bay has three large tidal inlets with maximum depths from 15 to 28 m. San Blas channel, 2.5 km wide and 12 km long, is the deepest tidal inlet and provides an example of a tidally dominated barrier lagoon with complex subsurface dune fields. Near the town of Bahía San Blas on the south side of this channel, the marshes are dominated by the southern and smooth cordgrasses, which grow year-round, forming dense monospecific stands across the entire intertidal region. These marshes commonly host very high densities of the crab *Neohelice* (*Chasmagnathus*) *granulata* – often around 60 crabs/m^2 in areas of better-oxygenated sandier sediments aerated by the tidal flow (Daleo and Iribarne, 2009). Experimental studies of the salt marshes at Peninsula Valdez, ~42.42° S, show that here the presence of pickleweed prevents smooth cordgrass from colonizing the high marsh, but the succulent forb cannot survive the frequent submersion by turbid seawater when transplanted to the cordgrass-dominated low marsh zone (Idaszkin *et al.*, 2010).

8.2.5 Bahia Bustamante, near Commodore Rivadavia

Bahia Bustamante on Golfo San Jorge is one of the less-disturbed salt marsh regions of Argentina. Here the tidal range is also high and the extensive high marshes are sparsely covered by widely spaced shrubs of *Sarcocornia perennis*, *Atriplex lampa*, seablites (*Suaeda fruticosa*, *Suaeda patagonica*) and Patagonian seaheath (*Frankenia patagonica*). The succulent, shrubby salt marsh vegetation of this semi-desert region visibly resembles that found around San Quintin Bay in Baja California, in western North America, although the species at San Quintin are different (*Arthrocnemon subterminale*, *Atriplex watsonii*, *Suaeda californica* and *Frankenia grandifolia*). The salt marsh at Bahia Bustamante is one of a vital chain of feeding areas for migratory waterfowl that spend the austral summer (Northern Hemisphere winter) in Tierra del Fuego and migrate to northern Canada for breeding in the northern summer. The birds include the Hudsonian godwit, *Limosa haemastica*, which is the smallest and least well-known of the four species of *Limosa* shorebirds. There are major difficulties in studying this godwit because it spends much time in very remote South American wetlands and then migrates over 8000 km to its nesting areas in the Hudson–James Bay region and Mackenzie–Anderson Deltas of northern Canada (Senner, 2007). At Bahia Bustamante, penguins also feed in salt marshes (Figure 8.9) and flocks of hot-pink flamingos are protected on a huge private sheep farm in Argentine Patagonia that is the same size as the large Hawaiian island of Kauai.

8.3 South American West Coast temperate region

There are few salt marshes on the West Coast of South America because of the steep coastline and the presence of the arid Atacama Desert, which does not have any permanent rivers or estuaries. There are only very small seasonal marshes along the 4200 km extent of Chilean coastline, mainly around Valdivia near the mouth of the only large estuary, and at the heads of embayments to the south. The coast here is cooled by the Humboldt Current flowing northward from Antarctica. The average summer and winter temperatures at Valdivia are

17 °C and 8.5 °C in summer and winter, respectively, and average rainfall is ~178 cm/yr. Southwards, these temperatures decrease to 13.6 °C and 4.7 °C at Puerto Aisén near Chiloe Island farther south, where the average rainfall varies from ~206 mm/month in summer to 300 mm in winter. The entire coastline is prone to earthquake shocks: during the twentieth century, Chile was struck by 28 major earthquakes, all with a force greater than 6.9 on the Richter scale (see Box 8.2 Measuring earthquake magnitude). The strongest of these earthquakes was at Valdivia in 1960 (reaching 9.5) and offshore of Central Chile in 2010 (estimated as 8.8 on the Richter scale).

There are four Ramsar coastal wetlands along the Chilean coast that protect brackish lagoons important for birds such as Andean gull (*Larus serranus*), the Andean flamingo (*Phoenicoparrus andinus*), James's flamingo (*Phoenicoparrus jamesi*), the horned coot (*Fulica cornuta*) and the black-necked swan (*Cygnus melanocoyphus*). The northernmost national park (Complejo Lacustre Laguna del Negro Francisco y Laguna Santa Rosa, ~27.3° S) also protects the vicuña (*Vicugna vicugna*), which has vulnerable status in Chile. At ~32° S, there is a small (34 ha) private nature reserve, Santuario de la Naturaleza Laguna Conchalí, which includes a brackish coastal lagoon typical of the central Chilean wetlands. Here the ecosystems of the Atacama–Sechura Desert and the Chilean–Matorral ecoregions meet, and the lagoon is a key staging area for migratory birds along the central Chilean coast. A creek feeds freshwater to the lagoon, and during high rainfall, the barrier island is flooded and the lagoon becomes a tidal estuary. The salt marshes here are dominated by salt grass, alkali seaheath (*Frankenia salina*) and Peruvian pickleweed *Sarcocornia peruviana*. The Coscoroba swan (*Coscoroba coscoroba)*, white-faced ibis (*Plegadis chihi*), the endemic Chilean mockingbird (*Mimus thenca)* and tropidurid lizard (*Liolaemus zapallarensis*) are noteworthy species and there are five endemic fishes, including *Odontesthes brevianalis* and *Mugil sp.* (Ramsar Site no. 1374; RIS information: 2004).

The Chilean salt marshes further south are dominated by *Spartina densiflora*, which commonly occurs from ~36−41.5° S and may be spreading both northwards and southwards (Bortolus, 2006). Tidal marshes of southern Central Chile are located in the estuaries of small rivers within sheltered bays, e.g. Bahia Quillaipe near Puerto Montt, and on Chiloe Island. Here there are six typical salt marsh communities: pickleweed and smooth cordgrass in the low marsh, arrowweed and the grass *Puccinellia glaucescens* in the middle marsh, and more diverse herbs, grasses and sedges (*Cotula coronopifolia, Anagallis alternifolia, Distichlis spicata, Selliera radicans* and *Scirpus* spp.) in the high marsh. Brackish swamps are dominated by arrowweed and sedge, *Scirpus californicus*. The EHW margin includes dry saline prairies with Baltic spikerush *Juncus balticus*, herbs *Lotus corniculatus* and *Centella* (a buttercup) and *Agrostis* grasses in wetter areas (San Martin *et al.*, 1992).

8.3.1 Valdivia, Chile

The largest earthquake ever measured occurred in coastal Chile north of a small marsh in the Valdivia River Estuary, where the trace of the 9.5 M_L quake occurred in 1960 (Jennings *et al.*, 1995, see Section 12.2). The sedimentological and microfossil sequences associated with this event are similar to those recorded for the 1992 Alaskan earthquake (Section 7.2). The data needed for a paleoseismological study of the Chilean earthquake came from surveys of the

Measuring magnitude of earthquakes

The *moment magnitude scale* (abbreviated as MMS; denoted as M_W) has now replaced the 'Richter scale' (M_L), which was used to indicate earthquake size from ~1930 to 1970. The new scale is used by seismologists to measure the size of earthquakes in terms of the energy released. The magnitude of M_W is based on the seismic moment of the earthquake, which is equal to the rigidity of the Earth multiplied by the average amount of slip on the fault and the size of the area that slipped. Although the old and new formulae for calculating earthquake size are different, the new scale retains the same continuum of magnitude values defined by the Richter scale.

tidal marshes in the Valdivia Estuary within 25 years after the earthquake, when the salinities in the river-dominated estuary were found to be <10 psu most of the year, but higher in summer. The main species at the seaward end of the estuary in 1989 were the spike rush *Juncus balticus*, the sedge, *Scirpus californicus*, and arrowgrass *Triglochin maritima*, in addition to the widespread pickleweed and salt grass. Sedges, the fern *Blechnum rasschilensis* and raspberry *Rubus constrictus* were the dominant species higher in the marshes and further inland. In a follow-up study carried out in 2008, Reinhardt *et al.* (2010) found that although the Great Chilean Earthquake of 1960 caused 1–2 m of land subsidence far upstream in the estuary, sedimentation has now filled in the river bank channels to about half of their pre-1960 depth. However, it is estimated that along this arid, microtidal coastline, the floodplains will require one to two centuries to recover completely, in contrast to a shorter time of one to two decades estimated for macrotidal inlets in northern Alaska. It is also notable that about 33% of the 80 plant species that invaded the newly formed mudflats are exotic invasives (Costa *et al.*, 2009). One of these invaders, the submergent Brazilian waterweed (*Egeria densa*) then became the main food source for black-necked swans, grebes and coots, and many of these birds died when the weed was killed by toxic waste from local industries. The question of whether or not to restore the favoured invasive weed is a difficult one to answer and is still being debated.

8.3.2 Bahia Quillaipe: a beautiful bay full of sea anemones and worms

Bahia Quillaipe 41.5° S is at the head of an extensive drowned bay system between the Chilean Coastal Range on the west and the glacier-covered Andes on the east (Figure 8.10). This is one of few areas in Chile where the coastal wetland's macrofaunas and microfaunas have been studied in detail. Reise (1991) found that the upper mudflats of this bay are uniquely characterized by abundant (1 per cm^2) sea anemones *Bunodactis hermaphroditica* that co-exist with symbiotic green algae. In contrast, the sandflats contain a more diverse fauna of burrowing ghost shrimp (*Callianassa brachyopthalma*), molluscs and worms. The polychaete worms include 19 species, of which 8 are endemic or restricted to cold waters of South America (Hartmann-Schröder, 1991). Around MLW there are banks of Chilean mussels *Mytilus chilensis*, below which there are fields of the red alga *Gracilaria* (Irish moss) that are periodically harvested commercially; there are no seagrasses in this estuary.

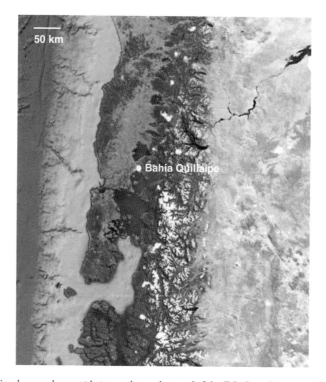

Figure 8.10 Bahia Quillaipe is in a large embayment between the southern end of the Chile Coastal Range and the icefields (white areas) of the southern Andes (from Google Earth). The whole southern fiord region is called Patagonia.

A study of the vertical distribution of foraminifera (Fernandez and Zapata, 2010) reveals nine calcareous and nine agglutinated foraminiferal species in two zones: Zone I is only in the salt marsh and has agglutinated species (mostly *Haplophragmoides manilaensis*) living in low salinities (~18.7); Zone II occurs on the tidal flat and is a calcareous-agglutinated community (mainly *Ammonia beccarii*) associated with higher salinities (~32.8 psu). At Bahia Quillaipe, an agglutinated association of *Trochamminita salsa–Jadammina macrescens* is restricted to the top of the salt marsh, where salinity and pH values are low, in contrast to the high marsh at Valdivia where only *Trochamminita salsa* is domitant.

8.4 South American subarctic salt marshes: oil spill and ozone damage

8.4.1 Bahía Lomas, Tierra del Fuego, Antarctic Chile

The long coastline of Chile ends in the subarctic region of northern Tierra del Fuego Island, where the Bahía Lomas wetland (58 946 ha) is Ramsar's (2007) second southernmost global site. Bahía Lomas is on the Chilean–Argentinian border, facing the Magellan Strait; it has the

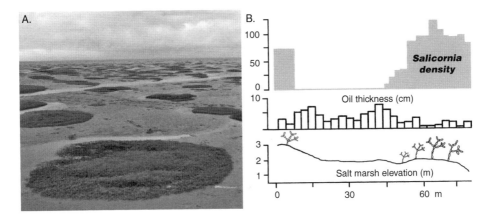

Figure 8.11 Subarctic marshes, northern Tierra del Fuego. A: *Salicornia* and *Suaeda* in the mudflats at Bahia Lomas, located at 52.63° S (Ramsar, 2007); B: diagram showing oil thickness and extent of *Salicornia ambigua* regrowth in the marsh east of Puerto Esporo 17 years after a super-tanker oil spill (data from Baker *et al.*, 1993).

largest intertidal flats in Chile, including a 69 km-long beach and several salt marshes (Figure 8.11A). The bay supports high concentrations of migratory shorebirds from October to March: >41 000 red knots *Calidris canutus* (over 88% of the Americas population), 23% of the global population of Hudsonian godwits, many white-rumped sandpipers *Calidris fuscicollis* and some near-threatened species e.g. the Magellanic plover *Pluvianellus socialis* and Chilean flamingo *Phoenicopterus chilensis*. Precipitation is low and vegetation is typical of the Patagonian steppe, dominated by the grasses *Festuca pallescens* and *F. gracillima*. Large cetaceans are frequently stranded on intertidal flats: 21 species have been recorded. The human population is scarce – mostly sheep farmers – but an oil extraction industry operates from two drill platforms within the flats and pollution from oil spills by large tankers and/or the platforms is a concern, despite the vigorous current circulation in the strait.

The impact of oil spills in this region is demonstrated by the example of the Puerto Espora marshes in the eastern Strait of Magellan, where the super-tanker VLCC *Metula* went aground and spilled ~47 000 tonnes of light Arabian crude and 3000–4000 tonnes of Bunker C oil in August 1974 (Baker *et al.*, 1993). The mousse-like petroleum layer was up to 30 cm thick, killing >4000 birds, including cormorants and penguins, and smothering the salt marshes containing shallow-rooted succulents, *Salicornia ambigua* and *Suaeda argentinensis*. The area was simply left to 'long-term natural clean up'. Field studies 17 years later showed that thick mousse-like oil deposits were still visible at the surface, and oil beneath this thick skin remained fresh, preventing any vegetation recovery. However, other areas with a thinner mousse layer had begun to weather and were being recolonized by *Salicornia*. The relationship between oil thickness and plant regrowth is shown in Figure 8.11B. It appears that marsh recovery rate depends in part on the salt marsh plant growth characteristics: a very long 'natural' recovery time is required for the shallow-rooted Puerto Esporo glasswort and seablite marsh species compared to a shorter time of 15 years to full recovery in a British marsh dominated by the deep-rooted perennials, spikerush and arrowgrass.

8.4.2 Rio Chico, Tierra del Fuego, Argentina

The high sea cliffs along the coasts of southern Patagonia and Tierra del Fuego mean that there is little area available for salt marsh development (West, 1977), and sediment supply on the east coast of this macrotidal region is limited by wave erosion of Tertiary sedimentary rocks or reworking of Quaternary glacial deposits (Isla *et al.*, 2005). The entire coastline is marked by high-energy waves and intense westerly winds; the southern coast is also subject to periodic tectonic activity along the boundary of the Scotian and South American Plates (Bujalesky, 2007). However, where streams cut the coastline, there are small marshes in wave-protected areas. These marshes are dominated by salt shrubs, including *Sarcocornia*, *Suaeda patagonica*, *Atriplex macrostyla*, in the diurnally inundated lower marsh. Endemic perennial herbs *Lepidophyllum cupressifolium*, *Limonium patagonicum* and *Frankenia microphyllum* occur in drier, more saline high marshes. The microfauna here is also sparse – usually 100 specimens per 10 cm^2 compared to several 1000s in temperate and tropical marshes of South America. Studies of the microfossils (Figure 8.12) and fossil molluscs in raised beach deposits from Rio Chico south to Beagle Channel ~55° S are interesting. This is because the Atlantic and Pacific waters meet here, on the margin of Drake Passage, separating Tierra del Fuego from western Antarctica (Gordillo *et al.*, 2005). Fossil records show that the fauna expanded somewhat during a slightly warmer period about 5000 yr BP, but there has been no major molluscan faunal change since the last glaciation.

Today the main concern in this area is the search for petroleum reserves in the neighbouring Falkland Islands – and the associated need for baseline environmental assessment

Figure 8.12   Marshes at Rio Chico 53.64° S are among the most southerly in the world. Vegetation is sparse and salt marshes are restricted to a narrow ledge below the 5000 yr BP high-stand elevation and above the present channel at MLW. The former marsh where Marcelo stands is no longer intertidal. On the horizon, high beach ridges probably mark the limit of the last RSL maximum during MIS 5e, ~120 000 years ago (photo by D. B. Scott).

studies onshore. There was almost no information on the microorganisms of these marshes before 1986. At that time, scientists from Mar del Plata at a meeting in Ushuaia near the southern tip of Tierra del Fuego obtained surface samples from the Rio Chico marshes that allowed the first detailed microfossil examination of these marshes (Scott *et al.*, 1990). The Tierra del Fuego marshes are now also being monitored for impacts of increased exposure to solar UV-B radiation resulting from the expanding Antarctic ozone hole. It appears that growth of both *Salicornia* and *Puccinellia* are inhibited by higher exposure to irradiation (Bianciotto *et al.*, 2003).

There are unpublished reports of very small marshes possibly located on the Antarctic Peninsula at 65° S, but it is difficult to obtain images to verify these reports, and no salt marsh peat was found in a study of Holocene beaches of eastern Antarctica (Baroni and Orombelli, 1991).

9

Africa: selected marsh and mangrove areas

Key points

Africa's pantropical location bridges Atlantic and Indian Ocean mangrove forests; microtides support extensive estuarine and lagoonal wetlands on low coastal plains; population growth drives wetlands degradation and escalates mangrove loss from global warming and sea level rise; now many African countries are conserving and restoring mangroves; the case history for the Gambia River, Africa's longest estuary, illustrates problems of protecting mangroves for services to local peoples; the Nile Delta on the world's longest river records near-disappearance of the wetlands over the past ~2000 years; in South Africa, a 220 km-long UNESCO World Heritage Site spanning St. Lucia Bay (iSisangaliso) marks growing awareness of estuarine values and conservation needs; smaller East African sites have escaped major changes and new studies show caution is needed when interpreting wetland changes from grey mangrove pollen archives.

9.1 Location and biodiversity: introduction to Africa as a pantropical bridge

The continent of Africa effectively is a link between the New World continents of North and South America (Chapters 7–8) and the Old World regions of Europe and Asia (Chapter 10), with its southern satellites, Australia and New Zealand (Chapter 11). The African continent spans the Equator roughly equally north and south (36° N to ~35° S), covering warm-temperate and tropical regions, but not extending into any cool climate or polar region (Figure 9.1). Most of Africa lies in the Afrotropical biodiversity region of the Ramsar Convention (1977). However, the West Coast wetlands border the Atlantic–East Pacific biogeographical region, while the East Coast wetlands border the Indian Ocean and belong to the Indo-West Pacific region. Hence, Africa, the least industrially developed continent in the world, is a transition zone between the long-settled Old World and the rapidly settled and exploited New World.

All the shores of Africa have microtidal regimes with tidal ranges less than 2 m (Eisma, 1997), and there are extensive low-lying intertidal areas behind sandy barrier beaches that support about one-fifth of the world's mangroves. Approximately 70% of the mangals are found in 19 countries of central West Africa (Corcoran *et al.*, 2007). About three-quarters of the commercial fish species in the West African Marine Eco-Region (from Guinea to Mauritania) depend on these mangroves for reproduction and sustenance during early-stage

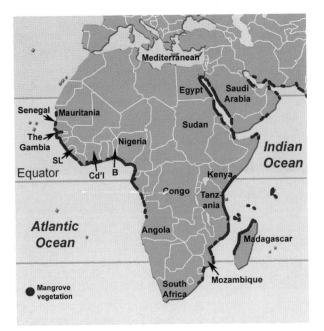

Figure 9.1   Map of Africa showing mangrove distributions in various countries according to Ramsar (2007). Exact boundaries of some countries and distributions may have changed in the past decade. SL = Sierra Leone; Cd'I = Côte d'Ivoire; B = Benin (modified from WRI/UNEP-WCMC).

growth, while the associated tidal flats may yield an average of 90 kg/ha of harvested fish and shellfish (Crow and Carney, 2013). Fourteen West African nations have ratified the Ramsar Convention and are attempting to protect and restore mangrove forests that have become degraded from overuse and pollution.

The northern limit of mangrove forests in West Africa is on the southern edge of the Sahel Desert in Mauritania (Figure 9.1) where stunted black mangroves (*Avicennia germinans*) survive dry summers and cold winters. Mangrove diversity and abundance increases progressively southwards to Nigeria and the Democratic Republic of Congo (formerly Zaire) which are near the Equator. In 1977, the Niger Delta, the largest in Africa, was covered by huge forests of red, white and black mangroves, with pandan shrubs and raffia palms at the inner border. These forests supported diverse animal life, including pelicans, terns, crocodile birds and fish eagles, two crocodile species, the Mona monkey and mudskipper fishes (Chapman, 1977). In contrast, the Congo (Zaire) River estuary, with the second largest water discharge in the world, has less-diverse estuarine mangrove vegetation and fauna, and southwards, wetlands with only three mangrove species reach their southern limit at ~12° S in the Namib Desert of Angola.

The West African mangrove tree species are similar to those of the Americas and share no species with the East African mangals. The vertical zonation of mangrove animals also shifts within these biogeographic regions (Box 9.1 Panmangal 3-D zonation). Por (1984) and Twilley and Day (2013) list the main genera of mangrove trees and their associated trunk/ root and mudflat faunas that are found in Pantropical, Atlantic-East Pacific and Indo-Pacific

Box 9.1   **Panmangal 3-D zonation**

Sketch (not to scale) of changes in vertical zonation patterns of mangrove biota in Atlantic and Indo-Pacific regions compared with the 'Panmangal' zonation of Por (1984) expected on a global scale. In mangrove communities, there is a 3-D zonation: a vertical zonation on the trees and another zonation for the 2-D shoreline elevation zones. Note that the warmer, more saline, tropical Indo-Pacific waters support fewer endemic species in the high water zone than found in the Atlantic region, but there are many parallels in the zonations of the lower water and mudflat zones. b–g = blue–green.

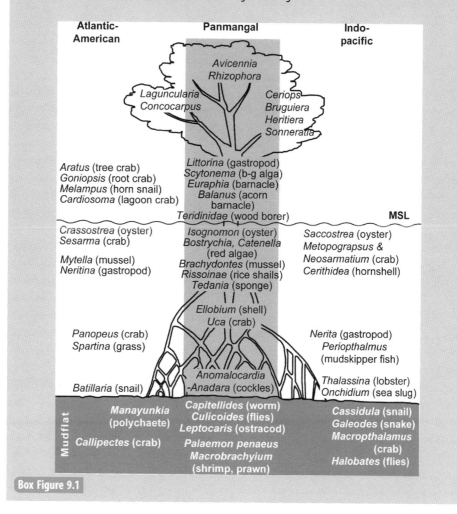

Box Figure 9.1

regions. Despite the natural biogeographic divisions, however, when transplanted, some mangroves such as the Indo-Pacific nypa palm, *Nypa fruticans*, have aggressively invaded the West African wetlands. In contrast, other introductions, such as the Amazon mangrove creeper *Brachypterys ovata* from Central America have not become invasive, being limited to parts of Sierra Leone and Guinea.

9.2 West Coast geomorphology: deltas, lagoons and anthropogenic changes

The main characteristics of the African coastal wetlands are outlined by Chapman (1977), Zahran (1977), Eisma (1997), and an up-to-date account is given in a richly illustrated World Atlas of Mangroves by Spalding *et al.* (2010). Here we focus on some selected African regions that either did not previously receive detailed coverage or which illustrate the dilemmas of preserving African wetlands in the face of awakening national interests in coastal development for industry, commerce and tourism. In effect, Central Africa previously offered protection to large-scale changes in coastal areas, but its development is now beginning to impose a heavy ecological footprint on the wetlands. This impact is particularly evident in the tropical mangrove forest of the West Coast where there are four types of mangrove forests (Saenger and Belan, 1995; Figure 9.2).

Most of the microtidal West African coast is tectonically stable and is backed by a relatively low-lying coastal plain drained by rivers of various sizes. From Cape Verde at the westernmost tip (15.1° N), to Liberia (6.4° N), the coast is notched by deep-water estuaries with muddy sandbanks, including those of major rivers in southern Senegal, The Gambia (see Section 9.3) and Sierra Leone. The mangroves here grow only on the deltas and tidal sections of river banks. The Gulf of Guinea coastline from Liberia to the Nigerian–Cameroon border is low and lined with barrier islands enclosing lagoons into which rivers flow with volumes that vary greatly with the monsoon rainfall. Here the mangroves grow in both the lagoons and on the deltas. Elsewhere along the coast, however, the shoreline is sandy and open, with relatively high wind exposure and wave energy. On this exposed coastline, such as Benin on the Gulf of Guinea, degraded remnants of once-extensive, tall mangrove forests survive only as shrubs inside lagoons where the water is increasing in salinity as a result of global warming trends.

9.2.1 Niger Delta: largest in Africa but strongly altered by petroleum industry

The Niger Delta is the largest delta in Africa and includes extensive barrier beaches, which in 1997 sheltered 9000 km^2 of mangrove-covered mudflats. At that time, tall mangrove trees (up to 45 m) generally stabilized the tidal channels, despite deep-burrowing giant mudflat crabs that caused erosion and shifting of the distributaries. In 1969, this delta supported 127 fish species (listed by Pillay, 1969) and an important fishing industry. Recently, however, drainage and road construction for petroleum exploitation and urban development caused a 25% loss of mangroves, lowering the water table and sediment supply and leading to a 50% decrease in aggradation (Syvitski *et al.*, 2009). Major efforts are now being made to restore the mangal vegetation by planting of red mangrove seedlings on this delta and on the lower Casamance River in Senegal (Figure 9.3). A Ramsar Convention preserve now protects about 20% of the mangroves on the 10 km-thick Nigerian Delta fan, which remains a major target for petroleum drilling.

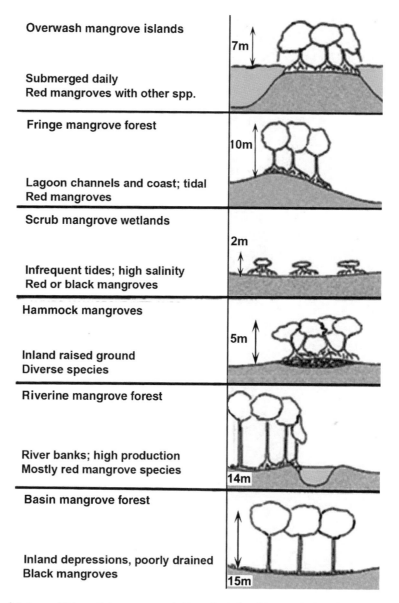

Overwash mangrove islands

7m

Submerged daily
Red mangroves with other spp.

Fringe mangrove forest

10m

Lagoon channels and coast; tidal
Red mangroves

Scrub mangrove wetlands

2m

Infrequent tides; high salinity
Red or black mangroves

Hammock mangroves

5m

Inland raised ground
Diverse species

Riverine mangrove forest

River banks; high production
Mostly red mangrove species

14m

Basin mangrove forest

Inland depressions, poorly drained
Black mangroves

15m

Figure 9.2 Sketches of six types of Paleoarctic-type mangrove habitats also in found in tropical Africa (from Twilley, 1985).

Pollen studies of drill cores from the Niger Delta (Adebayo *et al.*, 2012) show that both the red and black mangroves have been part of the coastal vegetation for at least 45 million years. Farther south, on the Congo River fan and offshore, studies of pollen and taraxerol (a geochemical biomarker for red mangrove) show that mangroves have persisted through both glacial and interglacial climate intervals during the past 1.25 million years (Scourse *et al.*, 2005). Large fluctuations in abundance of the pollen and taraxerol primarily record climate-driven cycles of sea level regression and transgression, and secondarily, they record

Figure 9.3 Young red mangroves near Zinguinchor, Casamance River, Senegal, where 10 000 seedlings/ha were planted in 2008–2009 (Photo by Hilton Whittle, February 2013). Similar restoration projects are ongoing in the Niger Delta and in Benin.

the megaflood history of the Congo River, which starts in East Africa and has the world's largest discharge volume after the Amazon. Despite this long history of West African mangrove survival, two-thirds of the mainland mangroves in the Cameroon have recently died back at rates up to 3 m/yr over the past 30 years, and the island mangroves have decreased by up to 89%. This dieback probably reflects reduced sedimentation supply and increased erosion resulting from anthropogenic shoreline changes and rapidly rising sea level (Ellison and Zouh, 2012).

9.2.2 Sierra Leone–Côte d'Ivoire barrier lagoons: prehistory and industrial impacts

The mangroves of the Sierra Leone Estuary and the lagoons of Côte d'Ivoire (Figure 9.4) illustrate other problems involved in protecting the West African coastal wetlands, including the fluctuating African monsoon rains and urbanization. The tropical climate here receives warm-season monsoon rains from May through to October, and the rest of the year is relatively dry. Intense thunderstorms mark the beginning and end of the rainy season. During the cool season from November through February, the humidity is reduced by dry Harmattan winds from the Sahara in the coastal region; inland, however, this wind produces severe dust storms and extreme heat. The average annual temperature range for Freetown, the capital of Sierra Leone, is 21–31 °C; annual rainfall is ~300–400 cm.

The coastal wetlands in the complex of rivers forming the Sierra Leone Estuary east of Freetown (Figure 9.4) mostly have three species of red mangroves (*Rhizophora racemosa*,

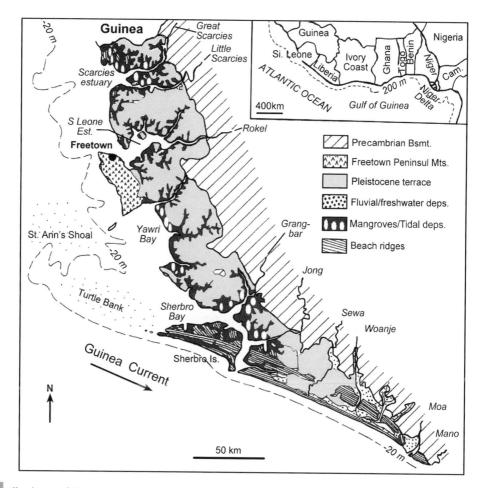

Figure 9.4 Sketch map of Sierra Leone, showing the complex coastal geomorphology and shallow, sandy inner continental shelf of varying width, with multiple rows of beach ridges in the south where terrane is steeper and the inner shelf is narrow (modified from Anthony, 1996). Inset shows locations of West African countries around the Gulf of Guinea. Si = Sierra; Cam = Cameroon.

R. harrisonii and *R. mangle*), and white, black and buttonwood mangrove trees up to 5 m tall. Where there are large amounts of freshwater seepage and sewage, however, only *Rhizophora* is present (Saenger and Belan, 1995). The Sierra Leone mangroves are an important resource for migratory waterfowl, including globally important populations of ringed and Kentish plovers, sanderlings, curlew sandpipers and redshanks (Spalding *et al.*, 2010). There are small areas of salt marshes near EHW, with the main species being the succulent sea purslane (*Sesuvium portulacastrum*), tropical grasses (*Paspalum vaginatum*, *Sporobolus robustus*) or sedges and rushes (*Scirpus maritimus*; *Eleocharis geniculata*, *E. mutata*). The succulent-leaf amaranth *Philoxerus vermicularis* replaces the typical temperate-region chenopods such as samphire and pickleweed that prevail in North and South America. Eisma (1997) describes the distribution of the marshes and tidal flats in

Aberdeen Creek, south of Freetown, and the activity of the diverse burrowing fauna that causes high bioturbation of the silt and sandy mudflats.

Anthony (1996) gives details on the complex geomorphological evolution of Sierra Leone's estuaries since the mid-Holocene when sea level reached a highstand about 5500 years ago, flooding the shallow, sandy continental shelf and drowning estuaries (Figure 9.4). A possible earlier (10 000 yr BP) establishment of the modern West African mangroves is questioned by Lézine (1997) who used pollen of *Rhizophora* to study the history of mangrove vegetation in estuaries of Sierra Leone and Senegal. In the Senegalese Sine–Saloume Delta, about 10 km inland and 1.5–2 m above the present shoreline, a layer of fossil oyster shells (a *Crassostrea gasar* biostrome) marks a slow-down of sea level rise from 5440–4850 yr BP after the early Holocene transgression (Demarq and Demarq, 1992). Microfossil foraminifera (mostly *Ammonia beccarii, A. tepida* and 10% *Elphidium* spp.) and Ostracoda (*Kroemelbeina ebbuttamatense, Neomonoceratina idoense, Miocyprideis leybarensis*) show that the biostrome bed salinity was lower than the present hypersaline condition, while the mangrove pollen indicates a more humid climate, with greater river discharge. The biostrome growth ended very abruptly, apparently because of a combination of sea level regression and drier climate. Today, most of the estuary–lagoon complexes are prograding onto the sandy shelf and are partly infilled by mudflats aggrading over peat layers that mark former mangrove swamps. However, the more wave-protected estuary at Sherbro Bay in Sierra Leone remains largely unfilled, illustrating the complexity of sediment dynamics in this coastal region, which makes it difficult to predict the impacts of future global sea level rise. The complex topography and salinity regimes of the tropical West African region are also reflected in the variable shoreline and root-dependant vertical distributions of the mollusc fauna (Plaziat, 1984).

Although extensive mangrove forests remain along the northern border adjoining Guinea and around Sherbro Island, the mangroves of Sierra Leone have been heavily altered by rapid population growth, land clearance for rice plantations and intensive harvesting of wood (Figure 9.5). High demand for the land and wood, coupled with a previous lack of community participation in resource management led to less mangrove regrowth and more small trees in many areas. Other damage is from siltation and pollution, including untreated waste from Freetown industries and associated harmful algal blooms, and oil spillage from tankers in the deep-water port on the Sierra Leone Estuary. Even inside nature reserves, vegetation clearance and unsustainable fishing threaten the mangrove ecosystem. This was especially intense during the civil war in the 1990s when regional forestry employees often were unpaid, but logging and massive deforestation occurred in forest reserves (Corcoran *et al.*, 2007). Secret contracts between African governments and Chinese companies are also reported as a source of overexploitation of the fish resources (Pala, 2013).

The neighbouring country of Côte d'Ivoire (Ivory Coast) has fewer mangrove areas, with smaller trees in the coastal lagoons and larger (>20 m) trees on deltas near the Liberian border. The Côte d'Ivoire has designated four Ramsar Convention mangrove preserves totalling >700 km^2. The Sassandra–Dagbego Complex on the Sassandra River estuary harbours the best-preserved mangrove stands in the country (Ramsar database, 2013). Primates, reptiles, tortoises, sea turtles, bats and over 208 bird species, especially waterfowl,

Figure 9.5 West African mangrove lagoons are important sources of fish and hardwood needed for fuel and construction by local people (Photo by Emily Corcoran, UNEP, 2006).

herons and gulls, are found here, and Ebrie Lagoon supports over 150 species of fish (Spalding *et al.*, 2010). The mangroves of Grand Bassam lagoon are an important habitat of the chimpanzee, the lesser white-nosed monkey and the sooty mangabey monkey, and it provides a spawning and nursery site for molluscs, fish and crustaceans. The rare forest elephant and pygmy hippopotamus also use the mangrove forest, although this is not their main habitat.

In Côte d'Ivoire, as in Sierra Leone, the main drivers of change are population growth, urban development (pollution and clearing of mangroves), and increased political instability up to March, 2007, which sparked civil war over demands for land and citizenship reforms. With the explosion of coastal development, large areas of mangroves were cleared for urban development and their survival in estuaries is also threatened by construction of dams and reservoirs upstream. Industrial and domestic pollution of the Ebrie Lagoon near the capital city of Abidjan (= Yamoussoukro; population 3.5 million) caused the disappearance of mangroves in many areas (Corcoran *et al.*, 2007).

9.3 Tropical West Coast estuary: The Gambia case history

9.3.1 Background history shapes the future: from Mandinka Empire to small British colony

The mangals of The Gambia are of special interest because they are located near the northern limit of mangrove survival, where increased drought and salinization would prevent their full growth in the absence of water from the 1500 km-long Gambia River (Figure 9.6). The Gambia and its surrounding countries are amongst the 49 least-developed countries on the world and are thus considered particularly vulnerable to the impacts of future climate changes. The Gambia River is the life-force of The Gambia, which today is the smallest

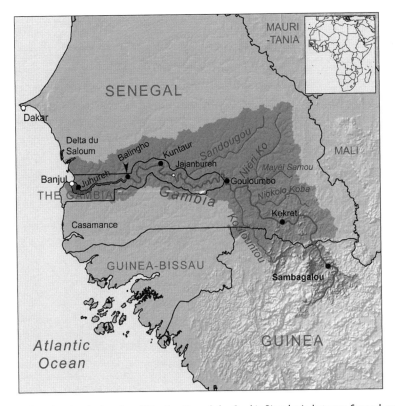

Figure 9.6   Map of West Africa showing the location of The Gambia and the Gambia River basin between Senegal and Guinea-Bissau, north of Freetown in Sierra Leone, and west of the Sub-Saharan desert regions Mauritania and Mali (from Jason Florio, 2012, by Karl Musser).

country in Africa (about the size of Jamaica), but was once part of the vast Mali (a.k.a Mandinka) Empire, stretching from Senegal to east of Timbuktu. During the era of French and British colonization in 1881, the land around the lower Gambia River, from the coast to 320 km upstream, and for ~16 km north and south of the river, was designated as British Gambia.

When the Republic of Gambia became an independent nation in 1965, it was reduced to its present small, narrow shape bordering the river from latitude ~13°–14° N, and longitude ~13°–17° W, with a population of about 1.5 million people. Despite this small size, The Gambia still has some of the most pristine mangrove vegetation in Africa (Corcoran *et al.*, 2007), which sustains vitally important shrimp and fisheries on the West African coast (Darboe, 2002). The Gambia River is also one of the last aquatic ecosystems in West Africa not yet affected by major environmental changes and human disturbances (Simier *et al.*, 2006). However, the man-made boundaries of The Gambia mean that the river headwaters are now controlled by Senegal and Guinea. Here, the construction of a giant dam at Sambagalou (Figure 9.6) is planned to supply hydropower and irrigation water to the countries surrounding The Gambia, thereby threatening the survival of the Gambian mangrove vegetation (Verkerk and van Rens, 2005).

9.3.2 The shared Gambia River: longest estuary in Africa

The Gambia River originates at 1500 m altitude in the wet, evergreen forest of the Fouta d'Jallon Plateau in Guinea, south of Sambangalou (Figure 9.6) and it cascades westwards to eastern Senegal where the slope drops to less than 1%, then remains very low down to Banjul at the mouth of the estuary. Consequently, the Gambia River is seasonally weakly tidal almost up to the eastern boundary of The Gambia, 550 km inland, making it probably the longest estuary in Africa. The deep river is the principal communication highway of the country, being navigable by ocean vessels from Bajul to Kuntaur and by boat to Georgetown. The regional climate is Sub-Guinean type within the Tropical Sudanese Zone (Twilley, 1985), being alternately dominated by humid African summer monsoons (June–October) and dry Harmattan air during autumn and winter. The geographical location, near the Tropic of Cancer, means that the rainfall volumes vary greatly from year to year. In addition, there are prolonged droughts, such as the continent-wide >30 year-long Sahel Drought that progressively lowered the river discharge from 460 to 90 m^3/yr between ~1968 and 2000.

Despite this reduction in rainfall and runoff, because of its origin in the Guinean highlands, the Gambia River has remained a strongly positive estuarine system (see Box Figure 2.2), with each semi-diurnal tide during the dry season producing a salinity gradient of 39 at the mouth to about 1 psu ~220–248 km inland. The mean tidal range is 1.2 m, with spring high water of 1.4 m at Banjul increasing to 1.7 m at Balingho 188 km upstream, where the estuary narrows to 1 km wide. The Gambia River is presently one of the last aquatic ecosystems in West Africa not yet irreversibly altered by recent environmental change and human disturbances. In contrast, in the neighbouring Casamance and Sine–Saloum estuaries in Senegal, dam construction has converted the 'normal' positive salt-wedge estuarine circulation into a hypersaline 'inverse' circulation, where saltwater flows *over* the fresher riverwater. In the 1980s, however, a plan was made for a saltwater barrage on the Gambia River to halt the upstream movement of saltwater and make a freshwater reservoir for year-round rice irrigation above a bridge to be built where the river is presently crossed by boat. Subsequent environmental impact studies (EIS) revealed that a barrage would expand the flooded freshwater area and increase risks of potentially lethal waterborne diseases such as malaria and bilharzia; it would reduce the mangrove area and nutrient transfer to coastal fisheries. A barrage could also result in coastal erosion by obstructing sediment transport to the ocean. Funding was thus withdrawn for the project, as well as for a dam upstream at Kekreti in Senegal, which the EIS revealed was within a wildlife preserve and breeding area of the endangered African elephant (Degeorges and Reilly, 2007).

9.3.3 The mangrove vegetation and its services

The mangrove vegetation of The Gambia, like that of Guinea-Bissau, Cameroon, Gabon and DR Congo (formerly Zaire), is essentially confined to river estuaries where the seasonally high rainfall and river flow provide sediment for extensive mudflat development (Corcoran

Mangrove type		Riverine	Fringe	Basin	Scrub/dwarf
Forest physiognomy & productivity	Profile 				
	Canopy height (m)	12.64±1.43	7.65±0.94	12.14±1.29	0.83±V0.09
	Litter production (MT/ha/yr)	12±1.0	9.0±0.7	6.6±0.7	1.9±0.6
Hydrology	Water source	Ocean tides & stream flow	Ocean tides	Ocean tides & saline groundwater	Ocean tides & saline groundwater
	Hydroperiod — Duration	Hours-days	Hours	Days-months	Perennial
	Hydroperiod — Frequency	Daily or seasonal	Daily	Seasonal	Continuous
	Hydroperiod — Depth	Shallow-deep	Shallow	Shallow	Shallow
Soil chemistry	Salinity (psu)	0/26	33/38	25/60	33/46
	Redox potential (mv)	−48/+116	−96/+103	+87/+279	−244/−105
	Sulfide (mM)	0.0/0.2	0.1/0.3	0.1/0.2	0.9/2.2

Figure 9.7 Profiles of the four mangrove formations in the Gambia River estuary (from Twilley, 1985), showing the corresponding productivity (as leaf litter production), tidal conditions and associated conditions of water salinity and soil oxidation.

et al., 2007). There are four mangrove community types in The Gambia (Figure 9.7) which cover 45 000 ha of forest and permanent tidal swamps, forming a semi-continuous belt along the lower estuary for about 200 km upstream. The dominant mangal type is *riverine mangrove wetland* that extends all along the shoreline except near the estuary mouth and its upper tidal creeks called 'bolons' (Twilley, 1985). Afrotropical West African riverine mangrove wetland has the same six mangrove species as the Neotropical eastern South America biogeographic region. These mangroves occupy the meso- to oligohaline parts of the Gambian estuary (see Box 2.2 for salinity terms). *Rhizophora racemosa* is the pioneer species, forming tall riverine/gallery forests up to 40 m high along a thin strip of the riverbank that is tidally inundated twice a day (Figure 9.8A). Other red mangroves, *R. harrisonii* (~3−6 m high) and the smaller *R. mangle*, grow together, either at the inland boundary mixed in with black mangrove (*Avicennia germinans* = *A. africana*) around

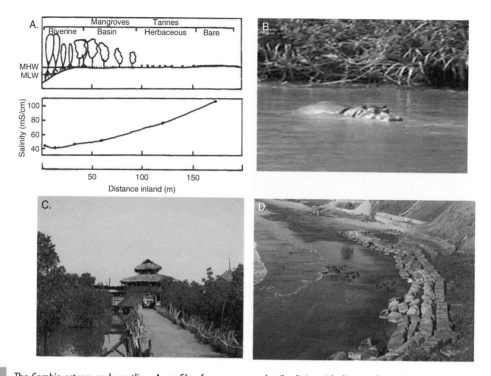

Figure 9.8 The Gambia estuary and coastline. A: profile of mangroves and soil salinity with distance from the river (after Twilley, 1985); B: hippo next to fringe mangrove forest; C: Lamin Lodge, a popular restaurant in cleared mangrove forest near Bajul; D: Banjul outer coast, showing the boulder rip-rap to try and stop the cliff erosion after sand removal for construction of cliff-top tourist facilities (photos by P. J. Mudie).

MHW, or on fibrous clay after removal/dieback of *R. racemosa. R. mangle* also grows in the drier, more saline landward edge of the riverine wetland inundated only once daily. Black mangroves (7–20 m-high) occupy higher elevation fibrous clay or sandier soils. White mangrove (*Laguncularia racemosa*) is uncommon, but may form shrubs <3 m high along with low-growing red or black mangroves. Mangrove associates in low-salinity areas commonly include wild yellow hibiscus (*Hibiscus tiliaceus*), buttonwood (*Conocarpus erectus*), bitter oil tree (*Carapa procera*), the legume trees (*Drepanocarpus lunatus, Dalbergia ecastaphyllum, Oxystigma mannii* and *Pterocarpus*), the pandan (*Pandanus candelabrum*), date palm (*Phoenix reclinata*) and the fern (*Acrostichum aureum*).

Basin mangrove wetland is found inland of the mixed *Rhizophora–Avicennia* mangroves, above MHW where there is a jump in soil salinity because of infrequent (fortnightly) tidal inundation on spring tides and high evapotranspiration (Figure 9.8A). This wetland often has monospecific growths of black mangroves which decrease in height and density landwards from the tidal channels, grading into *scrub mangrove wetlands* (Figure 9.7). The scrub mangrove wetlands are dominated by black mangrove, with minor amounts of dwarf *R. mangle* and *Laguncularia* less than 2 m tall. Inland there are unvegetated hypersaline saltflats called bare tannes (see Chapter 13, Figure 13.2). The highest areas (Figure 9.8A), which are flooded only on storm tides are called tannes (= saltflats or 'salinas'). These tannes have patches of herbaceous marsh vegetation, with the succulents *Sesuvium portulacastrum* (sea fig family) and *Philoxerus vermicularis* (amaranth family), grasses (*Paspalum vaginatum*) and spikerushes (*Eleocharis* spp.) interspersed with bare salt-covered areas. At the upper margins of the saltflats, there are scattered groves of palms, hibiscus or pandans. *Fringe mangrove wetlands* are found only near the river mouth on shores of bays and lagoons where salinity is fairly constant year round. These mangroves are dominated by *Rhizophora mangle* trees about 5–7 m high, forming forests like those of the bays in southern Florida (Figure 7.32). Inland of the fringe mangroves and in areas covered only by storm tides, there may be scrub mangrove wetlands dominated by black mangrove, with minor amounts of dwarf *R. mangle* and *Laguncularia* less than 2 m tall.

Services provided by mangroves in The Gambia, as elsewhere in tropical West Africa, primarily support essential livelihood needs in providing durable, termite-resistant timber for dwellings, cooking and heat. Red mangroves are preferred because they are rich in tannin, and burn almost smokelessly, giving grilled food and shellfish a pleasant taste. In some areas, salt is produced by villagers by boiling brackish water in clay bowls over fire made from *Avicennia*, but it takes 7 tonnes of wood to produce 1 tonne of salt and the harvest places a heavy demand on the mangroves. *Rhizophora*, *Avicennia* and *Conocarpus* have various traditional medicinal uses, as listed by Corcoran *et al.* (2007), and in some regions, the mangroves also have rich cultural and spiritual values amongst the local people. Indirect mangrove services are leaf litter (Figure 9.7) produced by the riverine forest wetland that contributes up to 85% of the total carbon input to the Gambia Estuary and provides a nutrient-rich nursery habitat for many fish species and oyster fisheries (see Section 9.3.5). Mangroves also protect the exposed sandy coastline from erosion and storm surges and shelter many rare and endangered species, including the African manatee, the clawless otter and the bottlenose dolphin.

9.3.4 Gambian mangrove and estuarine animals

The mangroves and waters of the Gambia River sustain large bird (Table 9.1) and fish populations (Table 9.2), which are fished throughout the country. There is also an important estuarine and offshore prawn fishery that relies on the leaf detritus of the mangrove swamps for nutrients and shelter. The Nile crocodile (*Crocodylus niloticus* Figure 4.3) and the hippo (*Hippopotamus amphibious* Figure 9.8B) still live in the brackish reed swamps and inter-island channels above Kuntaur; although they have been hunted to near-extinction and are now endangered in Gambia, they are still a hazard to be carefully avoided by the local canoe traffic. The African manatee (*Trichechus senegalensis*), clawless otter and green turtle are also now endangered Gambian estuarine species (Table 9.2). The green turtle nests at a few sites on the sandy beaches at the river mouth where development of tourist facilities has been high, including large-scale sand removal for construction (Figure 9.8D). Expansion and intensification of agriculture and the destruction of riverine wetlands upstream have resulted in increased erosion and siltation, requiring dredging to keep the river channel open throughout most of its upper course. However, four national parks and several nature preserves have been created to protect the river fauna, including the Baboon Island

Table 9.1 List of major birds found in the Bao Bolong Wetlands Reserve on the north shore of the Gambia River, according to the BirdLife International census in 2012.

Common name	Scientific name
Ducks	
Comb duck	*Sarkidiornis melanotos*
Spur-winged goose	*Plectropterus gambensis*
White-faced whistling duck	*Dendrocygna viduata*
Garganey	*Anas querquedula*
Northern pintail	*Anas acuta*
African pygmy goose	*Nettapus auritus*
Northern shoveler	*Anas clypeata*
Waders	
Western reef heron	*Egretta gularis*
Little egret	*Egretta garzetta*
Woolly-necked stork	*Ciconia episcopus*
Hammerkop	*Scopus umbretta*
Squacco heron	*Ardeola ralloides*
Little bittern	*Ixobrychus minutes*
Goliath heron	*Ardea goliath*
Great egret	*Egretta alba*
Marabou	*Leptoptilos crumeniferus*
Night heron	*Nycticorax nycticorax*
Grey heron	*Ardea cinerea*
Cattle egret	*Bubulcus ibis*
Sacred Ibis	*Threskiornis aethiopicus*
Birds of prey	
Pels fishing owl	*Scotopelia peli*
Fish eagle	*Haliaeetus vocifera*
Osprey	*Pandion haliaetus*
Woodland kingfisher	*Halycion senegalensis*
Pied kingfisher	*Ceryle rudis*
Swallow-tailed bee-eater	*Merops hirundineus*
Giant kingfisher	*Ceryle maxima*
Malachite kingfisher	*Alcedo cristata*
Blue-breasted kingfisher	*Halcyon malimbica*
Big birds	
Abyssinian ground-hornbill	*Bucorvus abyssinicus*
European roller	*Coracias garrulous*
Senegal parrot	*Poicephalus senegalus*
Rose-ringed parakeet	*Psittacula krameri*
Pink-backed pelican	*Pelecanus rufescens*
African darter	*Anhinga rufa*

Table 9.2 Selection of Gambian River animal diversity

Animal	Scientific name	Main location and notes
Mammals – 32 species known		
African clawless otter	*Aonyx capensis*	Estuary (endangered)
Atlantic humpback dolphin	*Sousa touszii*	Estuary
Bottlenose dolphin	*Tursiops truncatus*	Estuary and saline creeks
Hippo	*Hippopotamus amphibious*	River, reed swamp
Manatee	*Trichechus senegalensis*	River
Bushbuck	*Tragelaphus scriptus*	Floodplain-savannah
Duikers	*Cephalophinae*	Floodplain-savannah
Sitatunga antelope	*Tragelaphus spekeii*	Floodplain-savannah
Warthogs	*Phacochoerus africanus*	Floodplain and swamps
Leopard	*Panthera pardus*	Floodplain predator
Spotted hyenas	*Crocuta crocuta*	Floodplain predator
Chimpanzee	*Pan troglodytes*	Riparian forest preserves
Green monkeys	*Chlorocebus*	Riverine mangroves
Guinea baboons	*Papio papio*	Riverine mangroves (rare)
Marmosets	*Callithrix*	Riverine mangroves (rare)
Patas monkey	*Erythrocebus patas*	Riverine mangroves
Red colobus monkey	*Piliocolobus*	Riverine mangroves
Senegal bushbaby monkeys	*Galago senegalensis*	Bolons (rare)
Reptiles (Barnett and Ems, 2005)		
Nile crocodile	*Trichechus senegalensis*	River
Green turtle	*Chelonia mydas*	Coast: sandy beaches
West African mud turtle	*Pelusios castneaus castneus*	River banks
African rock python	*Python sebae*	River
Sand boa	*Gongylophis muelleri*	Bao Bolon wetlands
Fish – 80+ species known (Darboe, 2002; Pillay, 1969)		
Cichlid	*Tilapia nilotica*	Freshwater upper river
African walleye	*Clarias luzerra*	Freshwater upper river
Bonga shad	*Ethmalosa fimbriata*	Mid-estuary – coastal; 80% of catch
Cornish jackfish	*Mormyrus anguillaris*	Uppermost estuary – low salinity
Silversides	*Alestes baremose*	Uppermost estuary – low salinity
Threadfins	*Polynemidae* sp.	Brackish river
Catfish	*Chrysicthys maurus C. nigrodigitatus*	Upper estuary
Angel-fish	*Ilisha africana*	Mid-estuary – brackish
Marine catfish	*Arius* sp.	Brackish – coastal
Solefish – red and black	*Cynoglossidae* sp.	Coast – estuary
African bonytongue	*Heterotis niloticus*	Freshwater upper river
Flat tonguefish	*Cynoglossus senegalensis*	Upper estuary

Table 9.2 (cont.)		
Animal	Scientific name	Main location and notes
Gambian/upside-down catfish	*Synodontis gambensis*	Mid-estuary – euryhaline
Bagrid catfish	*Auchenoglanis occidentalis*	Upper estuary – river
African carp	*Labeo senegalensis*	River – low salinity
Big-eye catfish	*Chrysichthus furcayus*	Estuary – ocean
Croakers	*Sciaenidae*	Upper estuary
Grunts	*Pomadasyidae*	Outer estuary
Mullet	*Mugilidae*	Outer estuary, coast
Mackerel	*Scomber japonicus*	Offshore – commercial importance
Sardines	*Sardinella aurita, S. maderiensis*	Offshore and lower river
Jack-fish	*Carangoides* sp.	Offshore – commercial importance
Barracuda	*Sphyraena* spp.	Coast/estuary – commercial importance
Invertebrates (shellfish)		
Blue crab	*Callinectes latimanus*	Upper estuary
Cockle	*Anadara gambiensis*	Mangrove mudflats
Gambian murex	*Purpurellus gambiensis*	Outer estuary/coast
Cymbio/top-shells	*Cymbium pepo, Cymbium* spp.	Outer estuary/coast
Gambian oyster	*Crassostraea gasar*	Red mangrove roots
Black striped shrimp	*Parapenaeus longirostris*	Exotic, farmed
Gambian/pink shrimp	*Penaeus (Farfantepenaeus) notialis*	Lower estuary and bolons – coast
West African prawn	*Metapenaeus* sp.	Ocean

rehabilitation centre for orphan baby chimpanzees rescued from DR Congo, where around 450 adult 'chimps' are slaughtered illegally for bushmeat each year. There is a proposal to protect some mangroves and lagoons in Kiangs National Park on the south bank of the Gambia River, where wetlands have been extensively cleared, destabilizing the mudflats and diminishing the once-extensive oyster colonies, and where tourist industries are expanding into the mangroves (Figure 9.8 C, D).

The Bao Bolong Wetlands Reserve between Jufureh and Balingho includes 268 species of birds in 62 families and the Tanbi wetlands near Banjul support 362 bird species (Spalding *et al.*, 2010). Typical wetland birds (Table 9.1) include various ducks, herons, storks and egrets, pelicans, the fishing owl and fish eagle, as well as colourful kingfishers, roller birds and parakeets (BirdLife International, 2012). Here there are over 32 species of mammals (Table 9.2) including the clawless otter, sitatunga antelope, bushbuck, duiker deer and warthogs. Predators include the noisy spotted hyenas and stealthy leopards. The lion and striped hyena have been hunted to near extinction. Primates include the common red colobus, Patas and green monkeys, and rare Guinea baboons, marmosets and Senegal bushbaby monkeys.

The Gambia's continental shelf covers an area of 3855 km^2 and may be one of the richest fishing grounds in the West African subregion (UNEP/FAO/PAP, 2002). In addition to its marine resources, the Gambia River and its tributaries contain significant freshwater fish resources. The demersal fauna is extremely varied, with the most numerous species near the estuary being croakers, grunts, thread fins and mullet. The most common fish offshore are small pelagic sardines, and closer inshore, bonga fish are abundant. Commercially important pelagic fish are mackerels, jacks and barracudas. The maximum sustainable yield of pelagics is estimated at about 60 000 tonnes per year.

A total of over 80 fish and shellfish species (in >40 families) live or reproduce within the Gambia estuarine system (Darboe, 2002; Table 9.2). About 80% of artisanal river catches are bonga shad and other marine fish, threadfins, marine catfish and solefish. Also abundant are African bonytongue, upside-down/Gambian catfish, bagrid catfish, African carp and the big-eye catfish (Guillard *et al.*, 2004). In The Gambia, occurrence of species at different estuarine zones and times appears to be salinity-driven. During a high flow interval in December 2000, about 28 species were recorded between Bajul and Balingho in the upper estuary, the most widespread species being the croaker fish and blue crab, flat tonguefish and two catfish species. The euryhaline Gambian catfish and bonga shad have much narrower distributions in a mid-estuarine zone of optimum salinity. During a very low discharge period in June 2001, the number of species in the whole estuary dropped by about 50%, indicating the high sensitivity to fluvial-driven salinity changes. During a high-discharge rainy season in September 2001 that lowered the salinity to 0–5.5 psu in the middle estuary, the distribution of several species changed, including bonga shad, angelfish, pink shrimp, catfish and sardines. These fish moved towards the lower estuary around Banjul and oligohaline 'freshwater' species like the Gambian catfish also moved downstream. In contrast, other environmentally sensitive low-salinity species such as Cornish jackfish and silversides remained in the upper estuary.

Increased salinity at the mouth of the Gambia Estuary during lower freshwater flow may impede the entry of larvae and juveniles of marine species into the estuary so they cannot complete their early-stage growth. For example, salinities lower than ocean values are needed by the larvae and juveniles of the Gambian shrimp to sustain an osmoregulatory balance during three months of early development. After completing metamorphosis in the estuary, the pre-adult shrimp then need higher salinity for osmoregulation and they move towards the ocean to feed and grow. Increased salinity in the upper estuary could reverse this movement and shift the fisheries upstream, meaning that an oceanic industrial shrimp fishery would replace the estuarine artisanal fishery. This change could significantly alter the economy because Gambian shrimp are one of the main foreign exchange commodities. Reversal of the estuarine salinity as a result of dam construction would also impact the Gambian finfish populations, as happened in the neighbouring Casamance when a long drought resulted in heavy estuarine fish mortality and altered species distributions. Unlike the Gambia River salinity regime, however, the estuarine temperature regime does not vary much vertically or horizontally. Consequently, the bottom water is usually oxygen- and nutrient-rich, except after a large-scale mangrove die-off when the bottom becomes anoxic from decomposing mangrove leaf litter.

Figure 9.9   Oysters are a major source of household support in the lower Gambia River. A: red mangroves at low tide, with festoons of oysters on the roots; B: traditional oyster harvesting by the Jola women of the region (photos by Anita Whittle). See also colour plates section.

9.3.5 Gambian oyster industry

Oyster harvesting is an important occupation along both the north and south banks of the Gambia River (Njie and Drammah, 2011). Wild oysters (*Crassostrea gasar* = *Crassostrea tulipa*) are gathered at low tide, mainly from the roots of the red mangrove trees (Figure 9.9). Traditionally, most oyster collectors are women, who depend on the harvest for their livelihoods – selling the smoked or steamed oysters (Crow and Carney, 2013). The shucked shells (see Figure 13.2) are saved for making white lime and gravel for roads (primarily the job of older men). The oyster women often cannot swim and have to complete their harvest by foot or small dug-out canoe (Figure 9.9C) during the ~6–7 hours around the low tides of the harvest period, which is traditionally from December to June in the Tanbi wetlands. These women also collect cockles, murex and cymbio (top-shell) from muddy or sandy banks and areas between mangrove tree roots. Traditionally, a stick is used to remove the oysters. Because of the short time window for harvesting and the dire need to collect enough oysters for survival, however, some harvesters started to remove the oysters with a machete (large cutlass knife) that injured or cut through the roots, killing the plants and leading to overharvesting of the remaining healthy plants. A decline in size and number of oysters harvested in areas of development around Banjul led many local collectors to move to the North Bank, where the oysters are less exploited. An education programme is now in place in The Gambia and Senegal to teach that more careful harvesting produces higher yields, and the harvest season has been reduced to December–March because oysters spawn during the low-salinity rainy season. A nine-month summer growing season may also increase oyster production and sustainability. In the past, the processing and marketing techniques have been simple – shucking of the shell, boiling or grilling and local roadside or market sales. Plans for expansion of this market include better management of wild harvests and assessment of seafood sanitation issues, including the potential of contamination from human pathogens and need for post-harvest sanitary processing practices.

9.3.6 What to expect in the future?

Predicted scenarios of extreme changes accompanying continued global warming and Sahel droughts makes The Gambia one of the countries listed as most vulnerable to global climate change (Jallow *et al.*, 1996). A 1–2 °C rise in global air temperature, accompanied by a 10% reduction in precipitation can cause a 40–70% drop in mean annual river discharge. Under this scenario, there could be a complete change in the hydrological and salinity balance of the Gambia Estuary, which would change the fish abundance, species composition and distribution, and possibly reach the threshold levels for other aquatic organisms, particularly if mangrove die-off progressed due to higher temperature and salinity. Oysters are particularly vulnerable to increased salinity and temperature. The oyster fishery has already partially collapsed from deteriorating environmental conditions and substrate loss because of mangrove clearance and cutting of the roots during the oyster harvesting. Thus, the vital role of mangroves in curtailing erosion has been impaired and saltwater is intruding into rice fields, vegetable gardens and water wells. Overall, the salt-covered tannes are expanding, especially where wetland margins have been cleared for farming and road construction. Crow and Carney (2013) also point out that the efforts to protect and improve the south shore oyster industry focus on production but fail to consider fully the implications of expanded value chains for oysters products; the changes may actually accelerate regional degradation of Gambian mangrove ecosystem.

9.4 Nile River Delta case history: early civilizations and recent destruction

9.4.1 Background environment and historical geology

The coastal wetlands of the Nile River Delta at ~31.5° N, 31° E in Northeast Africa provide a big contrast to those of the wet tropical estuaries of the West African coast (Figure 9.10). The Nile Delta lies in the driest part of the Mediterranean Sea – a hyperarid hot desert region where temperatures average ~30 °C from June to September (July maximum is ~35 °C) and mean annual precipitation is less than 100–200 mm. The high evapotranspiration rates, which control how much water remains in the soil, range from ~600 to 1100 mm/yr and consequently there is a large net deficit in the annual moisture supply. Winter temperatures average ~11 °C, but January and February minima are ~8.5 °C, preventing the growth of mangroves on the delta, although grey mangrove shrubs (*Avicennia marina*) can grow at

Figure 9.10 Aerial view of the Nile Delta (courtesy of S.E.A Kholeif, National Institute of Oceanography and Fisheries, Alexandria). Most of this large delta is farmland or urban development, criss-crossed by narrow drainage channels; little salt marsh remains.

~27.3° N on the Suez Gulf just east of the Delta (Zahran, 1977) and at 28.21° N on the Sinai Peninsula (see Section 9.6). The tidal range of the Nile Delta is of a mixed, mainly semi-diurnal type and a typical microtidal Mediterranean Sea mean range: <26 cm at Rosetta, with highest storm levels only ~100 cm (AbdAllah *et al.*, 2006).

The Nile Delta has a classical fan shape, being built out into the Mediterranean Sea by sediments deposited at the mouth of the Nile River – the longest river in the world, extending >6650 km from the highlands of Kenya to the Egyptian coast. The Nile River has two major tributaries that meet in the Sudan, where there are huge freshwater papyrus (*Cyperus papyrus*), grass and bulrush (*Typha latifolia*) swamps called 'Sudd' (hence the name Sudan), forming the world's largest tropical freshwater wetlands. The shorter Blue Nile tributary arises in the mist forests of Ethiopia, and fossil pollen and plankton records from the Nile Cone show that for over 12 000 years, this has been the source of most of the water, soil and nutrients transported to the Nile Delta (Kholeif and Mudie, 2009). Older geological records from areas south of Cairo show that the extinct mangrove tree fern (*Weichselia reticulata*) grew there in the late Cretaceous (~80 million years ago), along with large dinosaurs, reptiles and turtles. In Oligocene time (~34–25 million years ago), a rich mangrove flora was present on the shores of the Tethys Sea, together with ancient precursors of whales, crocodiles and elephants (Spalding *et al.*, 2010).

Today, however, from the Sudan to the Egyptian coast, the Nile flows through barren desert, and for more than four millennia, most of the region's people have depended on the river for water. North of Cairo, the Nile splits into two major branches (distributaries) that cross the delta and supply freshwater to two large tidal lagoons: the Rosetta Branch flows to Borullus (= El-Brolus) Lagoon near Alexandria to the west; the Damietta Branch flows to the brackish El-Manzala Lagoon near Port Said on the Suez Canal in the east. The delta plain covers ~22 000 km², bounded on the north by a wave-cut arcuate coast ~225 km long. In

1998, >90% of Egypt's population relied on the Nile for survival, and ~40 million people (locally >1000 persons/km^2) crowded onto the Nile Delta (Stanley and Warne, 1998).

9.4.2 Holocene and historical records of Nile Delta change

Radiocarbon ages from delta cores show that from ~7500 years BP to ~1900 AD, the shape and elevation of the delta plain were in balance between sea level rise and sediment accumulation (Figure 9.11). Archeological records show that early farmers began to settle

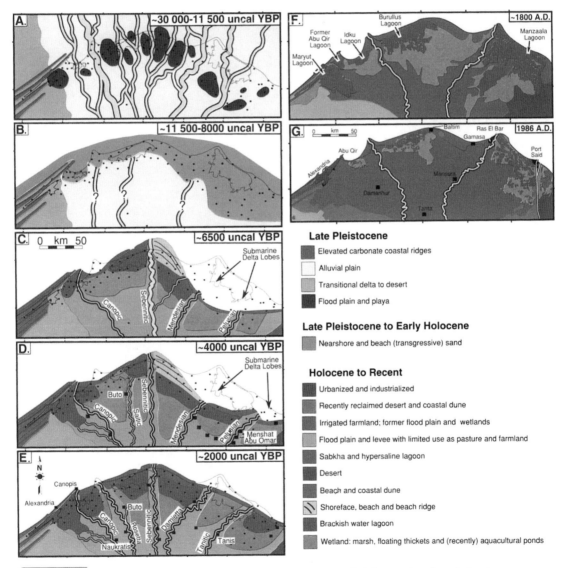

 Figure 9.11 Paleogeographic maps showing the evolution of the northern Nile Delta at various times during the past ~30 000 years (Stanley and Warne, 1998).

the Holocene Nile Delta shortly after its formation ~8000 years ago, when flooding of the
Nile was much greater than recent (pre-1960s) time, discharging water as far north as the
Aegean Sea east of Greece (Kholeif and Mudie, 2009). The early civilizations used natural
flood basins on the delta to capture river water and nutrient-rich soil in late summer and
early fall. By the Dynastic period ~5050 years BP, however, the climate in Egypt became
hyperarid and Nile flow lessened. The early farmers therefore made levees and canals to
capture more floodwater and irrigate their crops when flow was low. During the First
Dynasty ~4900 yr BP, deliberate flooding and draining of fields by sluice gates began,
and three centuries later, only the outermost delta wetlands were not directly affected
by irrigation structures. Climate-driven aridification increased to its present level by
~4500 years BP, leading to more levee building, enlargement of natural overflow channels
and use of mechanical lift irrigation by the shaduf at the start of the Pharaohs' rule after
~3550 years BP. The summer irrigation allowed planting of several crops a year, even during
the dry season, but farms began to suffer from increased salt build-up.

Herodotus, the famous Greek historian (unfortunately, also a notorious liar) travelled
to Egypt ~ 2450 years ago and wrote that the Nile Delta was actively prograding seawards
(de Selincourt, 1972). As farming and delta settlement expanded, however, salinization of the
delta became acute, especially at times of low Nile discharge, when agricultural productivity
decreased. This situation worsened during the Ptolemaic period (~2300−2030 yr BP),
when political centres shifted to the delta and agriculture intensified. Biblical stories of
catastrophic droughts and famines were likely based on factual records, such as
Nileometers, which measured heights of the annual floods, although it is difficult to calibrate
these data precisely because of changeable riverbed heights (Ellenblum, 2012). When the
original seven main delta distributaries were artificially reduced to two ~1000 years ago
(Figure 9.12), salinization became a serious problem, particularly when year-round irrigation became widespread after 1800 AD. By the mid nineteenth century, a dense irrigation
system began to drain the delta, although annual floods still deposited a thin blanket of silt
and washed salts and excess nutrients to the sea. In 1964, however, construction of the

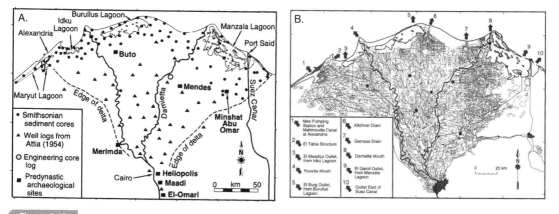

 Figure 9.12   Changes in extent of Nile Delta marshes. A: tidal marshes and brackish water vegetation before ~1800 AD; B: reduction in
wetland vegetation and expansion of irrigation canal systems in the past 200 years (Stanley and Warne, 1998).

Aswan High Dam reduced the delta sediment load to 2% of its original amount, and hundreds of million tonnes of sediment were trapped above the dam, forming a new Nile Delta in the southern Lake Nasser reservoir. Since 1964, therefore, the lower Nile River has carried much less sediment and no river-transported sediment reaches the coast, which is now eroding. An equivalent volume of less than 10% of the pre-dam river sediment load, however, is now pumped into the Mediterranean from coastal lagoons that would otherwise fill in rapidly. These lagoons and their delta marshes are sustained by water rerouted from north of Cairo to the delta coastal marshes; consequently, sedimentation rates in the coastal wetlands have nearly doubled since 1964. Barrages near the mouths of the Rosetta and Damietta distributaries also trap water on the delta, where a 10 000 km-long network of canals distributes tiny amounts of sediment to farmlands (Figure 9.12B). This high-density canal network, combined with inadequate drainage, have increased soil waterlogging and salt build-up on the delta (Stanley and Warne, 1998). Consequently, very little remains of the once extensive Nile Delta salt marsh vegetation; however, Chapman (1974) reports that behind the sand dunes west of Alexandria, there is a zonation of succulent halophytic shrub plant associations (Table 9.3). These Egyptian salt marshes represent a transition from the

Table 9.3 Nile Delta salt marsh vegetation zones and species (data sources: Chapman, 1974; Ayyad *et al.*, 1992).

Zonation	Common plants	Notes
Behind sand dunes west of Alexandria (Chapman, 1974)		
Lowest zones	Pure stands of desert species *Halocnemon strobilaceum*	Mostly shrub plant associations
~50 cm ASL	Desert species + *Arthrocnemum glaucum* and *Salicornia fruiticosa* (now *Sarcocornia fruticosa*) OR *Arthrocnemum glaucum–Limonium articulatum* associations with herbs *Frankenia levis, Cressa cretica, Mesembryanthemum nodiflorum* and grass *Parapholis incurva*	
Higher zones (arid and saline)	*Zygophyllum album* *Atriplex halimus*	Shrubs
Higher zones (wetter, less saline)	Rushes (*Juncus subulatus*) Sedges (*Scirpus maritimus*)	
Other sites of the central delta (Ayyad et al., 1992)		
Levees	Shrubs and trees (*Tamarix tetragyna*)	Salt-tolerant species
Canal banks	Trees *Casuarina* + *Eucalyptus*	Introduced species from Australia
Inner edges of coastal lagoons	Bulrush (*Typha domingensis*), spiny rush (*Juncus acutis*), sedges (*Cyperus laevigatus* and *Scirpus tuberosus*)	Reed swamp species

western Mediterranean communities to those of the Middle East, as far as the Caucusus Mountains, where similar salt marsh vegetation is found on the shores of the northern Black Sea (see Chapter 10).

9.4.3 The Nile Delta in the future

The human modification of the lower Nile system during the past 150 years has almost halted the seaward expansion of the coastline and the elevation increase required for delta growth with rising sea level. The classic deltoid (fan-shaped) outline of the Nile Delta appears to have been altered more by anthropogenic activity than most of the other inhabited world deltas, mainly because of the rapid twentieth-century population increase (Stanley and Warne, 1998). The primary consequences of the overpopulation, combined with changes from dam construction and draining of the delta, include: (1) the incursion of the sea and saline groundwater into the northern delta, (2) erosion of the coastline and (3) diminished capacity of the delta to regenerate itself through accumulation of river sediment that is now trapped in lakes behind the dams. The Nile Delta is currently listed as the fourth most imperiled of the world's deltas (Syvitski *et al.*, 2009) because relative sea level rise here is almost 5 mm/yr due to the reduced sediment accumulation and shrinkage of the now-drained wetland peats. Ellenblum (2012) believes that severe droughts throughout Egypt and the Middle East during the last warm climatic period (Mediaeval Warm Period, ~1000–1100 AD) caused widespread starvation and subsequent invasion of Turks from the northwest and North African Bedouin tribes from the south, thereby irreversibly changing the ethnic and religious characteristics of the region. Continued global warming may represent a similar dilemma for the future of Egypt.

9.5 South African wetlands: the Indian Ocean influence

The South African coastline extends over ~3000 km from Kosi Bay near the eastern Mozambique border to the Orange (Gariep) River at the Namibian border in the west (Figure 9.13). Some 300 rivers reach this coastline, forming estuaries ranging from small coastal ponds, occasionally connected to the ocean, to large, permanently open estuaries or barrier lagoons linked to the sea by shifting channels in the sand ridges (Whitfield, 1992; Eisma, 1997; Allanson and Baird, 1999). As in other countries, the estuaries are popular sites for both human activity and development, and the durable wood of mangrove trees is commonly harvested. Recreational uses of the estuaries include bait collection, bird watching, boating, fishing and swimming, and there are five protected estuarine wildlife reserves, including a UNESCO World Heritage site from St. Lucia to Kosi bays. However, many areas around estuaries have also been targeted for urban and tourist developments. Therefore, in KwaZulu–Natal province alone, mangroves have been completely removed from three estuaries near Durban (Adams *et al.*, 2004), and one of the largest mangals in South Africa that occupied over 200 ha of inner Durban Bay in the 1940s is now just a small

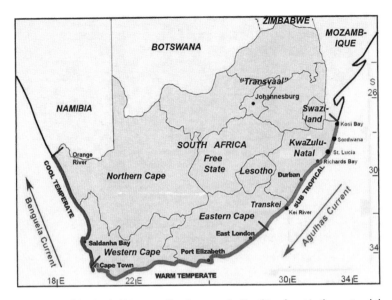

Figure 9.13 Biogeographical regions of the South African coastline, between the Namibian desert in the west and the mangrove swamps of the Mozambique in the east. Map from Allanson and Baird (1999).

20 ha grove on Bayhead Island. In the Eastern Cape Province, 17 estuaries contained mangroves in 1982, but the trees disappeared from three of these areas over the following two decades.

In an earlier summary of the coastal wetlands in Africa, Chapman (1977) comments that 'little is known of the vegetation south of Angola-Namibia until South Africa is reached' and that temperatures on the West Coast and around Cape Horn are too cold for mangrove growth. The distribution of different types of estuarine vegetation in southern Africa (Figure 9.13) is determined primarily by ocean currents. The West Coast is bathed by the cold waters of the Benguela Current, a northern branch of the icy Antarctic current, while the East Coast is warmed by the Agulhas Current of tropical origin. The wetland vegetation from the Tropic of Capricorn southwards to ~35° S is salt marsh. The salt marshes in both the cool western and the warmer eastern regions to just beyond East London superficially resemble those of North America – but in South Africa, there are different species and genera. Another cordgrass species, *Spartina maritima*, and Cape eelgrass (*Zostera capensis*) dominate the low marsh, followed by the succulent halophyte *Arthrocnemum perenne*, then high marsh with *Limonium linifolium* and several succulents, including seablites (*Suaeda* spp.), the endemic suss-bossie (= salt bush, Afrikaans), *Chenolea diffusa* and *Crassula maritima*, a South African species related to the ornamental jade plant. Around EHW there are other shrubby halophytes – *Arthrocnemum africanum* and *A. pillansii* – and the spiny, salt-excreting salt couchgrass, *Sporobolus virginicus*. In South African salt marshes, mudprawns (*Upogebia africana*), crabs and barnacles grow on the cordgrass, while gastropod snails *Ceridithea* and *Assiminea* live on the *Arthrocnemum*. Eisma (1997) provides details of the salt marshes and the mudflat biota and sediment dynamics of Langebaan Lagoon, Saldanha Bay,

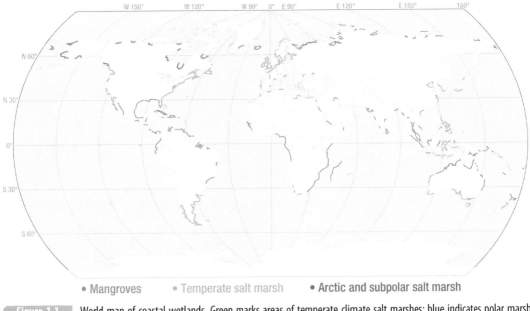

 Figure 1.1   World map of coastal wetlands. Green marks areas of temperate climate salt marshes; blue indicates polar marshes; red shows areas of tropical mangroves.

Figure 3.1   Mudflats are stabilized by gelatinous surface-growing microfloras. A: diatoms on a sandflat at Puerto Peñasco, Mexico; B: blue-green algae bind loose sand between stolons ('runners') of colonizing cinquefoil at Taylor Bay, Alaska (photos by P. J. Mudie).

Figure 3.5 Root systems of salt marsh and mangrove plants. Top left: white roots of salt grass (a) have air passages and include shallow vertical anchor roots and spreading horizontal roots (b); sea lavender has one thick black tap root that grows deep to obtain subsurface low-salinity water. Top right: tropical *Scaevola* mangrove shrub also has a deep anchor root and shallower surface roots absorbing water from rain. Bottom left: black mangrove has abundant pencil-like pneumatophores (foreground) protruding above the mud; red mangroves have branching stilt roots keeping them high in the water (photos by P. J. Mudie). Bottom right: lung roots of the apple mangrove at Tamil, Indonesia (photo by Patrik Nilsson).

Figure 4.2 Examples of birds commonly found in coastal wetlands. Top: bald eagle (photo by P. J. Mudie); roseate spoonbill (photo by Harold Wagle/Flickr); great blue heron (photo by Dori Chandler). Bottom: greater yellowlegs (photo by J.F. Gauthier); clapper rail (photo by USFWS/Flickr); song sparrow (photo by Dori Chandler).

Figure 5.3   Melting of subsurface permafrost (white cliff area) in the Mackenzie Delta causes massive collapse of the shoreline, resulting in mudslides and new colonization of salt marsh vegetation (green areas on the mud). Note the polygonal cracks in the bluff vegetation, which are typical of tundra soils and aid in release of methane gas on melting of permafrost. Jorgenson and Brown (2005) estimate an average of 0.18 million metric tonnes of carbon are released annually from erosion of the Alaskan-Beaufort sea coast (photo by P. J. Mudie).

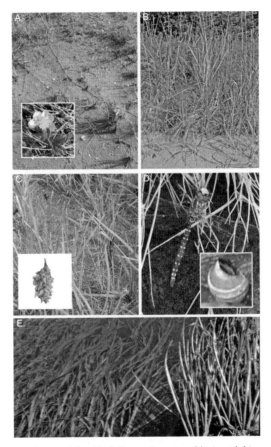

Figure 6.2   Arctic salt marsh plants are mostly perennials with shallow main roots and horizontal rhizomes on the surface or near-subsurface that allow them to spread over the unstable soil. If broken by frost heave or ice, the plants can regenerate from rhizome fragments, as well as seeds. A–D: low marsh: sedge (*Carex*) with V-shaped leaves, grass (*Poa phryganoides*) and arrow-grass with grey leaves, and cinquefoil (*Potentilla*, with inset yellow flower) near Anchorage, Alaska; B: high marsh, with taller lyme grass (*Elymus*); C: sedge and plantain with characteristically orange-pigmented leaves for protection from excessive UV and PAR; D: blue darner dragonfly feeding on lyme grass, above dense mats of blue-green algae grazed by gastropods (inset); E: *Zostera marina* and arrowroot (*Trichlochin maritima*) in a subtidal channel (photos by P. J. Mudie).

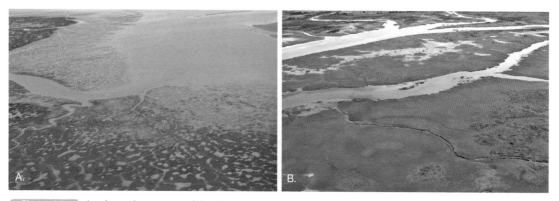

Figure 6.3 A: salt marsh vegetation of the Russian Arctic tundra estuaries at Baydaratskaya Bay (from Morozova and Ektova, 2013); B: *Suaeda* and *Spergularia* in Alaskan marshes have reddish leaves; also note the large driftwood logs (photos by P. J. Mudie).

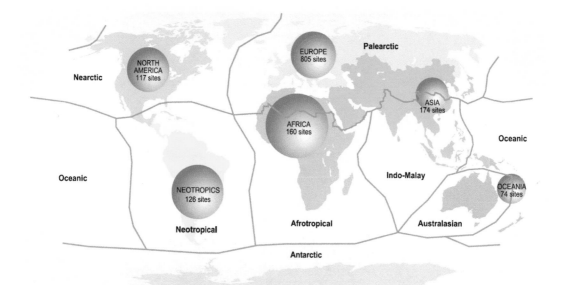

Figure 6.5 Distribution and biogeographic realms of wetlands listed as Internationally Important Ramsar Sites (from MEA, 2005). The amounts shown include all Ramsar wetland types; circle sizes range from 8–37 million ha, from smallest to largest (estimated in 2005). Only about 1–10% of the Ramsar temperate region wetlands are tidal salt marshes and about 25% of the tropical-subtropical sites are mangrove wetlands.

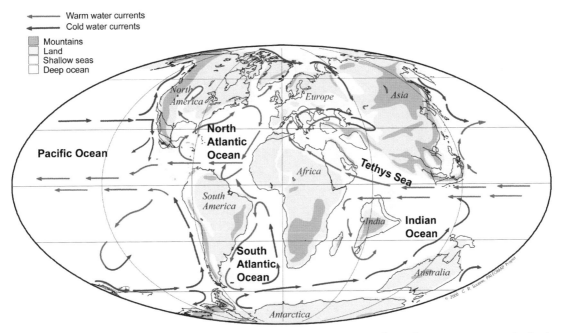

Figure 6.6 Late Cretaceous (66 Ma) ocean circulation allowed interocean migration of water-borne mangrove propagules, but it separated the northern and southern continents, which therefore developed different species diversity and regional adaptations (adapted from C. Scotese, PALEOMAP project).

Figure 7.5 Alaskan marshes north of Glacier Bay where many Alaskan coastal wetlands are at the entrances of inlets with glaciers calving into the headwaters. Left: Calving glaciers and debris-laden ice at Lake Alsek near the head of Dry Bay. Right: Aerial view of the entrance to Lituya Bay (50 km south of Dry Bay), with extensive sedge marsh inside the beach barrier and a large plume of silt marking the outflow of glacial sediment to the Pacific Ocean (photos by P. J. Mudie).

Figure 7.12 Oyster farming in Pacific salt marshes. Left: lines of oyster shell ('seed') are used to provide a firm surface for the seeded oyster larvae ('spat') in Drake's Estero, California, 1986 (photo by P. J. Mudie); Right: in unusually warm El Niño years, closure of oyster farms in Washington State is needed because of outbreaks of toxic red tide algae; in 1995, the Grays Harbor oyster beds were also closed for harvest 10 times from pollution by a sewage treatment plant and pulpmill effluent (photo by Kai Schumann).

Figure 7.27 Inner Chesapeake Bay salt marshes and tidal channels. Straight lines radiating from meandering channels are spoil-lined ditches dug for mosquito control, which reduces invertebrate and fish populations and allows invasions of shrubs like marsh elder *(Iva)* (image courtesy of Chesapeake Bay Program).

Figure 7.17 Excessive plant growth associated with effluent discharge and low salinity in Los Peñasquitos lagoon when the channel entrance is closed. Left: surface scum of *Enteromorpha* algae beginning to decay in the main channel at onset of summer; Right: strands of ditchgrass mixed with masses of blue-green algae grazed by *Melampus* snails; white bubbles are hydrogen sulfide from black masses of decaying plants (photos by P. J. Mudie).

Figure 7.35 These two images show brown marsh in Louisiana marking the dieback of huge areas of cordgrass and meadow salt grass in subtropical salt marshes, which is associated with rising soil water salinity because of increased heat and lowered river water levels since 2000. Green areas are black mangroves and salt grass, which are more salt-tolerant species (USGS: Gabrielle Bodin images).

Figure 7.39 Salt marsh habitats of Northwest Mexico. Left: characteristic *Arthrocnemon–Allenrolfea–Monanthochloe* salt marsh at San Quintin, with the succulent subtropical-tropical species *Sesuvium portulacoides* in the foreground; Top right: meadow of purple sea lavender in *Sarcocornia* marsh between salt flats at San Quintin; Bottom right: dead tule reeds on the banks of a tidal channel at low tide, Colorado River Delta near El Golfo, Sonoran Mexico, 1971 (photos by P. J. Mudie).

Figure 7.42 Tourism has been both a major attraction to the Neotropical mangroves of North America and the basis of their destruction. Left: permanently submerged stilt-roots of red mangrove in the Virgin Islands (photo by Caroline Rogers/USGS Southeast Ecological Science Center); Right top: dead mangroves outside La Paz in 1973, following construction of a new highway; Right bottom: tidal inlet near St. Thomas, US Virgin Islands, with *Rhizophora* roots in silt-laden water (photos by P. J. Mudie).

Figure 9.9 Oysters are a major source of household support in the lower Gambia River. A: red mangroves at low tide, with festoons of oysters on the roots; B: traditional oyster harvesting by the Jola women of the region (photos by Anita Whittle).

Figure 9.15 'Lung' roots of the apple mangrove (*Sonneratia alba*) heavily colonized by oysters (white crust and rings), showing some large scars where careless harvesting has cut the pneumatophores (photo by Patrik Nilsson, taken at Tamil). The portions of dead root may be tunneled and used as brood pouches by the mangrove-borer isopod *Sphaeroma terebrans* (see Woitchik *et al.*, 1993). Originally from India, this crustacean spread both east and west with early shipping, and now has a global distribution in the tropics.

Box Figure 10.1 b View of farmland that has replaced the coastal wetlands at the ancient city of Troy; photo taken from archeological excavations at the modern village of Troy (photo by P. J. Mudie).

Figure 11.5 Stromatolites in Western Australia. A: Large concentration of living stromatolites in the intertidal zone at Shark Bay (photo by D. B. Scott); B: a slice through one stromatolite shows layers of organic cyanobacteria and other microorganisms alternating with calcite mineral deposits (photo from University of Wisconsin, Board of Regents); C: another large assemblage of living stromatolites is found in Hamelin Pool at the southeast end of Shark Bay; the carbonate rock 'stools' look dried out here, but when covered by tides, they will expand and become slimy (photo by D. B. Scott, 2002).

Figure 10.4 Left: the Rhône Delta near Marseilles, showing the main tributary and islands of the Camargue marshes in the saline Vaccares lagoon (CNES Spot image); Right: the Giraud salt flats with flamingos in front of violet 'fleur-de-sel' ponds; white 'mountains' of harvested sea salt are in the background (photo courtesy of Lorenzo Borghi, artborghi.wordpress. com).

Figure 10.9 Mud volcanoes and solanchak on the Taman Peninsula, Black Sea coast of southwest Russia. Left: large, mature (inactive) volcano encircled by salt marsh plants (*Salicornia* and *Suaeda*); electricity for the village in the background is provided by the heat and methane gas – at the risk of occasional breccia explosions; Centre: young, active volcano with hot mud venting methane gas; Right: halophytic salt marsh plants on the white, salt-encrusted soil of the once-submerged beach terrace (photos by P. J. Mudie).

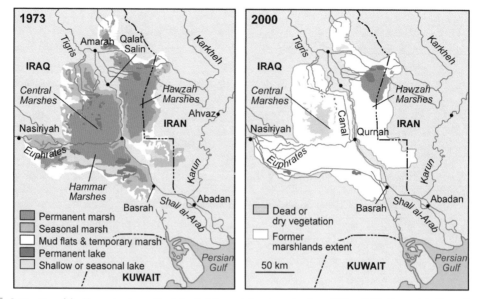

Figure 10.12 Destruction of the Mesopotamian wetlands by draining and river damming from 1973 to 2000. Maps originally made from satellite images by H. Partow (UNEP, 2001).

Figure 10.18 The Irrawaddy Delta and impact of Cyclone Nargis. Left: SRTM altimetry of the delta, showing flooded areas in red, based in MODIS imaging (from Syvitski *et al.*, 2009 and permission from Macmillan Publishers Ltd); Right: aftermath of Cyclone Nargis (courtesy of Dr George Pararas-Carayannis, www.drgeorgepc.com).

Figure 10.19 Top: floating markets on the Mekong River sell produce raised on delta farms; Bottom: traditional fishing net on the delta at high tide (photos by P.W. Mudie).

Figure 10.22 High water areas of tropical mangrove forests in the equatorial region include a wide diversity of shrubs with colourful flowers and large trees. Left: buttress roots (up to 1 m high) of a fig tree, *Ficus macrophylla*; Right: red wild hibiscus with exerted stamens pollinated by birds (top) and the trumpet-shaped yellow flowers of sea hibiscus (*Hibiscus tileaceus*) (bottom) pollinated by butterflies (photos by P. J. Mudie).

Figure 10.23 Relatively little remains of China's tidal original wetlands, but large areas of salt marsh still cover parts of the Yellow and Yangtze River deltas. A: meadows of native sedges fringed by taller cordgrass behind the heavily barricaded coast of Nanhui, near Shanghai, with fishing towers on the South Channel; B: rapidly aggrading mudflat east of Chongming Island, central delta, where cordgrass is colonizing the Dongtan Nature Reserve; industrial developments line the North Channel (top centre); C: cordgrass invading more diverse native marsh on Changxing mud island, South Channel; D: layers of peat and silt in the rapidly aggrading channel at Jiuduansha, a new island formed since 1950; note peat layers in rapidly aggrading reed marsh (photos courtesy of Li Yuang, E. China Normal University, Shanghai).

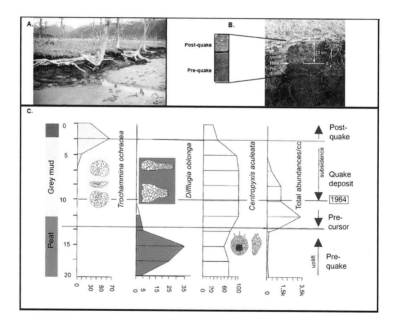

Figure 12.3 Geological record of the megaquake at Girdwood Flats, Alaska, in 1996. A: dead spruce trees drowned by seawater when the earth subsided after the 1964 quake; B: sediment section from the marsh showing an abrupt transition from pre-quake forest peat to tidal mudflat before 1964; grey sediment is mud deposited when the coast subsided before and during the earthquake; C: simplified diagram to show the microfossil distributions in the section just before 1964, where the freshwater thecamoebian (*Difflugia*) and forest peat disappear and are replaced by abundant brackish water *Centropyxis*. The earthquake is marked by the grey mud with fewer *Centropyxis*, and after the 1964 earthquake, the salt marsh foraminiferan *Trochammina* appears (from Scott *et al.*, 2001).

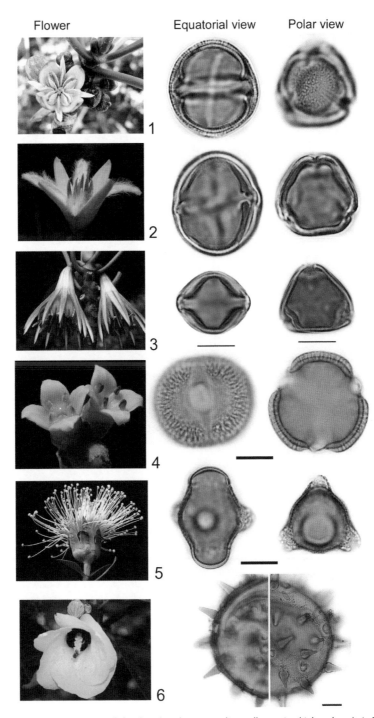

Figure 12.11 Flowers of tropical mangrove trees (left column) and corresponding pollen grains (right column). 1. *Rhizophora apiculata*, 2. *Rhizophora mucronata*, 3. *Bruguiera sexangula*, 4. *Avicennia marina*, 5. *Sonneratia caseolaris*, 6. *Hibiscus tiliaceus*. Scale bar for pollen: 1–4: 10 microns; 5, 6: 20 microns (data from Mao Limi, adapted from Mao *et al.*, 2012).

Figure 13.3 Comparison of flowers in *Salicornia europaea* (left), and *Crithmum maritimum* (right). The *Salicornia* flowers have no petals and are buried inside the fleshy stem and leaves. The *Crithmum* flowers are bright yellow and easy to see (photos by P. J. Mudie).

Figure 13.5 Genetic experiments with halophytic plants. Top left: salt-tolerant Galapagos Island tomatoes with tiny yellow fruits have been hybridized with the commercial tomato species to produce an intermediate-sized 'cherry' tomato (Läuchli and Epstein, 1984); Top right: a selection of salt-tolerant cultivars available today (photo by Shawna Henderson); Bottom left: field trial of salt-resistant hybrid rice (photo by IRRI); Bottom right: ancestors of the new salt-resistant forage crop, tall wheatgrass (*Thinopyrum ponticum*), come from the remnant salt marshes on the Turkish shores of the Black and Marmara Sea (photo by P. J. Mudie).

northwest of Cape Town; other marsh ecosystems are described in Allanson and Baird (1999).

On the warm, high wave-energy southeast coast of South Africa, there are distinctive temperate mangrove communities with *Rhizophora mucronata* (a red mangrove) and *Bruguieria gymnorrhiza* (large-leaf orange mangrove) in protected estuaries from Mozambique southward to Kei River, 32.63° S, in the Transkei. Chapman (1977) reported that the southern limit of the most common South African mangrove *Avicennia marina* was at ~32.5° S, but Steinke (1999) reports it near East London, ~33° S. This southern boundary probably varies over time because periodic die-off of *Avicennia* groves is associated with drowning of the 'breathing roots' after the closing of a tidal estuary by sand bars, although orange mangroves apparently survive such flooding (Adams *et al.*, 2011). Temperate mangroves at the limit of their growth are particularly vulnerable to increased environmental stress from climate, disease or predation. The additional stress is because in colder climates, mangrove wood vessels (water-transporting tissue) are narrower, restricting the water regulation required in fluctuating salinities, and the maturation of their reproductive propagules is slower (Morrisey *et al.*, 2010). Tropical cyclones making landfall at Kosi, St. Lucia and Sodwana bays (Figure 9.14) often result in lagoon closure and high lagoon water levels, causing extensive mangrove tree dieback because the pneumatophores cannot function when permanently submerged. If the water in the closed lagoons freshens, there is danger of infestations of schistosomiasis (= bilharzia), which is a chronic parasitic disease of humans caused by liverflukes (*Schistosoma* spp., a flatworm) hosted by freshwater snail vectors. Schistosomiasis, which is also widespread in southern Asia and Brazil, is the second most socioeconomically devastating parasitic disease after malaria (Kupferschmidt, 2013). Like malaria, bilharzia is proving to be very difficult to control and eradicate by development of a vaccine, and it accounts for 200 million infestations worldwide and 300 000 deaths/yr in sub-Saharan Africa.

The largest estuarine system in Africa is The Greater St. Lucia (*iSisangaliso* in Zulu) Wetland Park which stretches for 220 km from Kosi Bay (~27° S) in the southern Mozambique coastal plain to St. Lucia at ~29° S, near the eastern border of the famous Hluhluwe Game Reserve in KwaZulu–Natal Province (Figure 9.14). The park consists of 13 adjoining protected areas and it extends offshore to include the southernmost coral reefs on the African continent at Sodwana Bay where coelacanth (*Latimeria chalumnae*), the world's most ancient and rarest fish, is occasionally found. Climatically, this region is a subtropical zone transitional between the southern temperate coastal wetlands and the tropical mangrove forests to the north. Thus the park encompasses an exceptionally large biodiversity, including many endemic plant and animal species. The protected area is home to the largest African population (800) of hippopotamus and about 1000 crocodiles, as well as a wealth of plant and animal life. The park shelters two endangered mammals: the humpback whale and the African elephant, which was re-introduced into the wetlands in 2001 to help restore the ecosystem to its former natural condition. In addition, it also shelters the endangered Gaboon adder (*Bitis gabonica*), that has the longest fangs and the highest venom yield of any snake in the world. The sandy outer beaches of the park are important nesting areas for large numbers of leatherback and loggerhead turtles. Other animals include a large variety of snakes, including the deadly black mamba, boom-slang and puff-adder,

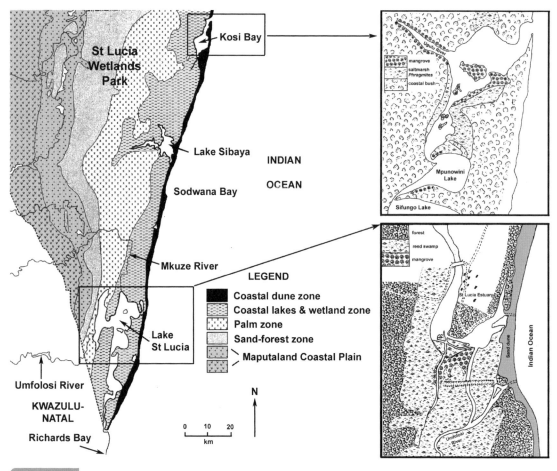

 Figure 9.14 Left: the largest mangrove swamps in South Africa are in the St. Lucia Wetlands Park (from Rajkaran, 2011; after Watkeys *et al.*, 1993); Right: mangroves and extensive reed swamps occur at St. Lucia and Kosi bays (Macnae, 1963).

multitudes of frogs in the freshwater swamps and crabs on the mangroves, and about 520 species of birds, including large breeding colonies of pelicans, storks, herons and terns.

St. Lucia and Kosi Bay are the main areas with coastal wetland vegetation. These wetlands consist of estuarine lakes with a total water area >1200 ha that occupy drowned river valleys separated from the sea by vast tree- and grass-covered fossil sand dune systems. The semi-diurnal spring tidal range in the region is ~1.8 m, but the tidal prism of the estuaries is small and the temperature of the water in the shallow lakes (0.9 m at St. Lucia) fluctuates seasonally from ~11 to 22 °C in winter and from 23 to 29 °C in summer (Rajkaran, 2011). Shifting dunes can close the tidal channels and isolate the estuarine lakes from the sea. The salinity of the estuarine lakes therefore is highly variable, ranging from brackish (~8 psu) to hypersaline (>80 psu), depending on fluctuations in rainfall (normally ~1 m/yr) and seasonal evaporation.

The vegetation of the Wetland Park includes four major plant associations (Figure 9.14), with extensive areas of coastal lakes and wetlands behind a coastal dune ecozone, seaward of a palm zone and a sand-forest tree zone (MacNae, 1963; Watkeys *et al.*, 1993). The wetlands include freshwater swamps with diverse tall grasses, including millet (*Sorghum bicolor*), beds of reeds (*Phragmites australis* and *P. mauritanius*) and sedges (*Scirpus littoralis*), and areas of mangroves. A salient feature of the whole park mangrove region is the absence of the red mangrove, except occasionally at Kosi Bay. This anomaly may be because of periodic prolonged freshwater flooding of the estuarine lakes, although laboratory studies show a wide range of salinity and inundation tolerance for *R. mucronata* (Hoppe-Speer *et al.*, 2011). At St. Lucia, there are only two mangrove trees, the black and the orange mangroves, as also found in the Transkei; at St. Lucia, however, the trees are up to 7 m high compared to <4 m in the south. The mangroves occur from MHW to EHW, intermingled with reeds, and at the upper edge of the marsh, they are joined by the mangrove fern *Acrostichum aureum*, the mangrove hibiscus shrub and the powder-puff mangrove, *Barringtonia racemosa*. This tree has stilt roots like those of the red mangrove, but it can also grow in dry conditions.

At Kosi Bay, where winter minima and summer maxima are ~1° C higher than at St. Lucia, there are more extensive mangrove forests in the intertidal zone, greater species diversity and two other Indo-Pacific shrubby mangroves. These additional species are the widespread tropical yellow mangrove (*Ceriops tagal*; Rhizophoraceae), closely related to *Rhizophora*, and the high marsh Pacific black mangrove (*Lumnitzera racemosa*, Combretaceae), which is in the same family as the white/button mangrove *Laguncularia racemosa*.

9.6 Other important East African coastal wetlands and some cautionary notes

The mangroves from Kosi Bay to northern Kenya (Figure 9.1), including the west coast of Madagascar, have encountered a smaller amount of change (~8%) relative to the west and north coasts (Spalding *et al.*, 2010). Small-scale restoration efforts have begun in Kenya and Tanzania. The mangrove flora and fauna of this region are similar to that at Kosi Bay, with the addition of *Sonneratia alba* (mangrove apple) and 5–20 associated shrubs and salt marsh species that are widely distributed in the tropical Indo-Pacific region. *Heritieria littoralis* (looking-glass mangrove) and *Xylocarpus granatum* (cannonball mangrove) are two other common Indo-Pacific mangroves that also grow in this region. Gang and Agatsiva (1992) describe the mangroves (7 spp.) and associated tropical shrubs and salt marsh plants (32 spp.) in lagoons of the Kenyan coast, where there was a total of 52 980 ha in 1981. This mangrove area apparently expanded to 60 950 by 2010, despite clearance for sugar plantations and damage from oil pollution around Mombasa. The best-protected mangrove forest is in northern Tanzania and on the nearby islands of Zanzibar. Offshore, extensive seagrass beds play an important role in stabilizing sediments and protecting the coral reefs from siltation and erosion.

Por (1984) considers that there is also another type of mangrove ecosystem in East Africa which is also found on the Low Isles of the Australian Great Barrier Reef. This variant, called the 'hard bottom mangrove', characterizes dry tropical and subtropical climates and low-lying coral atolls, where salinity is high (47 psu). It is marked by low-growing, shrubby black or red mangrove and a rich development of algae, seagrass and sessile animals on the roots (Box 9.2 Sinai Desert 'hard bottom' mangroves). Nine species of seagrasses in the genera *Halophila*, *Halodule*, *Thalassia*, *Thallassiodendron* and

---

Box 9.2   **Sinai Desert 'hard bottom' mangroves**

Detailed studies have been made of the mangroves on the south coast of the Sinai Peninsula because, unlike the typical tropical coastal wetlands with high rainfall year-round and soft, river-transported muds filling deltas and lagoons, the Sinai and Red Sea mangroves are in hot, dry deserts with little rain and with strong sunlight year-round. Salinity and temperature fluctuations are among the highest in the world, and only grey mangrove *Avicennia marina* grows here. The sketch (using data from Por and Dor, 1984) shows the typical vertical zonation of sessile animals on a pencil-root pneumatophore (~0.3 m high) of the Sinai 'hard bottom' mangal. In addition to the animals, 46 species of nitrogen-fixing blue-green algae grow on different parts of the root, the tip of which has chlorophyll pigment and grows upwards towards the light like a green shoot, not downwards or sideways like the normal primary and cable roots that anchor the trees below the mud surface. Other algae form stromatolite mats on the hard mudflat surface between the pneumatophores. No foraminifera were found on the mangrove pencil roots, but there were 17 species in surface sediments and on leaves of four seagrass genera.

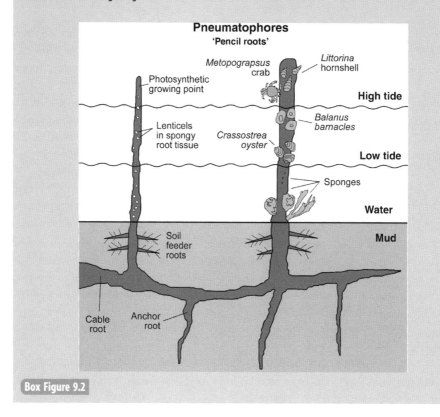

Box Figure 9.2

Cymodocea are found within Gazi Bay (Kenya), and 20 kinds of seaweeds grow on the mangrove roots; another 15 species of epiphytic algae also grow on the shoots of sickle-leaf cymodocea (*Thallassiodendron ciliata*) and 279 species of benthic diatoms were recorded in a survey of the wetlands by Schrijvers *et al.* (1993). This list for Gazi Bay illustrates the great biological diversity associated with unpolluted tropical mangroves in regions where sufficient light can penetrate the tree canopy and where sediments are sandy, so that the water contains little silt and clay to reduce light penetration and clog the pores of sponges.

Although the mangrove degradation in the East African region is lower than in most countries, a vigorous timber and fuelwood trade with the Middle East extends back at least 2200 years and illegal harvesting continued into the last century, when 0.5–0.7 million mangrove poles were exported each year. *Sonneratia* timber is relatively light and this may explain its successful survival because it is less suitable for construction and it does not last as long as denser mangrove timber. However, the air-filled 'lung roots' (pneumatophores) of the apple mangrove are harvested for fishing floats, and careless oyster harvesting may result in large scars (Figure 9.15).

Punwong *et al.* (2013) studied the pollen and charcoal in four radiocarbon-dated cores from the mangroves on a Zanzibar Island near Unguja Uuku and used these proxy data to reconstruct the mangrove ecosystem dynamics during the Holocene. They found that *Rhizophora mucronata* and other mangroves were well established by about 7000 cal BP

Figure 9.15 'Lung' roots of the apple mangrove (*Sonneratia alba*) heavily colonized by oysters (white crust and rings), showing some large scars where careless harvesting has cut the pneumatophores (photo by Patrik Nilsson, taken at Tamil). The portions of dead root may be tunneled and used as brood pouches by the mangrove-borer isopod *Sphaeroma terebrans* (see Woitchik *et al.*, 1993). Originally from India, this crustacean spread both east and west with early shipping, and now has a global distribution in the tropics. See also colour plates section.

and migrated landward during a sea level highstand. After the mid-Holocene, however, there was a decrease in apple mangrove, and increases in associated mangrove species, salt marsh grasses and charcoal, indicating drier climatic conditions, at the same time as the first habitation of Zanzibar. Increased charcoal after ~1400 cal yr BP may also reflect early human settlement at Unguja Ukuu. Recent anthropogenic impacts are marked by a reduction in mangroves, possibly associated with the mangrove pole trade and fuel wood consumption in Zanzibar, particularly after ~530 cal yr BP.

North of Kenya are the deserts of Ethiopia, Eritrea, Sudan and Egypt, with patchy stands of mangroves around sheltered bays, where they have escaped heavy browsing by camels and degradation from prolonged Sahelian droughts. The mangroves reach their northern limit on the Sinai Peninsula of the Red Sea, which is part of the tectonic fault system of the African Rift Valley. Mangroves on the gravelly carbonate shore of the Red Sea are mostly shrubs < 0.5 m high (Figure 9.16; Kholeif, 2007) although they may be trees 4 m high, where they escape grazing and/or there is some freshwater seepage. An important study of the pollen in sediments of the black mangrove area at Abu Ghosan (22.36° N, ~31.30° E) shows that the only species, grey mangrove (*Avicennia marina*), does not produce much pollen at its distribution limit, and this pollen does not travel far beyond the pneumatophores. This means that caution is needed in interpreting absence of this pollen in a sediment core as indicating the mangroves were not either not yet established (i.e. sea level transgression had

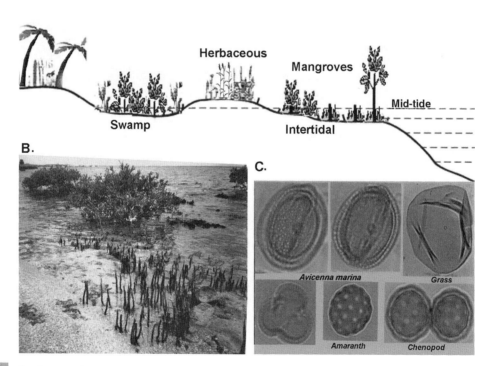

Figure 9.16 Abu Ghosan study area, Red Sea. A: shoreline vegetation zones; B: *Avicennia* shrubs and pneumatophores; C: indicator pollen of mangrove and salt marsh vegetation (courtesy of Kholeif, 2012).

not occurred) or that the mangroves had died out from anthropogenic overexploitation or pollution damage. This work on perhaps the smallest stand of mangroves in Africa has profound consequences for interpretation of Holocene and recent mangrove pollen data on a global scale, since *A. marina* is the most widespread mangrove of the Old-World Indo-Pacific region, and the same principles of pollen dispersal may pertain in the Atlantic and Pacific regions of the New World.

10   Europe and Asia: a view of what remains

Key points

Eurasia is the largest continent, but its coastal wetlands have long histories of human occupation and few tidal wetlands remain; temperate salt marshes dominated by grasses and pickleweed occur in Western Europe and the northern Mediterranean; these wetlands are grazed by livestock or used for salt production; extensive microtidal Danube Delta marshes survive as wildlife refugia because of wild pigs and malaria mosquitoes; Mesopotamian and Yangtze basin marshes are highly imperiled by upstream dam construction; Southeast Asia has 10 densely populated delta wetlands in high to extreme danger of survival; tsunamis and tropical storms damage fringe mangroves on coastlines long changed by farming and aquaculture; monsoon floods and tidal surges drown millions of people in China despite giant dams; efforts are being made to replace mangals destroyed for wood, and rice or shrimp farms, and to stabilize mudflats with cordgrass.

Background

The continents of Europe and Asia (Eurasia) make up about one-third of the world's land area, but the combined acreage of their coastal wetlands is less than in the much smaller Nearctic and Afrotropical landmasses (Ramsar, 2007). Coastal wetlands of both Europe and northern Asia belong to the Palearctic biogeographic province (see Figure 6.5). Eurasian temperate marshes therefore share many common characteristics, particularly large-scale anthropogenic alteration over thousands of years, beginning in the Middle East, Europe and China, which have been ancient centres of agriculture and the locations of first cities since about 9000 yr BP (Roberts, 1998; Riehl *et al.*, 2013). In this temperate region, there are now only two major estuaries: the Rhine Estuary (Netherlands) and the Shatt al-Arab (Iraq); both are seriously endangered by rising global sea level and global warming trends. Further south in Eurasia, the tropical-subtropical wetlands lie in the Indo-Malayan and Polynesian sectors of the Oceanic biogeographic region. Here, ten vast, densely populated estuaries are under moderate to extreme threat of destruction from rising global sea level. The tectonically active Southeast Asian region on the western 'Pacific Ring of Fire' is further subject to frequent earthquakes and tsunamis, as well as intense river flooding during the monsoon rain season.

Because of the great size of the Eurasian landmass and large interregional diversity among its coastal wetlands, we can only present a few case histories highlighting representative regional ecosystems, land-use histories and salient problems in this book. We have chosen to document regions less-covered in other wetlands texts, but excluding the vast Arctic coastline, which has very little available published information and is partially covered in Section 6.1.

10.1 European temperate regions: waddens, estuaries and the Low Countries

There, two times in each period of a day and a night, the ocean with a fast tide submerges an immense plain, thereby hiding the secular fight of the Nature whether the area is sea or land.
(From Oxford Dictionary re 'wadden')

The tectonically stable coastline of Western Europe (Figure 10.1) is located on a broad continental shelf with partly glaciated shorelines that are variable in both tidal range and in coastal geomorphology. This coast includes macrotidal estuaries in the Seine River (France), and the Severn, Solway and Dee Estuaries (England, UK), extensive mesotidal inlets in the Wadden Sea–Rhine Delta areas of Holland and Germany, and the microtidal Dovey (Dyfi) Estuary of Wales. Eisma (1997) and Pratolongo *et al.* (2009) give summary descriptions of this coastal geomorphology and the marshlands. Books by Chapman (1974; 1976) give details of the English wetlands, while Davy *et al.* (2009) provide a comprehensive account of human modifications of the western European salt marshes and recent conservation efforts of the European Union. Here we present a detailed example of the history of the largest and most altered of the western European estuaries – the delta tributary complex in the vicinity of the mouth of the Rhine River in the Netherlands (neder = low, Dutch), including the Zuiderzee.

Temperate tidal marshes occur throughout the region, from Spain to Denmark (Figure 10.1), and many of the earliest classical ecological studies of salt marshes were made here (see Chapman, 1976; Adam, 1995). Application of pollen studies for understanding the Holocene salt marsh histories also began here. The largest of the salt mash–mudflat complexes, called waddens (wad = mudflat, Dutch), are those behind the sand barriers along the coast from Mount St. Michel Bay in France to the Baltic Sea entrance. These wadden marshes developed behind shifting coastal sand-dune barriers and were hazardous environments for the earliest inhabitants, as described by the Roman magistrate, Pliny in ~80 CE, who considered the marsh people a miserable race living in a 'no-man's land' between land and sea. The Rhine Delta, at the confluence of the Rhine, Meuse and Scheldt Rivers, lies between the major cities of Amsterdam and Rotterdam. This delta is one of the largest in Western Europe and is now in moderate danger of destruction by rising sea level (Syvitski *et al.*, 2009).

Construction of protective dikes to reclaim the Dutch and German waddens probably began about 300 BCE, and it was well underway around 1000 CE (Figure 10.2), when many towns/cities were present throughout the area. Draining of the peat bogs for urbanization and farming lowered the marsh surface, which was then extensively flooded by the disastrous St. Lucia storm surge in December 1287. The name Zuiderzee (= Southern Sea) was given to

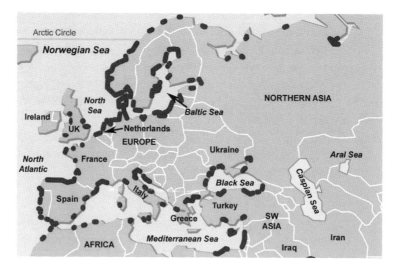

 Figure 10.1   Map of Europe and western Asia, showing distributions of coastal wetlands (thick black lines) and main locations of countries mentioned in the text (modified from WRI; UNEP-WCMC).

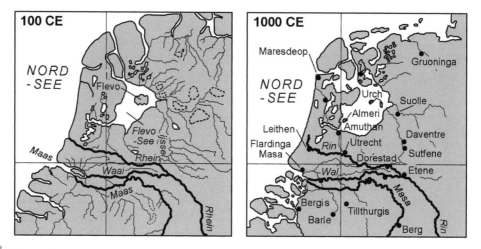

Figure 10.2   Old maps showing the early development of the Zuiderzee in the Netherlands (including modern Belgium south of the Rhine = Rhein/Rin River, and Holland to the north). Left: tidal channels connecting the Flevo See and North Sea (Nord-See) ~100 CE; Right: larger Almeri Sea about 1000 CE, with the developing cities of Utrecht in Belgium and Amuthan (old Amsterdam) in Holland. Note the decrease in small rivers, the increase in towns and expansion of the tidal lagoon between the second and eleventh centuries (redrawn from Otto Humlum, with permission).

the 120 km-long area of the North Sea coast drowned by the St. Lucia flood, which was the fifth largest in history and killed 50 000–80 000 people. In 1362, another winter hurricane called Grote Mandrenke (= Great Drowning of Men, Dutch) caused a second massive flooding, which further eroded the coastline. As a result, the originally small, semi-enclosed Flevo Sea barrier lagoon of early Roman time became fully connected to the North Sea.

Because of dike construction and drainage, by the late Middle Ages, the waddens were transformed to farmland, and by the seventeenth century, the dikes were moved seawards to allow more land reclamation. This construction phase peaked in the nineteenth and twentieth centuries, when the cities of Amsterdam and Utrecht were built on former peat- and marshlands (Feddes, 2012).

Compared to North America, where sustained farming only began after ~1590–1680, the coast of Western Europe was farmed at least two millenia earlier, and by ~800 years later, about 70% (133 km^2) of the salt marshes in western France alone were converted to agricultural land (Pratolongo *et al.*, 2009). Today, the salt marshes in most of the remaining wetlands, from Belgium to Denmark (~390 km^2, including mudflats) are partially drained and grazed by sheep. Therefore, not much is known about the original salt marsh vegetation, despite centuries of detailed studies of the surviving marshland, using the species-association method (Chapman, 1977; see Section 6.2). A series of historical maps (Dijkema, 1987) show how the tidal salt marshes of the Wadden–Zuiderzee area were changed from 1660 to 1826 CE by construction of embankments, infilling and draining for agricultural and military defense purposes. The largest change was in 1932, when construction of the Afsluitdijk (= Cut-off dike) was completed across the outer Zuiderzee. This massive dike converted the former Flevo estuary into the freshwater Lake Ijsselmeer, after which much of the inner lagoon was reclaimed for urban development and farmland (Figure 10.3).

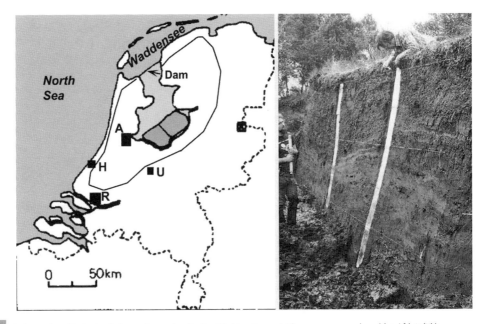

Figure 10.3 Left: modern Netherlands has salt marshes in the Wadden Zee outside an enormous dam (the Afsluitdijk), and the freshwater Ijsselmeer is now largely reclaimed by and dredge infill (dark grey shading). The polygon marks land that would flood if the dam broke, drowning cities marked by squares. Right: soil profile of reclaimed wadden, showing about 1 m of newly formed dark peat layers grown over the light-grey tidal mudflat sediments since diking (slides by B. van Geel, University of Amsterdam, with permission).

Today, the typical zonation of Netherlands and other European temperate salt marshes starts with a seaward low marsh zone sparsely covered by patches of tall cordgrass (*Spartina anglica*), interspersed with annual glassworts (*Salicornia dolichostachya, S. fragilis*) and on the landward edge, small patches of salt marsh grass (*Puccinellia maritima*). The middle marsh has more plant cover and diversity, including the halophytes *Salicornia ramosissima, Suaeda maritima, Halimione portulacoides, Atriplex portulacoides* and salt marsh daisy *Aster tripolium*. High marshes are dominated by spikerush and the grasses *Festuca rubra* and *Elymus athericus* (wild rye-grass). In northern Europe, wild rye-grass is invading the middle marsh, forming monospecific growths that replace *Atriplex* marshes; this change may reflect heightened soil nitrogen levels from agricultural runoff. Elsewhere, the hybrid cordgrass *Spartina* × *townsendii* (native *Sp. anglica* × North American *Sp. alterniflora*) are invading the low marsh, but in some areas, this hybrid is undergoing dieback and decline after 30 years of aggressive spreading (Pratolongo *et al.*, 2009).

10.2 European Mediterranean: vanishing oceans and sinking cities

The Mediterranean Sea basin is a large depression caused by the tectonic African Plate sliding under the Eurasian Plate. From ~45 to 12 million years ago, it was part of the wide Tethys Seaway linking together the Atlantic and Indo-Pacific Oceans (see Figure 6.6). During the warm Messinian period ~7−5 million years ago, a near-desiccation of the Mediterranean dropped the water level below Gibraltar Strait, and the ancestral Rhône and Po rivers carved huge valleys across the exposed shelves of the Tyrrhenian and Adriatic seas as the Alpine ranges pushed upwards, triggering earthquakes and giant landslides. At the end of the Messinian, the Atlantic Ocean broke through Gibraltar Strait and refilled the Mediterranean Sea, flooding the river valleys and filling them with marine sediments before the Pleistocene Ice Age. Cycles of interglacial marine transgressions and glacial-stage regression are recorded in sediments of the Po valley southwards to Sicily. The final transgression began after the last glaciation, ~20 000 yr BP, submerging the shoreline until about 5500 yr BP, after which large deltas and coastal wetlands began to grow seawards. An entertaining account of the Mediterranean Sea evolution is given by Dorrik Stow (2010) in *The Vanished Ocean*. The impact of Holocene climate change in the region east of Italy is dramatized in a book by Ronnie Ellenblum (2012), *The Collapse of the Eastern Mediterranean*. Ellenblum reconstructs drought cycles from Syrian and Egyptian crop records and shows how they are linked with the rise and fall of early empires and cities.

10.2.1 Rhône Delta and Camargue Marshes, France: violet salt and pink flamingos

The Rhône is a major European river that rises in the Swiss Alps and flows to the western Mediterranean Sea through southeastern France. At Arles, the river divides into two branches, forming the large delta of the Camargue wetlands region west of Marseilles

Figure 10.4   Left: the Rhône Delta near Marseilles, showing the main tributary and islands of the Camargue marshes in the saline Vaccares lagoon (CNES Spot image); Right: the Giraud salt flats with flamingos in front of violet 'fleur-de-sel' ponds; white 'mountains' of harvested sea salt are in the background (photo courtesy of Lorenzo Borghi, artborghi.wordpress.com). See also colour plates section.

(Figure 10.4). Despite the construction of dikes and embankments, the boundaries of the Camargue still change periodically as the river annually transports about 20 million m³ of silty mud downstream. Although constrained by sea dikes, the delta coastline continues growing seawards, producing major landscape changes on a century scale. For example, the town of Aigues Mortes (Figure 10.4), built as a seaport in the thirteenth century, is now ~5 km inland.

In 1927, about 85 000 ha of the Camargue delta were designated as a nature reserve to try and maintain a balance between the wetlands and human activities (mainly tourism, hunting, agriculture and industry). The northern salt marshes are reclaimed for rice, cereal farming and vineyards, but other areas remain brackish–saline ecosystems, with meadows of sea lavender and glasswort flourishing between levees covered in reeds and invasive tamarisks. In spring, the low-lying salt plains (sansouires) are covered with glasswort, which is grazed by the little endemic Camargue bulls and Camargue horses before the salt flats dry out in summer. The marshes are submerged in winter, and in spring the lagoons are flooded up to 2 m, becoming important feeding areas for coots, diving ducks and fishing birds such as grebes, terns, seagulls and pink flamingos. The ponds also support migratory and sedentary birds, including egrets, night herons, bitterns, mallards and wagtails. Here, economic development is not at odds with wildlife. The flamingos actually depend on a sustainable salt industry, as evident at the Etang du Fangassier lagoon refuge established in 1969 (Figure 10.4). This lagoon was filled with seawater by the salt company and for 30 years, it supported a flourishing colony of flamingos that were protected from predators, fed on brine shrimp and nested on an artificial island in the Rhône Delta. When a partial strike at the saltworks stopped the pumping of seawater in December, 2007, the lagoon dried up and the flamingos moved to areas with food and better-protected nesting sites.

Salt harvesting in the lower Grand Rhône valley dates back to Roman times and includes one of the biggest saltworks in the world, producing up to 15 000 tonnes/day in summer.

Some salt is bulk-harvested for use as table salt, but the prime Fleur De Sel de Camargue ('flowers of salt') is hand-raked and harvested for a medicinal product (Figure 10.4R). For this product, only the top layer of the salt bed is used; the name Fleur De Sel comes from the aroma of violet-scented flowers that develops as the salt dries. There was once a vast network of special salt roads to transport Camargue salt to France via the mountain passes and to Italy by sea. *Salicornia* plants and salt crystals were also incinerated together to produce soda-salt (sodium carbonate and potassium carbonate) for soap making and glass-making, but this was replaced by industrial soda by the end of the nineteenth century. Tellinas ('sunset shells', *Donax trunculus*) are small cockles from the sandy shores of the Camargue that are traditionally cooked with either onions, herbs and white wine, or with chopped parsley and garlic.

10.2.2 The Po Estuary and sinking of Venice Lagoon into the Adriatic Sea

For twelve hundred years, the city [of Venice] has performed a magical balancing act. Not quite land, not quite sea, Venice seems to float in its own world.
(NOVA, Sinking City of Venice, 19 November 2002)

The Po River is the longest river in Italy, with about one third of its length in the Alps and two-thirds on the northern Italian Plain, where it ends in a large delta on the Adriatic Sea (Figure 10.5L). The delta wetlands contain some of the largest marshes in the Adriatic Sea, and although many areas have been reclaimed for farmland, there are still healthy marshes and a local eel fishery. The wetlands are protected by a UNESCO World Heritage Site, which contains 53 653 ha of wetlands, coastal forest, dunes and salt pans, with a high biodiversity, including over 1000 plant species and 374 vertebrate species (mostly birds).

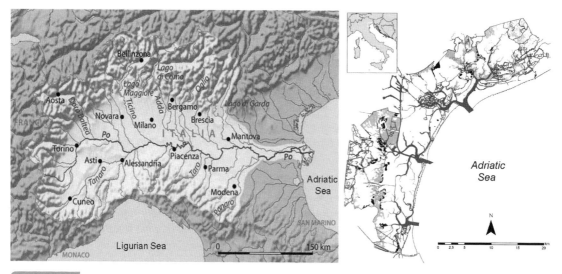

 Figure 10.5 Po River watershed and Delta at the Adriatic Sea, south of Venice Lagoon. Left: map of the Po Valley in northern Italy (photo by Nord Nord West, CC); Right: details of Venice Lagoon with three tidal channels in the lagoon barrier, remnant salt marshes on the inland borders (dark grey), surrounded by upland and urban areas (light grey and white).

Over the past 5000 years, the Po River Delta prograded seawards, with an estimated rate of ~4 m/yr from 1000 BCE to 1200 CE. By the mid twentieth century, however, anthropogenic changes had changed the entire coastline of the northern Adriatic Sea, including the Po Delta wetlands inside Venice Lagoon, which is the largest in Italy and one of the largest in the Mediterranean (Day *et al.*, 1999). The city of Venice, originally built on islands off the coast, is now sinking and is at high risk of flooding because of continued subsidence and rising sea level. The recent changes are mainly because of reduced sedimentation following dam construction upstream and excavation of river-sand for industry. Furthermore, agricultural use of the river water is high, causing the drying-up of freshwater aquifers and subsurface flow of saltwater into the coastal water table. The lower Po Valley is now one of the most imperiled of the world's major estuaries (Syvitski *et al.*, 2009). Water pollution is also a problem, particularly during low streamflow in the hot, dry Mediterranean summers. Sewage is discharged into the river; as result, in 2005 the water contained 'staggering' amounts of the chemical ecognine benzoate, which is excreted in the urine of cocaine users. On 24 February 2010, the Po River was contaminated by an oil spill of about 600 000 litres from a refinery near Milan in western Italy.

The modern part of the Po Delta, with active tidal channels, began forming in 1604 when the city of Venice diverted the main Po tributary southwards to bypass flow into Venice lagoon, which was filling with river sediment. The mean depth of the lagoon is now 1.1 m and has a microtidal range of 60–100 cm, controlled by huge floodgates (Figure 10.5R). Most of the tidal lagoon (about 370 km^2) is bordered by intertidal salt marshes. Extensive tidal flats below MLW are partially vegetated by macroalgae and the seagrasses *Zostera marina*, *Zostera noltii* and *Cymodocea nodosa*. The salt marsh plant associations range from a low marsh with annual *Salicornia* spp. from 5–10 cm above MSL to a *Limonium serotinum–Spartina maritima* association to 20 cm above MSL, then a middle marsh *Limonium–Puccinellia* association (from 15–30 cm MSL) and a high marsh *Puccinellia palustris–Arthrocnemum fruticosum* association at 25–40 cm above MSL.

The Venetian marshes remain important because of their size and high productivity, although the area has shrunk from about 12 000 ha around 1900 AD to about 4000 ha in 1992 because of reclamation, erosion, pollution and land subsidence (Day *et al.*, 1999). Artificial islands created from dredge soil in Venice lagoon now support successful breeding of 13 species of waterfowl and shorebirds (up to 10% of the Italian populations) in areas where intense human recreational use otherwise keeps birds from their natural beach nesting areas (Scarton *et al.*, 2013).

10.3 Southeast Europe and Southwest Asia: Mediterranean–Indian Ocean transition

Figure 10.1 shows the Eastern European and Southwest Asian subregions of Eurasia, with the locations of salt marsh sites, several of which have been studied for their histories of delta formation and sea level change after the last glaciations (Mudie *et al.*, 2011). Some studies are covered in the Nile Delta section of the previous chapter (Section 9.2). However,

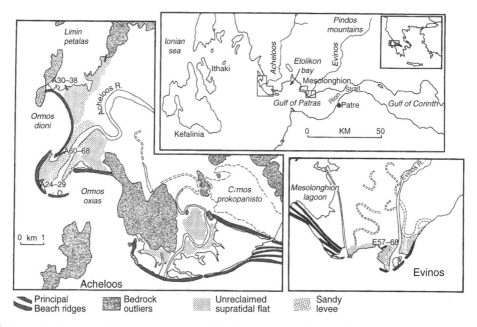

Figure 10.6 Maps of the Acheloos and Evinos marshes on the Gulf of Patras, southern Greece, where foraminifera were studied in 1979 by Scott, Piper and Panagos. Here the salt marshes occupy a narrow coastal belt on the microtidal shoreline of the Gulf of Patras, where the hinterland landscape is intensely overgrazed (see Figure 10.7).

most of the marshes in the summer-dry Mediterranean region are very small and have not been examined in detail; others are large tidal freshwater marshlands, including the Ukrainian Danube and Dniester Deltas and the Shatt al-Arab in Iraq. Some information is available for salt marshes in southern Greece, first studied by Scott *et al.* (1979) at Patras off the Ionian Sea (Figure 10.6), and more recently, investigated at Korphos on the Aegean Sea by Nixon *et al.* (2009). The foraminifera and thecamoebians from three Korphos salt marsh cores show there were five apparent sea level advances (transgression events) after the end of the mid-Holocene global rise in RSL. Marsh accretion rates at Korphos were calculated from radiocarbon-dated peat and correlated to a series of submerged tidal notches. The combined data indicate that the transgressions were rapid and episodic, displaying a step-wise pattern of RSL change that is best explained by subsidence from earthquake fault displacement on the Hellenic subduction zone of the Northeastern Mediterranean.

Throughout the Northeastern Mediterranean and Black Sea regions, interpretation of salt marsh histories and rates of sea level change is complicated because of frequent volcanism and earthquakes associated with the African tectonic plate thrusting northwards into Eastern Europe and rotating eastwards towards central Asia. Further difficulty in understanding the native salt marshes arises from the 6000–9000-year-old history of agriculture, which began in the Iraq highlands and spread westwards along the coast into Europe. Consequences of early domestication of goats, sheep and horses in the summer-dry Mediterranean region are the intense grazing of the evergreen intertidal marshlands adjoining overgrazed hillside grasses and shrubs, and erosion of densely cultivated, terraced coastal slopes (Figure 10.7).

Figure 10.7 Overgrazing of hillsides and salt marshes in the Eastern Mediterranean. A: terraced hillside almost denuded of vegetation near the Gulf of Corinth, Greece in terrane similar to the Gulf of Corinth where the marsh foraminiferal studies were made; B: sheep graze on succulent *Atriplex* and *Suaeda* halophytes that continue to grow on a raised Holocene shoreline along the Strait of Dardanelles at the entrance to Marmara Sea, Turkey; C: wild horses (Ramsar 2007) and D: domestic donkeys graze in salt marshes throughout much of the region, from the Camargue in France (Left; Ramsar, 2007) to the Trikharti liman salt marshes west of Odessa in the Ukraine (A,B,D: photos by P. J. Mudie).

High rates of soil erosion have filled many major estuaries, such as the Karamenderes Valley north of Troy (Box 10.1 Ancient Troy, Trojan legends and salt marshes), and the Meanderes Valley east of Rhodos in Turkey. For example, boreholes, archeological studies and historical records for the Karamenderes (Scamander) Valley west of Troy show that from ~5500 to 5000 yr BP, the delta sediments aggraded ~3 km northwards, with a further ~3 km northward migration from 5000 to 3000 yr BP (Kraft *et al.*, 2003). At that time, the legendary city of Troy, now ~10 km inland (Box Figure 10.1), was on the shore of a harbour, and fish and shellfish were important foods in addition to pork, mutton and goat meat from domesticated animal herds. During the last 3000 years, however, the delta has migrated northwards another ~4 km and the bay is now filled and converted to farmland. This high rate of aggradation is largely from upstream erosion associated with deforestation and overgrazing, although tectonic uplift of the Dardanelles Strait shoreline may also be involved.

Box 10.1   **Ancient Troy, Trojan legends and salt marshes**

The Simoeis and Scamandar [Rivers] effect a confluence in the plain, and since they carry down a great quantity of silt they advance the coastline and create a blind mouth, and saltwater lagoons, and marshes.

(Strabo, 13.1.31, ~0 CE)

The history of the legendary city of Troy and its former harbour is recorded in works by several famous ancient Greeks – including the epic poem 'The Iliad' of Homer who lived around 800 BCE, and books by Strabo, the geographer during the reign of Cesar Augustus ~20 BCE–7 CE. Strabo describes the tidal salt marshes in the estuary of the Scamander (modern Karamenderes) River, which was not far from Troy at the time of Homer and the epic Battle of Troy (Box Figure 10.1). The word 'meander' which is used to describe the wave-like path of rivers in a floodplain comes from these Turkish rivers (Kara = black; menderes = winding stream). Molluscs in the fossil salt marsh peat indicate brackish to marine bay environments, where today only river sediment and farmland can be seen from the city ruins (below). Archeological studies also indicate that the Trojan marshlands were a breeding ground for malaria mosquitos that caused the death of a high percentage of Trojan children under the age of two years. The map shows the history of salt marsh infill associated with deforestation and livestock grazing: the river mouth is now located on the shore of the Dardanelles (Çanakkale) Strait, about 6 km north of the ruins of the legendary city (Box Figure 10.1) where the battle for Helen of Troy took place.

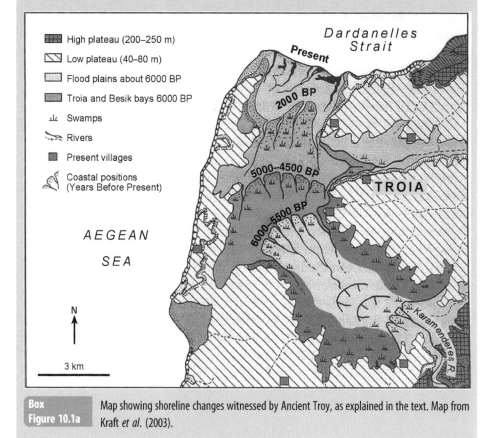

Box Figure 10.1a Map showing shoreline changes witnessed by Ancient Troy, as explained in the text. Map from Kraft *et al.* (2003).

Box Figure 10.1b View of farmland that has replaced the coastal wetlands at the ancient city of Troy; photo taken from archeological excavations at the modern village of Troy (photo by P. J. Mudie). See also colour plate section.

10.3.1 Northern Black Sea: tiny tides and giant mud volcanoes

Chapman (1974) considers that the salt marshes of Egypt in the southeast Mediterranean Sea (Figure 10.1) represent a transition from the western Mediterranean communities to those of the Middle East, as far eastwards as the Caucusus Mountains, between the Black and the Caspian Seas. Similar, but generally more prostrate, salt marsh vegetation, dominated by the succulent perennials *Halocnemon strobilaceum*, *Salicornia prostrata* and *Suaeda salsa*, are found on the shores of Ukraine and on the Kuban Peninsula, Russia, near the entrance to the Sea of Azov (Figure 10.8). Here the tides of the semi-enclosed seas are semi-diurnal or mixed semi-diurnal, with a microtidal average range of only 2−9 cm, but with larger fluctuations of ~45 cm corresponding to seasonal thermal expansion of the seawater and with additional interannual changes of up to 25 cm during freshwater floods (Sorokin, 2002). There are 19 coastal wetlands covering 635 000 ha along the Ukrainian coastline, in the deltas of large rivers and behind barrier lagoons (= limans). The surface salinity of the Black Sea is low − about 16 psu – and liman salinities range from 2 to 12 psu.

The zonation within the microtidal Black Sea marshes is barely visible (Figure 10.8A), but varies considerably with soil texture (sand versus mud) and with distance from large rivers, where the estuaries are often filled with the reed *Phragmites australis*, sedges and cat-tails. In silty soils around the Saskia liman in the Ukraine and in the Azov-Sivaś preserve, there is low

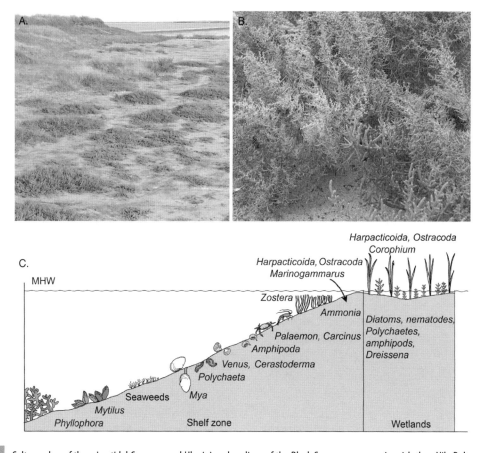

Salt marshes of the microtidal Caucuses and Ukrainian shorelines of the Black Sea are more species-rich than Nile Delta associations because of higher rainfall and lower surface water salinity. A: middle and high marsh in silty soil near Saskia liman, dominated by *Halocnemon strobilaceum*, with grasses, *Limonium* and *Artemisia*, in the high marsh; B: *Sarcocornia perennis* and *Suaeda* on sandy soil of Tuzla Spit, northeast Black Sea; C: profile of a sandy shoreline, northwest Black Sea, shows the pre-1980 zonation of plants and animals. After ~1980, however, invasions of the Japanese giant snail *Rapana thomasiana* and blue-green algae changed this succession in some areas, which are now creeping dead zones (photos by P. J. Mudie).

marsh with annual *Salicornia europaea* and *Suaeda maritima* below the perennial *Halocnemon* forbs. The high marsh contains *Atriplex* spp., several grasses (e.g. *Puccinellia fominii*; *Pholiurus pannonicus* and *Aleuropis littoralis*), *Limonium caspicum* and *Artemisia taurica* in a condensed zone around EHW (see Tonev *et al.*, 2008 for species lists). *Zostera marina* is the main seagrass in subtidal areas, together with the red alga *Phyllophora nervosa* in deeper water. Zaitsev and Alexandrov (1998) list the macrobenthos and algae associated with sandy and rocky shores of the northwestern Black Sea (Figure 10.8C). Characteristic meiobenthos distributions are given by Yanko-Hombach *et al.* (2013).

One highly unusual feature of the Black Sea area is the presence of mud volcanoes venting watery mud and methane gas from active pools of hot (~38 °C) bubbling mud cones

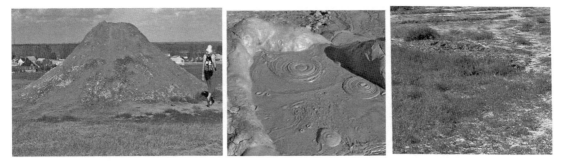

Figure 10.9   Mud volcanoes and solanchak on the Taman Peninsula, Black Sea coast of southwest Russia. Left: large, mature (inactive) volcano encircled by salt marsh plants (*Salicornia* and *Suaeda*); electricity for the village in the background is provided by the heat and methane gas – at the risk of occasional breccia explosions; Centre: young, active volcano with hot mud venting methane gas; Right: halophytic salt marsh plants on the white, salt-encrusted soil of the once-submerged beach terrace (photos by P. J. Mudie). See also colour plates section.

(Figure 10.9, centre). Lines of small active and larger 'quiescent' volcanoes are common on low terraces of the Kerch and Taman Peninsulas near the entrance to the Azov Sea, which were partly submerged during the last warm climatic interval ~125 ka BP. The origin of the mud volcanoes on the Taman Peninsula is variable and includes continental and marine sources (Shnyukov *et al.*, 2010). The gas and hot water probably originate from organic-rich sediments in ancient (mid-Tertiary) subsurface clay beds that are being squeezed by the weight of the overlying sediments and the gas pressure. The 'quiescent' volcanoes explosively erupt every ~50 yrs, spewing out mud, steam and mud balls (breccia), and endangering local villages, which use the methane for heating (Figure 10.9). The Taman mud volcanoes represent onshore examples for some submarine pingo-like features in the Beaufort Sea (see Section 7.1). The marine origin of some of the Taman mud volcanoes is clearly reflected in the presence of salt marsh plants around the mud cones (Figure 10.9 left) and halophytes (*Limonium suffruticosum*, *Halimione verrucifera*, *Artemisia santonica* and *Kochia prostrate*) in ancient salt marsh soils (solonchak = salt marsh, Russian) of the recently submerged terrace (Figure 10.9 right).

10.3.2 Danube Delta: the largest and best-preserved wetland in Central Europe

The Danube Delta is a lobate delta formed by shifting tributaries depositing large volumes of silt at the mouth of the Danube River, which is the longest river in Western Europe, being shared by nine nations (Pringle *et al.*, 1993; Giosan *et al.*, 2012). This river drains 70% of the subcontinent, and annually supplies about 33% of the freshwater in the Black Sea. The modern delta began to form about 6500 years ago, when Mediterranean water flowing from the Aegean Sea to the Black Sea reached the level of the modern shoreline. Today, some delta lobes aggrade into the Black Sea at a rate of 24 m/yr, and although numerous dams have reduced the sediment load by about 50% since ~1960, delta accretion is still greater than the rate of sea level rise (Syvitski *et al.*, 2009). The high volume of year-round

| Box 10.2 | Characteristics of tidal freshwater marshes (Pringle, 1993) |

Definition: Coastal wetlands influenced by tides, but little affected by saltwater
Distribution: Deltas and estuaries of large river systems with microtides
Salinity: Average salinity less than 1 psu, temporarily increasing during drought and/or large storm surges
Tidal range: May be higher than on open coast because of amplification in channels

Vegetation structure:

- mostly species restricted to freshwater or low salinities
- complex mosaics of perennial and annual species
- high percentage of emergent macrophytes (e.g. waterliles (*Nuphar, Nymphaea, Pontederia, Sagittaria*))
- many submersed species of pondweeds (e.g. *Potamogeton, Myriophyllum* spp., *Elodea* spp.)
- communities of annuals (*Bidens* spp., *Polygonum* spp.)
- often extensive stands of cattails (*Typha* spp.), wild rice (*Zizannia aquatica*), panic grass (*Panicum hemitomon*)
- 'floating marshes' buoyed by roots and rhizomes with airspaces
- large seed banks
- seed germination depends on river flooding
- no distinct shoreline zonation, but gradients from coast to delta/estuary head

Fauna:

- Low meiofaunal and invertebrate diversity
- High diversity of reptiles and amphibians
- Huge bird populations
- Aquatic mammals e.g. otter, mink, rice rats, nutria

freshwater flow, combined with the microtidal range of the Black Sea, produces tidal freshwater wetlands (Box 10.2 Characteristics of tidal freshwater wetlands) dominated by reeds, sedges, and floating and submerged pondweeds, with an outer fringe of barrier lagoons and salt marsh vegetation (Sârbu *et al.*, 2011). The delta wetlands include ~230 000 km^2 of river channels, lakes, ponds and low sandbars, stretching from southwestern Ukraine through eastern Romania, and it has the largest reedbeds in the world. These *Phragmites australis* reedbeds are fascinating because many of them form floating islands of vegetation (plavs/plaurs) which support layers of brackish water sedges, herbs and ferns, providing a safe haven for nesting birds and a refuge from the wild boars (*Sus scrofa*) in the wooded shorelines. Sârbu *et al.* (2011) document the main emergent and aquatic plants of the main delta channels in Romania, and Coops *et al.* (1999) list ten categories of submergent vegetation in the ponds between the grassy plaurs. A list of the aquatic macro- and microfaunas is given in Berczik *et al.* (2012).

Although the inner Danube Delta wetlands have been altered by construction of drains and canals for agriculture and transcontinental shipping (boats sail from Amsterdam to the Black Sea via the Rhine River and canals linked to the Danube), until the 1980s, much of

Figure 10.10 Danube Delta and its iconic animals. Top: Ukrainian section of the delta is sparsely inhabited by fisherman who also collect reeds for thatch and wall construction (photos by P. J. Mudie); Bottom: the UNESCO preserve protects flocks of white pelicans (photo by Ali Arsh, Flickr CC) and herds of wild boars, which have survived hunting since Neanderthal times by hiding in the swamps (photo by Marieke I. Jsendoorn-Kuijpers, Flickr CC).

the vast delta was largely unpopulated (<80 000 people) since its first colonization by the Scythians, Greeks and Romans, about 500 years BCE. Human occupation of the wetlands is still discouraged by the abundance of ferocious wild boars (Figure 10.10), wolves, wild horses, foxes and polecats, in addition to otters, minks, muskrats and hares. Tortoises, adders and other snakes live here. Furthermore, swarms of mosquitoes and biting flies infest the lowlands in summer, making habitation uncomfortable. Consequently, the Danube Delta provides a critical refuge for several globally threatened species, including 70% of the world's white pelican population, 60% of the world's pygmy cormorants, populations of glossy ibis and spoonbill, various species of egrets and herons and mute swans – a total of 330 bird species. The wetlands shelter about 110 species of fish, 36 of which live only in the delta, which is the source of half the freshwater fish production of Romania. The saltwater environment of the Razim–Sinoe lagoons includes perches, pike, mullets, mackerel and flounders in areas of preferred salinity. Marine waters in front of the delta shelter endemic Black and Caspian Sea sturgeons, while the brackish−low-salinity delta waters include the great sturgeon, common sturgeon, sevruga and red starlet, which produce the edible black fish-roe called caviar.

Other very large eastern European wetlands are located in the braided delta of the Volga River, northern Caspian Sea, where there are also extensive reed swamps and large lakes

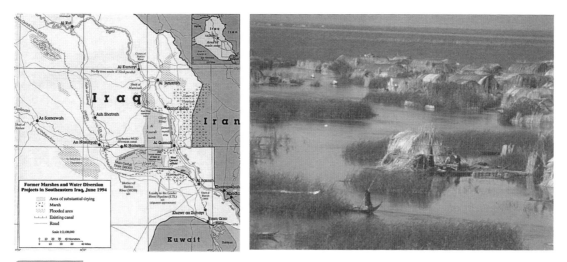

Figure 10.11 Mesopotamian wetlands. Left: Map showing Iraq, places mentioned in the text and the locations of adjoining countries involved in the politics of its water distribution (courtesy of Central Intelligence Agency). Right: Central Mesopotamian marshes, with Ma'dan reed house villages and wooden canoes (from UNEP, 2001).

covered by water lilies (Mitch and Gosselink, 2007). However, like the Aral Sea ecosystem, these inland seas do not have marine tides and are therefore not included in this book.

10.3.3 The Tigris–Euphrates Rivers case history: what happened to the Gardens of Babylon?

The Tigris–Euphrates Delta region of southern Iraq and Iran (Figure 10.11) was the largest wetland ecosystem of western Asia before 1950, covering about 20 000 km^2, almost half the total area of coastal salt marshes in the world today (Richardson and Hussain, 2006). The grassy Mesopotamian wetlands, surrounded by dry desert sands, were once the home of up to 500 000 indigenous Marsh Arabs (Ma'dan) who lived on islands in the channels, surviving by fishing, hunting wild pigs and birds, and raising water buffalo (Young, 1977).

From 1951 until the late 1970s, however, the marshes were drained to reclaim land for agriculture and to facilitate oil exploration. Remote-sensing image analysis shows that 63% of the marshes dried up (Figure 10.12) and lakes were replaced by evaporitic salt deposits from 1985 to 1992 (Brasington, 2001). The drainage programme accelerated when President Hussein tried to relocate the Ma'dan who had lived in the wetlands for about 5000 years, including the time of the biblical Gardens of Babylon, which were probably in this fertile region. By the end of Hussein's regime in 2003, the marshes were drained down to 10% of their original size, burned over and polluted by toxic gases. Fewer than 10 000 Ma'dan survived in the wetlands and the southwestern marshes were further damaged by acidic 'black rain' from the burning of oil well fields in Kuwait in 1991, although the magnitude of long-term oil damage has not yet been fully determined (Al-Otaibi *et al.*, 2006). Restoration of the marshlands is now being attempted by reflooding of dikelands, but droughts,

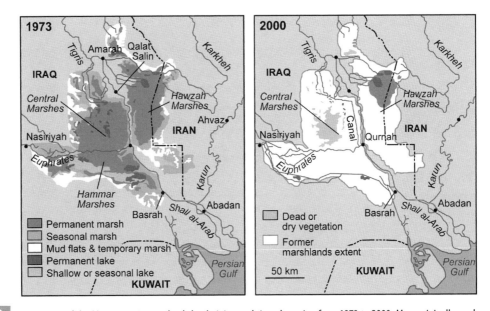

Figure 10.12 Destruction of the Mesopotamian wetlands by draining and river damming from 1973 to 2000. Maps originally made from satellite images by H. Partow (UNEP, 2001). See also colour plates section.

construction of 21 dams and international disputes regarding water rights to the vast Tigris and Euphrates river systems have hindered progress (Al-Yamani *et al.*, 2007).

Mesopotamia means 'in between rivers' and the Mesopotamian wetlands include three major areas sustained mainly by the Tigris and Euphrates Rivers, with additional contributions from the less-well-known Karun River (Figure 10.12). The marshlands stretch between two deltas: the main ancient inner delta produced by the Euphrates and Tigris, and a younger marine delta built by the Karun and Marunjerrahi Rivers at the head of the Persian Gulf (UNEP, 2001). Near the junction of the Tigris and Euphrates, the Mesopotamian plain narrows because of a huge gravel fan in the west, and the Karun delta complex in the east. The three rivers join to form the 100 km-long tidal Shatt al-Arab Estuary that drains into the Persian Gulf (= Gulf of Arabia) near Abadan. West of the Shatt al-Arab along the Gulf Coast are the Khor Zubayr tidal marshes, which are dissected by wide, deep subtidal channels with marine salinities (~32−40 psu) and a network of smaller channels draining large tidal flats. The Shatt al-Arab channel has relatively low salinities (1−1.5 psu) during most of the year, increasing to 8 psu during November and December. Because of the freshwater outflow from large rivers, the central Mesopotamian wetlands are primarily tidal freshwater ecosystems, with salinities of 0.5−2 psu during summer. However, salinities increased during the 1980s to ~2 psu, and there are now evaporitic lakes of 10−15 psu, surrounded by ponds and sabkhas (supratidal flats) with salinities up to 60 psu.

There are few accessible details on the geological evolution of the marshland, but it is clear that they developed from silt discharged into the Persian Gulf from the rivers after the slow-down of postglacial sea level rise ~6000 years ago (Lees and Falcon, 1952). Early geographers reported that before 4000 BCE, the head of the Persian Gulf reached Baghdad, which is now ~3000 km upstream (~150 km above Amarah), and that around 325 BCE, the coastline was ~180 km inland of its present-day position (Heyvaert and Baeteman, 2008). Before 1985, the largest marshes were in the floodplain between the Tigris and Euphrates Rivers, consisting of extensive reed beds surrounding several permanent lakes (UNEP, 2001). The dense reedbeds graded shorewards to meadows of low sedges, spikerush and bulrush (*Scirpus brachyceras*). Deeper permanent lakes supported diverse submerged aquatic vegetation, including hornwort (*Ceratophyllum demersum*), eel grass (*Vallisneria spiralis*), pondweeds (*Potamogeton lucens* and *P. pectinatus*), watermilfoil (*Myriophyllum* sp.), naiads (*Najas marina* and *N. armata*) and stoneworts (*Chara* spp.). In the smaller lakes and back swamps, there was a floating vegetation cover of water lilies (*Nymphoides peltata, N. indica, Nymphaea caerulea* and *Nuphar* spp.) and water soldier (*Pistia stratiotes*), and duckweed (*Lemna gibba*) was common (Rechinger, 1964; Scott, 1995). A total of 47 plant species were recorded for these wetlands in the 1970s (Hamdan *et al.*, 2010).

Until recently, the average depth of water in the main marshes during the flood season was about 1.0–1.5 m, up to about 2.0–3.5 m (rarely ~6 m). Today, the emergent vegetation is dominated by the common reed, narrowleaf cattail/reedmace (*Typha domingensis*), papyrus and occasionally the cane, *Arundo donax*. Reeds dominate the deeper areas, while *Typha* occupies the seasonally wet fringes, and bulrushes dominate temporarily flooded areas. When wetlands restoration began after 2003, emergent aquatic vegetation colonized parts of the shores, and there are now reed beds in the littoral zone. The sabkhas surrounding wetlands have a sparse vegetation of halophytic shrubs (*Haloxylon salicornicum*), alkali-tolerant shrubby perennials (*Achillea fragrantissima, Artemisia herb-alba* and *Rhanterium epapposum*) and scattered *Ziziphus* trees.

Seven major wetland types are found in the present-day wetlands that now cover about 60% of the former area: (1) permanent freshwater lakes with a rich growth of submergent aquatic vegetation and a marginal zone of floating aquatic vegetation, (2) permanent freshwater marshes dominated by tall stands of *Phragmites, Typha* and *Cyperus*, (3) rivers, streams, canals and irrigation channels, typically with little emergent vegetation and steep earth or muddy banks, (4) permanent ponds, mainly man-made irrigation ponds and duck-hunting ponds, typically with a pronounced drawdown in summer and little emergent vegetation, (5) seasonal freshwater marshes dominated by rushes and sedges, typically occurring as a broad belt around the edge of the permanent marshes, (6) seasonally flooded mudflats and semi-desert-like steppe, (7) shallow, brackish to saline lagoons, mostly seasonal and often with extensive areas of *Salicornia*. About 26 of the original 47 plant species have returned (Hamdan *et al.*, 2010), and there are ~12 new plant species, generally characteristic of dry subsaline soils, and including *Ruppia maritima* and *Zannichellia palustris* as pond weeds, and the invasive cane, *Arundo donax*. Along the Gulf shore, reed swamp and intertidal mudflats extend from the mouth of the Shatt al-Arab westwards along the Gulf shore for at least 50 km. Here the tidal amplitude is large (3 m or more) and the bare mudflats and swampy flats are backed by a belt of date palms and extensive bare silt flats. In the mid 1970s, the region included an estimated one-fifth of the world's 90 million palm trees. However, by 2002, warfare, salt and pests had wiped out more than 14 million palms and many of the remaining trees are in poor condition.

Table 10.1 Highly threatened birds of the Mesopotamian marshlands (Source: Scott and Evans, 1994; Scott, 1995)

Common Name	Scientific Name (notes)
Entire population supported by marshes:	
Iraq babbler	*Turdoides altirostris* (endemic)
Basrah reed warbler	*Acrocephalus griseldis* (endemic)
Dependent on marshes:	
Pygmy cormorant	*Phalacrocorax pygmaeus*
Dalmatian pelican	*Pelecanus crispus*
Imperial eagle	*Aquila heliaca*
White-tailed eagle	*Haliaeetus albicilla*
Marbled teal	*Marmaronetta angustirostris*
Slender-billed curlew	*Numenius tenuirostris*
Little grebe (subspecies)	*Tachybaptus ruficollis iraquensis* (endemic)
Breeders in the marshes:	
Goliath heron	*Ardea goliath*
African darter (subspecies)	*Anhinga rufa chantrei* (declining)
Sacred ibis	*Threskiornis aethiopicus*
Grey hypocolius	*Hypocolius ampelinus* (most of the breeding population)

The Mesopotamian marshlands are a major haven of regional and global biodiversity, still supporting significant populations and diversity of wildlife. Located on the intercontinental West Siberian–Caspian–Nile flyway of migratory birds, they are particularly important for birds (UNEP, 2001). The marshlands are a key wintering and staging areas for waterfowl travelling between breeding grounds in the Ob and Irtysh river basins in western Siberia to wintering quarters in the Caspian region, Middle East and northeast Africa. Two-thirds of West Asia's wintering wildfowl, estimated at several millions, may reside in the marshes of Al Hammar and Al Hawizeh. Incomplete surveys report 134 bird species in significant numbers from the area, of which at least 11 are globally threatened (Table 10.1).

Once abundant, the Mesopotamian marsh mammals have been under enormous pressure of elimination. Most lions were hunted and killed soon after the introduction of the rifle after World War I, the last lion being shot in 1945. Other mammals and reptiles and their statuses are listed in Table 10.2. The marshland ecosystem includes many fish, including several of economic and scientific importance. In 1990, the FAO estimated that 60% of Iraq's inland fish catch of 23 500 tonnes was caught in the marshes. A selection of fish species is listed in Table 10.2. Additionally, a wide range of marine fishes migrate upstream via the Shatt al-Arab, including occasional sharks. The penaeid shrimp (*Metapenaeus affinis*) migrates seasonally between the saline Arabian Gulf and low-salinity nursery grounds in the marshlands (Banister, 1994).

The Mesopotamian wetlands are famous for their rich archeological and historical legacy (UNEP, 2001). This was where the Sumerian civilization flourished about 1792 BCE (~4000 years ago), followed later by the Akkadian, Babylonian, Assyrian and

Table 10.2 Other animals in the Mesopotamian marshes (UNEP, 2001)	
Animal (Scientific name)	Notes or Status
Mammals	
Grey wolf	Globally threatened
Long-fingered bat (*Myotis capaccinii*)	
Smooth-coated otter (*Lutra perspeicillata**)	*subspecies *maxwelli* (endemic)
Honey badger	
Striped hyena	
Jungle cat	Rare in the 1980s; may be locally extinct
Goitered gazelle	
Indian-crested porcupine	
Wild boar (*Sus scrofa*)	Was once the most common mammal; declined due to agricultural draining of the marshes
Feral Asian water buffalo (*Bubalus bubalis*)	Probable ancestor of Indian water buffalo
Small Indian mongoose	
Asiatic jackal	
Red fox	Frequently sighted
Bandicoot rat (*Erthyronesokia bunnii*)	
Harrison's gerbil (*Gerbillus mesopotamicus*)	
Reptiles	
Caspian terrapin (*Rafetus euphraticus*)	
Soft-shelled turtle	
Variety of snakes	
Desert monitor (*Varanus griseus*)	Once common in desert regions bordering the marshes; overhunted and now rare
Fish	
Carp (*F. cyprinidae*)	Dominant species
Gunther (*Barbus sharpeyi*)	Known as *bunni*; endemic, commercially important
Giant catfish (*Silurus glanis* and *S. triostegus**)	*endemic
Hilsa shad (*Tenualosa ilisha*)	Use marshes for spawning
Pomphret (*Pampus argenteus*)	

Chaldean civilizations. At that time, the Gardens of Babylon, one of the classical Seven Wonders of the World, were built at the edge of the marshland when the Tigris–Euphrates Delta was near present-day Baghdad. Mesopotamia is an area of major significance in the history of Christianity, Islam and Judaism. Biblical scholars regard it as the likely original site of the legendary 'Garden of Eden' and the 'Great Flood'. Mounds (called tells) within the marshes were likely sites of ancient cities (Oppenheimer, 1998). The recent large-scale landscape changes associated with drainage, military activity and oil exploration hampers continued archeological research, yet the legacy of marshland drainage is not new. Archeological evidence shows that control of the Tigris and Euphrates through irrigation canals and flood protection has been vital to the settlement of lower Mesopotamia over the past 5000 years. Geoarcheological records also document several attempted irrigation agriculture failures due to the ensuing intrusion of salt.

No specific plans were made to drain the marshlands, however, until the twentieth century, when surveys for irrigation development were commissioned by the Ottoman Empire and the question of marshland damage was raised. In 1951, British engineer Fred Haigh provided the framework for the subsequent marshland reclamation (UNEP, 2001). Haigh's main objective was not marshland drainage, but to counter the problem of soil salinization caused by Mesopotamia's drained wetlands and irrigated farmlands. Haigh designed a drainage network to transfer saline waters from the farmland to the delta interfluves and to divert the Tigris waters from the Central and Eastern marshes (Figure 10.12). Construction of the huge Main Outfall Drain (MOD) to remove the saline drainage started in 1953, but in the 1980s, the focus shifted to marshland reclamation. In 1992, the Iraqi government completed the man-made Saddam River, which is a drainage canal extending from the Central Plain, tunnelling under the Euphrates, and finally emptying saline, nutrient- and pesticide-enriched water onto the northwest Gulf tidal flats. Other major water diversion schemes such as the 'Mother of Battles' River (Umm-al-Maarik) and the 'Crown of Battles River' (Tajj al-Maarik) finally reduced the wetlands to ~10% of the original size by ~1995. The Al Hammar Marshes were also divided into polders by dikes, resulting in soil desiccation. Ironically, because salt intrusion followed the water diversion, by 2000 CE most of the reclaimed farmlands were barren until the dikes were broken open in 2003.

The extraordinary scale and speed of land-cover change in the Mesopotamian marshlands is comparable only to the deforestation rates of Amazonia and desiccation of the Aral Sea (UNEP, 2001). In less than a decade, one of the world's largest and most significant wetland ecosystems has completely collapsed. The loss of the marshlands has had a catastrophic impact on wildlife and biodiversity well beyond Iraq's borders because the loss of wintering area for migratory birds on the Western Siberia–Caspian–Nile flyway is registered across thousands of miles from the Arctic to southern Africa. About 66 species of Mesopotamian marshland birds are at risk, in addition to the endemic mammals, and up to 40% of Kuwait's shrimp catch that originated from the marshes is threatened. Drying of the marshlands is also likely to impact the coastal fisheries in the Northern Gulf, with serious economic consequences. Desiccation of over 9000 km^3 of wetlands and lakes could also have changed the regional microclimate because lower humidity leads to increased aridity, higher temperatures and strong winds. The dry winds were previously buffered by the vast reed beds, but when unimpeded, they sweep over the salt crusts and dry mudflats, raising dust laced with pollutants and carrying these toxins well beyond Iraq's borders (Maltby, 1994; UNEP, 2001). Ecosystem degradation at this scale could seriously affect human health in several ways, including water scarcity and pollution, hyperthermia and toxic dust storms, on a scale comparable to those associated with the drying out of the Aral Sea.

After 2003, local inhabitants began to reflood some of the marshes, often in an uncontrolled and haphazard fashion. The breaching of levees and dams coincided with good rains in the following two years and the marshes superficially recovered and regained about 55%

of their former extent (CIMI, 2010). However, this recovery was short-lived and the marsh area shrank again after a drought in 2008–2009, only slightly recovering in winter 2009/10. It appears that the natural hydrological regime which underpins the ecosystem dynamics in the original Mesopotamian marshes cannot be fully restored. Nonetheless, the remnant marshes are still one of the largest wetlands in the Middle East, forming a unique island of wetland vegetation and biota in an otherwise extremely arid desert environment (UNEP, 2001).

10.4 Southern and Southeastern Asia: Indo-Pacific and Polynesian subregion of Oceania

In tropical systems it is possible that the effects of global climate change will be over-shadowed by other, larger disturbances such as deforestation and land-use changes.

(Ficke *et al.*, 2007)

Southern and Southeast Asia include all of the tropical-subtropical Indo-Malayan and the Polynesian sector of the Oceania biogeographic regions, encompassing ten of the most densely populated estuaries, which are under moderate to extreme threat of destruction from rising global sea level (see Table 5.1). The tectonically active southeastern region on the western 'Pacific Ring of Fire' is also prone to frequent earthquakes and tsunamis, in addition to intense monsoon river flooding that marks the entire region. Because of the large inter-regional diversity among the Eurasian coastal wetlands, only a few case histories high-lighting representative regional ecosystems, land-use histories and salient problems can be presented in this book. More than 80% of the Asian wetlands are in seven countries: India, China, Indonesia, Papua New Guinea, Bangladesh, Myanmar and Vietnam (= Malayasia/ Malay Archipelago). Of these, some of China's wetlands are covered by Mitsch and Gosselink (2007), so the main focus here is on the less well-known, but important, coastal wetland regions.

Overall, the mesotidal shores of India, Southeast Asia and Polynesia include a vast array of salt marshes and mangrove-covered coastal wetlands (Figure 10.13) in large delta estuaries and multiple coastal lagoons of the subtropical-tropical shorelines. Most of the region has a tropical climate (annual monthly temperature >18 °C) with high humidity and abundant rain (1800 mm), especially during the monsoon season of April–October. In 1997, about 40% of the world's mangroves were found in this region, including 62 of a total of 73 world species (Spalding *et al.*, 2010). However, rainfall is lower in the semi-arid regions bordering the Arabian Sea, so wetlands are not common along the West Coast of India, except in the Indus Delta of Pakistan and a few protected lagoons south of Mumbai and Goa (Ravishankar *et al.*, 2004). The microtidal, wave-dominated shores of Sri Lanka also support only a thin fringe of mangroves, and over 50% of the wetlands have been cleared for farmland or aquaculture (Figure 10.14).

The 3200 km-long Indus River has its headwaters in the Himalayan ranges of western Tibet, not far from the source of the Ganges River, which flows in the opposite direction to

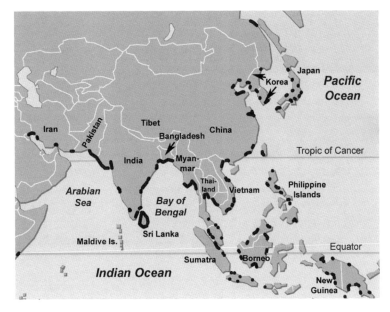

Figure 10.13 Distribution of coastal wetlands in the Indo-Pacific and Oceania biogeographic regions of Asia (modified from WRI; UNEP-WCMC).

Figure 10.14 The coast of Sri Lanka still supports 23 mangrove species, but many forests have been damaged by anthropogenic actions or by tsunamis. Left: undisturbed mangrove forest produces abundant leaf litter (photo by Steven Lutz, Blue Carbon Working Group, AGEDI: UNEP/GRID-Arendal); Right: terraced and densely farmed inner estuary near Colombo where the population grew from 80 000 in 1886 to ~0.6 million today (photo by Gary Sonnichsen, ~1995).

the East Coast of India. Sediments from the Indus form a wide delta south of Karachi, the capital city of Pakistan (Giosan *et al.*, 2006). The Indus used to be one of the most important sediment-producing rivers in the world, with an extensive delta and alluvial plains supporting the vast Harrapan civilization from about 3300–1300 BCE. Today, however, most of the river flow is diverted into the world's largest irrigation system, which was developed over the last 150 years by building large dams and barrages, leading to huge reductions (50–80%) in water and sediment discharge after 1950. In the early twenty-first century, the discharge was further reduced to ~two months/yr, and the survival of the delta and its irrigated farmland is now in question. Today, this delta is fronted by a series of almost bare, sandy barrier islands between Karachi and the river mouth, and huge tidal salt flats 40 km wide occur up to 6.5 km inland (Eisma, 1997). Mangroves are relatively rare: 95% are grey mangroves (*Avicennia marina*), with scattered red mangroves (*Rhizophora mucronata*), tagal (*Ceriops tagal*) and river mangrove (*Aegiceras corniculatum*) forming a fringe along the seaward channels, grading inland to thickets of tamarisk bushes.

The coastal wetlands of India cover ~500 000 ha of mangroves, including 19% in the Indian Andaman and Nicobar Islands west of Sumatra. Most of these mangal (~71%) are on the East Coast and in the Sundarbans of the Ganges Delta (Ravishankar *et al.*, 2004). Generally, the whole of Southeast Asia is characterized by vast estuarine wetlands supporting endangered mammals that are in high to extreme danger of survival. Earthquakes, tsunamis and tropical storms are also threats to the surviving thinned-out mangrove forests along shores long altered for farming or aquaculture. In the following sections, we present highlights of the vast Ganges, Irrawaddy, Mekong, Yellow and Yangtze delta regions and give some case histories for parts of the Polynesian islands. Details of the mangrove vegetation are provided by Spalding *et al.* (2010) and by Giesen *et al.* (2007); animal ecology is reviewed by Nagelkerken *et al.* (2008) and the management of the dwindling wetland wildlife is discussed by Belsare (1994); a review of the socio-economic importance of the mangals and their management is given by Walters *et al.* (2008). Mapping of the mangrove distributions by satellite imagery is reported by Giri *et al.* (2011) although the reader is cautioned that some areas of salt marsh without mangroves are included in these maps.

10.4.1 Ganges River Delta: The Green Delta and Beautiful Gardens

What we are doing to the forests of the world is but a mirror reflection of what we are doing to ourselves and to one another.

(Mahatma Ghandi)

The Ganges Delta, 22° N, 90° E, is one of the largest in the world. Spanning the borders of India and Bangladesh, this delta is one of the most densely populated areas on Earth. The Ganges–Brahmaputra Rivers, which together form this delta complex, carry a huge sediment load that supports rich farmland, but also creates frequent severe flooding. The Ganges Delta (also called the Ganges–Brahmaputra, the Sundarbans and the Bengalla Delta) stretches for 380 km across the northern Bay of Bengal from the Hooghly River

Figure 10.15   The Ganges Delta marshlands and mudflats. Left: Satellite image of the delta and distributaries fronted by mudflats (photo by Klein/NASA); Right: several hundred endangered Bengal tigers still live in the delta (photo by Bernard Landgraf, CC).

distributary in the west to the braided estuary of Meghna River in the east (Figure 10.15). The cities of Kolkata (formerly Calcutta) and Haldia in India, and Mongla and Chittagong in Bangladesh are large seaports on the delta. River outflow is primarily in the east where the Padma tributary of the Ganges joins the Jamuna River, the main distributary of the Brahmaputra River. After merging, these combined rivers join the Meghna River before reaching the sea in a huge complex of interconnected channels. Macrotidal influences dominate the western delta where semi-diurnal tides normally reach 5.6 m and can increase to 7 m above MSL during monsoons (Eisma, 1997). However, the worst floods in the Ganges Delta are not caused by monsoons but are associated with pre- and post-monsoonal tropical cyclones that are amplified by northward recurving of storms in the Bay of Bengal (Murthy and Flather, 1994). Severe storm surges are an additional hazard throughout this region and these can be linked to El Niño events, which are predicted to increase with continued global warming.

The Ganges Delta lies at the junction of three tectonic plates – the Indian, Eurasian and Burma Plates – where the Indian Plate is moving northwards and colliding with the Tibetan Plateau, forming the lofty Himalayan ranges. Since the Himalayan collision about 60 million years ago, enormous amounts of eroded river sediment has resulted in the seaward progradation of the delta and shelf edge by about 400 km. Today, the Ganges Delta is an 'arcuate' (arc-shaped) delta covering over 105 000 km^2. Although its location is Bangladesh and India, the delta is sustained by rivers with headwaters in other countries – Bhutan, China and Nepal – within and north of the Himalaya Mountains, where increased dam construction is planned because of shrinking glaciers. Most of the delta consists of fine-grained river-borne sediment that settles out when the currents slow down in the tidal channels. This alluvial soil is rich in plant minerals and nutrients, making it fertile and good for farming. The main delta area has a labyrinth of channels, swamps and lakes interspersed with bare flood plain sediments called chars.

Mangroves cover ~650 000 ha of the western delta, constituting one of the world's largest continuous mangals. These forests are naturally dominated by *Heritiera fomes* (= sundari) and *Excoecaria agallocha* (= Tbilla or Thillai), but since 1966, some areas have been replaced by plantations of *Avicennia officinalis* and *Sonneratia apetala* in an effort increase the buffer against storms (Spalding *et al.*, 2010). The usually ubiquitous tropical red mangrove is not present. This absence suggest that the tall stilt roots of red mangroves are more prone to storm damage than the pencil roots of the other mangroves, but field observations do not actually support this idea. Regional tectonic subsidence is shifting the river flow eastwards, which is reflected in lower soil salinity (~15–20) in the east versus 28–32 psu in the west (Blasco *et al.*, 1994). The iconic Sundarban mangroves grow on an older deltaic plain in the western delta, which was formed before the eastward shift of the river (Eisma, 1997). Channelization of the inner delta for agriculture has changed both the salinity and supply of sediments to the delta and has stopped tidal inflow to mangroves behind the highest artificial levees. Elsewhere, the spill-over of sand sheets above dikes and levees has smothered and killed the pioneer mangroves (Figure 10.16).

Between 125 and 143 million people live on the Ganges Delta, where most areas contain over 200 people/km^2, making it one of the most densely populated regions on Earth. Approximately two-thirds of the Bangladesh people are employed in agriculture on the fertile floodplains, and about four million people totally depend on the mangroves for their livelihood. This delta is reputedly one of the most fertile regions in the world, sometimes called 'The Green Delta'. The major crops are jute (hemp), tea and rice; fishing is also an important activity and a major food source for the artisans. Since the 1980s, a decline in the fisheries (particularly prawn fisheries), have been attributed to mangrove losses, and recently there has been an effort to improve fish farming methods. Ponds have been converted into viable fish farms; combined with an eightfold improvement in fish-raising methods, many people now live by raising and selling fish. Shrimp and salmon are also farmed in containers submerged in open water, but most of this harvest is exported. Important petroleum gas reserves have been discovered in the delta and several major oil companies have invested in exploration of the Ganges Delta.

Most of the Ganges Delta is in the tropical wet climate zone, and receives 1500−2000 mm of rainfall each year in the drier west and 2000−3000 mm in the wetter east. The area is prone to flooding by tropical cyclones, which hit the region four to eight times a year and are intensified by the funnel shape of the Bay of Bengal, so that waves surges are 5–8 m high (Blasco, 1977; Blasco *et al.*, 1994). In November 1970, the deadliest tropical cyclone of the twentieth century, Cyclone Bhola, hit the Ganges Delta region. This cyclone killed 500 000 people (official death toll), with another 100 000 missing. Guinness Book of World Records estimates the total loss of human life from the Bhola cyclone as one million. In 1991, another cyclone made landfall and killed about 138 000 people, and in 1998, a flood destroyed the entire rice crop. Although the most recent very severe tropical Cyclone Sidr took a smaller toll

Figure 10.16 In the Ganges Delta and other mangrove forests of eastern India, overharvesting of mangroves and marsh drainage has greatly reduced the original area of tidal wetlands. Top: large areas of mangroves are removed for construction of houses and for fishing poles (Ravishankar *et al.*, 2004); Bottom: pioneer mangroves with 'pencil root' aerial pneumatophores are killed by sheets of sediment deposited during floods or because of agricultural dike failure; the Asian water monitor is a large lizard common in the coastal wetlands of Southeast Asia (photo by P. W. Mudie).

of lives (~3500) because of improved forewarning and evacuation of 2.7 million people, the exfoliation and uprooting of mangroves in 5–25% of the area presents a new threat of shoreline erosion. Top-dying of the sundari trees has also taken a heavy toll; increasing salinity of the delta soils may contribute to this disease, which has killed an estimated 20% of this endemic mangrove. However, the greatest challenge that people living on the Ganges Delta face in the future is the threat of rising sea levels caused mostly by delta subsidence and magnified by accelerated climate warming and the accompanying sea level rise over the past 50 years. An increase of just 0.5 m could result in six million people losing their homes in Bangladesh.

There are three terrestrial eco-regions within the Ganges Delta: (1) The Lower Gangetic plains *moist deciduous forest eco-region* remains the largest eco-region, although much of the forest has been cleared for agriculture and thick stands of tall canebrake grass fill the wetter areas; (2) further seaward, the Sundarbans *freshwater swamp forest eco-region* is flooded with slightly brackish water during the dry season and with freshwater during the monsoon season. These forests have also been largely replaced by intensive agriculture, with only ~9% of the eco-region's 1 460 000 ha being protected; (3) along the Bay of Bengal and extending ~25 km inland, Sundarban's *tidal wetlands eco-region* forms the world's largest mangrove area, which was estimated as ~800 000 ha in 1977, making it almost four times larger than the second largest region, the Mekong Delta (Blasco, 1977).

The Sundarbans (Beautiful Garden) derive their name from the main mangrove species, *Heritiera fomes*, known either as *sunderi* (= beautiful) or *sundari* and *ban* (garden or forest). This mangrove, which is prized for its hard wood, grows only along the Bay of Bengal from Northeastern India to northwestern Malaysia, and is now threatened with extinction (IUCN Red List, 2010) because of habitat alteration for shrimp farming, palm tree plantations, the dieback disease and encroachment of its habitat by rising sea level. The shrub *Scyphiphora hydrophylacea* (=*narathanduga*) is also endemic to the Bengal coastline. In 1977, there were about 51 other salt-tolerant plant species in the Sundarban region; other mangroves include three garjans (*Rhizophora* spp.), and the Indo-West Pacific genera *Bruguiera*, *Ceriops*, *Sonneratia* and *Xylocarpus*, which are not found in the Atlantic region. Except for the near-endemic sundari mangrove, the biologically diverse forests of the Sundarban are similar to those of Papua New Guinea (see Figure 10.20). Other common trees in the Sundarbans are the nipa/mangrove palm (*Nypa fruticans*), and at the upper margin, mangrove date palm (*Phoenix paludosa*) and bamboo (*Bambusa*) species. The outer edges of the Sundarban mangroves are monopolized by light-demanding pioneer mangroves, such as *Avicennia* species and the poisonous milky mangrove, *Excoecaria agallocha* (Euphorbiaceae), which is also known as *Thillai* (blind-your-eye mangrove and river poison tree). The milky sap of this tree is extremely poisonous and skin contact can cause irritation and blistering; contact with eyes results in temporary blindness. Even dried and powdered leaves of the milky mangrove contain enough poison to kill fish quickly. Animals in the delta are listed in Table 10.3.

Other important wetlands on the Bay of Bengal include the protected lagoons at Orissa (Pattanaik *et al.*, 2008) and the Godavari River estuary (Ravishankar *et al.*, 2004) in eastern India, and wetlands of the Irrawady River Delta of Myanmar (= Burma). The macrotidal Irrawady (= Ayeyarwady) River Delta (Figure 10.17) includes about 530 000 ha of intertidal flats covered by sundari forest. Here the delta is aggrading seawards at a rate of 25−60 m/yr and is also accreting as a result of sediment deposited during floods (Eisma, 1997). The 32 species of mangroves in this delta located at ~16° N include taxa from both the Southeast Asian floral community of India and the Indo-Andaman community characterized by forests up to 25 m high, dominated by the stilt-rooted mangroves, *Rhizophora* and *Ceriops*.

In Myanmar (Burma), mangroves are disappearing faster than any neighbouring countries because of changes during the British colonial era (1852–1947), subsequent political turmoil, and large-scale conversion to agriculture (Spalding *et al.*, 2010). Furthermore, much of the remaining forest area is degraded by overharvesting for timber and fuel wood. Declines in the fisheries and increased bank erosion are also linked to loss of the mangrove

Table 10.3 Some animals in Sundarbans Ganges Delta (E = Endangered)

Mammals
Clouded leopard (*Neofelis nebulosa*)
Indian elephant (*Elephas maximus indicus*; Figure 10.17)
Bengal tigers (*Panthera tigris tigris*; Figure 10.16) – E; largest population found here
Chital/ Spotted deer (*Axis axis*)
Barking deer (*Muntiacus* sp.)
Macaque monkey (*Macaca mulatta*)
Indian fishing cat (Prionailurus viverrinus)
Irrawaddy dolphin (*Orcaella brevirostris*) – oceanic; enters from Bay of Bengal
Ganges river dolphin (*Platanista gangetica gangetica*) – true river dolphin; rare; E

Reptiles
Indian python (*Python molurus*)
Crocodiles (*Gavialis gangeticus*) – E: critical
Water monitor lizard (*Varanus salvator*)
River terrapin (batagur) (*Batagur baska*) – E: critical

Birds (315 species total)
Pallas's fishing eagle (*Haliaeetus leucoryphus*) – E
Masked finfoot (*Heliopais personatus*) – E
Great kingfisher (i.e. *Pelargopsis melanorhyncha*)
Shalik (*Acridotheres tristis*)
Swamp francolin (*Francolinus gularis*)
Doel (*Copsychus saularis*)

forests since 1984 (Oo, 2004). On 2 May 2008, the Irrawaddy Delta suffered a major disaster when Cyclone Nargis made landfall, with sustained winds over 210 km/h and gusts up to 260 km/h driving the storm surge 50 km inland, killing 138 000 and leaving ~2.5 million homeless (Figure 10.18). Nargis represents the deadliest tropical cyclone worldwide and one of the worst natural disasters; damage estimates of over US$10 billion also make Nargis the most destructive cyclone recorded for the Indian Ocean (Fritz *et al.*, 2009). More than 1 m of vertical erosion and 100 m of land loss were recorded along the coast and drinking-water wells were flooded with saltwater. At least part of this damage is the result of the long history of deforestation for charcoal and rice paddies, which left few primary growth mangroves to provide wave attenuation along a 30 km-wide coastal buffer-zone. In contrast, the coast of southern Myanmar and western Cambodia have been relatively unexploited and maintain a high diversity of mangals that have resisted various typhoons and the 2004 Sumatran tsunami (Spalding *et al.*, 2010).

10.4.2 The Mekong Delta: a biological treasure trove

The Mekong Delta ('Nine Dragon river delta'; Vietnamese) is at ~9° N, 105.1° E in south-western Vietnam where the Mekong River flows into the South China Sea (locally = East Sea) through nine major channels and a network of distributaries (Figure 10.19). The Mekong is

Figure 10.17 Top: people of the Irrawaddy Delta, Myanmar, live in and depend on mangroves for their livelihood; mangrove-wood poles are used for fish traps (photo by P. W. Mudie). Bottom: the endangered Indian elephant is now protected in various elephant farms of Southeast Asia like the Maesa Elephant Camp, Chiang Mai, Thailand (photo by S. J. Littlefair-Wallace).

Asia's third largest river and its delta plain is the third largest in the world, being exceeded only by the Amazon and Ganges–Brahmaputra Deltas (Xue *et al.*, 2010). The Mekong River arises on the northeast rim of the Tibetan Plateau of China, and flows 4160 km through or along the borders of China, Burma, Laos, Thailand, Kampuchea and Vietnam (ASEAN Centre for Biodiversity, 2013). Geomorphologically, the delta is puzzling because although the drainage basin is smaller than that of the Chinese Yangtze River (see Section 10.4.4), its sediment load is about the same at the Yangtze and its delta is two times bigger. The Mekong Delta wetlands started to form about 8000 years ago when sea level was higher and sedimentation

Figure 10.18 The Irrawaddy Delta and impact of Cyclone Nargis. Left: SRTM altimetry of the delta, showing flooded areas in red, based in MODIS imaging (from Syvitski *et al.*, 2009 with permission from Macmillan Publishers Ltd); Right: aftermath of Cyclone Nargis (courtesy of Dr George Pararas-Carayannis, www.drgeorgepc.com). See also colour plate section.

Figure 10.19 Top: floating markets on the Mekong River sell produce raised on delta farms; Bottom: traditional fishing net on the delta at high tide (photos by P. W. Mudie). See also colour plate section.

progressively filled in the former marine estuary, prograding seawards by ~250 km from the Cambodian border. Over the past 3000 years, most of the river sediment load (~160 million tonnes/yr) has been trapped in the delta, which has extended lopsidedly seawards by 16 −26 m/yr in response to variations in its shoreline, longshore currents and possibly a strengthened northeastern monsoon influence. The subaerial delta is additionally bordered by a subaqueous delta system up to 20 m thick.

Today, the Mekong Delta region covers a total of ~4 million ha, although the inundated area varies seasonally. The delta border extends from the Gulf of Thailand, where tides are diurnal (0.5−0.8 m) to the South China Sea where tides are semi-diurnal (3.5−4 m). Tidewater regularly penetrates up the branches of the Mekong for 20−65 km, supporting the famous floating markets (Figure 10.19). About 500 000 ha of land are affected by seawater intrusion up to 50 km inland during the dry season, but the large freshwater inflow keeps salinities along the eastern coast of the delta very low (<4 psu maximum), particularly during the flood season. At the end of the rainy season, the combination of floodwaters from the rivers, local rainfall and tidal inundation results in the flooding of the entire Vietnamese portion of the delta. During typhoon surges, flooding to depths of 60−80 cm can extend up to 70 km inland.

The tropical monsoonal climate produces a dry season from December to March/April and a distinct rainy season during the southwest monsoon from May to October/November. The average annual rainfall ranges from ~1500 mm in the centre and northwest to > 2350 mm in the south, with 70–80% occurring June−September. The mean annual temperature is ~26 °C with a small summer–winter range of about 5 °C, and high relative humidity throughout the year. Tropical cyclones are frequent − about 30 tropical cyclones annually, of which 11–12 tropical cyclones land in the South China Sea, and six to eight storms and tropical depressions impacted the territory from 1960 to 2005. Vietnam is considered one of the most disaster-prone countries in the world, and in some years is struck by ten or more typhoons; for example, in 1964, there were 18 typhoons, 1973 and 1978 had 12 typhoons in each year and 1989 had 10 typhoons (Nyugen, 2007).

The river valleys of Vietnam were one of the earliest centres of agriculture and the inner Mekong Delta was likely inhabited far back in prehistorical time because caves in neighboring Thailand have remains of stone-age hunters. Neolithic archeological sites date to at least ~7000 yr BP and older sites may lie beneath on the shallow water offshore (Oppenheimer, 1998). Archeological discoveries at Óc Eo (= Glass Canal) and other Roman- to Mediaeval-age sites, including the famous Ankor Wat temple, show that extensive human settlement may go back to the fourth century BCE and the delta was an important trading centre crossed by canals as early as the first century CE. After colonization by France from 1874 to 1954, the Mekong Delta was part of the Republic of Vietnam before becoming the modern country of Vietnam. In the 1970s, the Khmer Rouge regime attacked Vietnam in an attempt to recapture the delta region. This campaign precipitated the Vietnamese invasion of Cambodia and the subsequent downfall of the Khmer Rouge. Today, the Mekong Delta region is inhabited by approximately 17.4 million people, many of whom depend on it for their livelihood (Figure 10.19). The low-level plain less than 3 m above sea level is criss-crossed by a complex system of rivers, canals and diked ponds. The floating market of Cần Thơ operates in this famous rice-growing area, which produces about half of Vietnam's total rice output. The region also supports a large aquaculture industry of basa fish, Tra catfish and shrimp.

Although the Mekong was once covered by extensive mangals and flooded grassland, major loss of mangroves here began long before the Vietnam War (1959–1975). During the war years, however, ~100 000 ha were destroyed throughout Vietnam (Spalding *et al.*, 2010). The defoliant Agent Orange (2,4-dichlorophenoxyacetic acid, 2,4,5-trichlorophenoxyacetic acid) and MCPA (2-methyl-4-chlorophenoxyacetic acid) that had been tested successfully in Sierra Leone to eradicate *Avicennia africana* and *Rhizophora racemosa*, and the herbicides Dalapon and 3-(4-chlorophenyl)-1,1-dimethyl urea were sprayed on *Avicennia* pneumatophores to kill the trees directly (Walsh, 1974; Chapman, 1977). GIS image analysis from the period 1965−2001 shows that the total coverage of mangrove forests decreased by 50%, with the annual rate of destruction increasing from 0.2 to 13% from 1995 to 2001. After the deforestation by warfare, firewood collection and clearing for agriculture and shrimp farming became further drivers of wetland forest destruction (Thu and Populus, 2007). Loss of mangals accompanies aquaculture development during clearance for shrimp ponds, and also because soil deteriorates when shrimp farms fail from diseases. These failures stimulate more deforestation and pond construction, while the abandoned ponds dry out and became sterile saltflats with highly acidic soils. In Thailand, Vaiphasa *et al.* (2007) show that disposal of shrimp pond dredge material into the mangrove wetlands also prohibits the growth of several common mangroves, although the trees can tolerate shrimp biopesticide residues. Khoa and Roth-Nelson (1994) discuss the need for integrated delta wetland management of farm and mangrove ecosystems; they recommend the planting of red gum (*Eucalyptus tereticornus*) in raised beds to reduce soil acidity by lowering the water table and its associated sulfate accumulation zone, and by rotational cropping of various kinds of fruits (e.g. coconuts, bananas, papayas) with rice and cajeput trees (*Melaleuca*), as well as shrimp/fish farming and mangrove preservation in non-acidic soils.

The Mekong Delta is intensively cultivated and in addition to the mainstay rice industry, it produces vegetables, fruit, poultry and much of Vietnam's pork (50%) and fish (30%). Jute, coconut, sugar cane and other fruits are also farmed. Today only ~5.5% of the delta remains a tidal wetland, supporting about 280 000 ha of mangal and *Melaleuca* (cajeput) forest. The surviving mangrove forest consists of 46 plant species, 38 of which are of some economic importance. From the coast inland, the forest is dominated by *Avicennia marina*, *Rhizophora apiculata*, *Bruguiera parviflora*, *Ceriops tagal* and on the upper margin *Bruguiera gymnorrhiza* dominates, with *Nypa fruticans* in brackish areas (Chapman, 1977). Other mangrove trees are *Sonneratia caseolaris*, *Lumnitzera littorea*, *Xylocarpus granatum* and poison-milk, *Excoecaria agallocha*. The *Melaleuca* (cajeput) forest includes and additional 77 plant species, with *M. leucodendron* (bottlebrush tree) predominating throughout. The cajeput forests are a valuable source of honey from wild bees, amounting to five or six litres/ha of honey each year.

The lower Mekong River and Delta support one of the largest inland fisheries in the world. During the past decade, the Vietnamese part of the delta has yielded an annual harvest of about 400 000 metric tonnes of fish of which about 40% is from the brackish water and estuarine zone. In recent years, however, the fish production has declined as a result of overexploitation, forest destruction, use of herbicides and pesticides, and the wetland changes described above.

The Mekong Delta in Southern Vietnam is now an almost entirely human landscape although it has been dubbed a 'biological treasure trove', based on the finding of various previously unknown species for food/sale in marketplaces on the delta. Over 10 000 new species have been found in previously unexplored areas of this region, including an endangered species of rat, the Laotian rock rat (*Laonastes aenigmamus*), previously thought to live in only in limestone rock cliffs of Vietnam. It is not yet clear if the rat actually lives on the delta although it is sold there as a food delicacy. According to the ASEAN (2013) report, which is based largely on 1987 or earlier references, the Mekong Delta fauna includes 23 species of mammals, 386 species and subspecies of birds, 35 species of reptiles, 6 species of amphibians and 260 species of fishes. Five species of dolphins are present: two species, *Stenella malayana* and *Tursiops aduncus*, are confined to the estuarine zone, but the Irrawaddy dolphin *Orcaella brevirostris*, Chinese white dolphin *Sotalia chinensis* and black finless porpoise *Neophocaena phocanoides*, occur upstream into Kampuchea. Other wetland mammals are the crab-eating macaque *Macaca fascicularis*, smooth-coated otter *Lutra perspicillata* and fishing cat *Felis viverrina*. The clawless otter *Aonyx cinerea* may also occur. Belsare (1994) inventories the wetland wildlife of the entire Southeast Asian region and various plans for conservation management.

The bird fauna includes about 92 species of waterfowl and a variety of other species associated with wetlands, as listed in Table 10.4. The region is particularly important for its large populations of waders, which nest in huge colonies in the mangrove and *Melaleuca* forests. Populations of the storks have decreased notably in recent years and some are now very rare. The pelicans were common in the 1960s, but are now rare or absent. Cranes nested on the delta before 1964, but now are non-breeding visitors. It is also not clear if the endangered ibises and wood duck are still present (ASEAN, 2013).

Table 10.4 List of major birds found in Mekong Delta Wetlands according to ASEAN Regional Centre for Biodiversity Conservation (2013) R = Rare; E = Endangered

Common name	Scientific name
Raptors	
Osprey	*Pandion haliaetus*
Brahminy kite	*Haliastur indus*
White-bellied sea eagle	*Haliaeetus leucogaster*
Grey-headed fish eagle	*Ichthyophaga ichthyaetus*
Eastern marsh-harrier	*Circus spilonotus*
Kingfisher	Alcedinidae – eight species
Waders	
Cormorants	Three *Phalacrocorax* spp.
	Anhinga melanogaster
Heron	*Ardea sumatrana*

Table 10.4 (*cont.*)	
Common name	Scientific name
Ibis – black-headed	*Threskiornis melanocephalus*
Glossy ibis	*Plegadis falcinellus*
White-shouldered ibis	*Pseudibis davisoni* (E)
Giant ibis	*Thaumatibis gigantean*
Storks	*Leptoptilos dubius* (R)
	Mycteria cinerea (R)
	Mycteria leucocephala
	Anastomus oscitans
	Ciconia episcopus
	Ephippiorhynchus asiaticus
	Leptoptilos javanicus
	Threskiornis aethiopicus
Sarus crane	*Grus antigone sharpie* (E)
Pelicans	*Pelecanus onocrotalus*
	Pelecanus philippensis
	Grus antigone sharpie (E)
Ducks and Waterfowl	
White-winged wood-duck	*Cairina scutulaza,*
Little grebe	*Tachybaptus ruficollis*
Lesser whistling duck	*Dendrocygna javanica*
Pygmy goose	*Nettapus coromandelianus*
Spot-billed duck	*Anas poecilorhyncha*
	Anas querquedula – migratory
Slate-breasted rail	*Rallus striatus*
Common watercock	*Gallicrex cinerea*
Purple swamp-hen	*Porphyrio porphyrio*
Jacanas (lily trotters)	Two Jacanidae species
Shorebirds	
Beach stone-curlew	*Esacus magnirostris*
Oriental pratincole	*Glareola maldivarum*
Little pratincole	*Glareola lactea*
Greater painted snipe	*Rostratula (benghalensis) australis*
Masked finfoot	*Heliopais personata*
Indian skimmer	*Rhynchops albicollis*
Terns	
Asian dowitchers	
Various	20 other shorebirds

Mekong Delta reptiles include the monitor lizard, the python (*Python reticulates*), four water snakes (*Enhydris* spp.), the endangered river terrapin (*Batagur baska*) and the estuarine crocodile (*Crocodylus porosus*). Over 200 species of fishes contribute to the commercial fishery, along with many shellfish, mussels and clams (Mollusca), and prawns and shrimps, notably *Macrobrachium rosenbergii* and *Penaeus monodon* (ASEAN, 2013). In the brackish water and coastal zones of the estuary, the fish fauna is dominated by Clupeidae, Scombridae, Sciaenithe, Tachysauridae, Polynemidae, Tachysuridae and Cynoglossidae. Most of the species are migratory and the species of threadfins and marine catfish seasonally travel upstream to spawn in the oligohaline or freshwater zones of the estuary. The freshwater fish are mainly species of Cyprinidae, Siluridae, Clariidae, Schilbeidae, Bagridae, Sisoridae, Akysidae, Chanidae and Ophicephalidae.

10.4.3 Other Southeast Asian mangrove regions: more biodiversity

Chapman (1977) and Spalding *et al.* (2010) describe the great number of mangal species found in the abundant tropical forest communities of the Indo-Malaysian peninsula (including Malaya, Sumatra and Java) and on the nearby islands of Borneo, Celebes and Papua-Papua New Guinea. Here mangals over 25 m high extend ~100–320 km up the largest estuaries which support diverse communities (Figure 10.20), with the mangal composition varying with amount of coastal exposure and soil type. Average temperature is ~28 °C with almost no seasonal difference. Annual rainfall at Ao Nam Bor in Thailand is 2266 mm. The islands of Indonesia, from Sumatra to Papua New Guinea, are in a region of mesotidal to microtidal conditions containing a mixture of Pacific and Indian Ocean water with complex mixed or diurnal characteristics (Eisma, 1997) and spring tide ranges are between 2.0 and 2.8 m. This Indonesian region astride the Equator is outside of the locus of the tropical storms, therefore wave action is low. Many of the islands are of volcanic origin, with variable coastlines that include small river deltas, lagoons, low cliffs and coral reefs. The loose volcanic soils tend to erode easily and silt deposition is high, allowing mangrove growth in sheltered areas (see Figure 2.3 for examples). The volcanic islands of Sumatra and Java have steep slopes and narrow coastal shelves; here high erosion following deforestation has led to rapid delta formation in some areas – up to 70 m per year – and many river estuaries support large areas of mangroves. There are also extensive tidal flats with mangroves and supratidal salt marshes in Fiji and New Caledonia of the Pacific Islands. Details of the geomorphology of alluvial plains and deltas for other islands in the region are given by Eisma (1997).

On the semifluid mudflats of Malaysian mangrove marshes, mudskippers (*Scartelaeos viridus*) and Latreille's sentinel crab (*Macrophthalmus laterillei*) are common, but in the more stable substrates of the forested zones, there are other mudskippers (*Boleophthalmus boddaerti*), the mudcrab *Mtaplax crenulatus* and fiddler crab species (Figure 10.21). Landwards, there are mudskippers that climb the trees as the tide rises. Water snakes (Figure 10.21) hunt crabs down their burrows and various snails, including *Ceridithea*, graze on thin surface algal mats. Mangrove kingfishers are the most common resident birds,

Papua New Guinea, 6 S, 147 E

Figure 10.20 Top: height and diversity of mangroves trees and shrubs is greatest in the Southeast Asian region, as shown for a Papua New Guinea tidal zonation. A = Acrostichum fern, A1=*Avicennia marina;* Ac = *Aegiceras corniculatum;* B1, B2, B3 = *Bruguieria* spp., H= *Heritiera littoralis;* M = Metroxylon; P = *Pemphis acidula;* Ra and Rs = *Rhizophora apiculata* and *R. stylosa;* X = *Xylocarpus granatum* (from Saintilan *et al.*, 2009). Bottom: upper tidal channel near Sibu, Borneo (photo by P. W. Mudie)

and various visitors feed in the mangroves, such as cormorants, herons and the 'sea eagle' (*Ichthyophaga ichthyaetus*). Mammals include macaque monkeys (*Macaca irus*), and the leaf-nosed proboscis monkey (*Nasalis larvatus* spp.; Figure 10.21). Crocodiles and water monitors (Figure 10.16) hunt in the channels, and there are two pit vipers in Malaya and Sumatra (Figure 10.21). Mosquitos species abound, with larvae thriving in pools of water up to 13 psu salinity and in water-filled tree holes. Nests of weaver/tailor ants (*Oecophylla* spp.) are common in some mangrove trees – *Bruguiera, Sonneratia caseolaris* and *Ceriops.* In Thailand, the ant eggs and larvae are harvested for favourite dishes of either red-ant soup or crispy ant egg-larvae salad. Details of the interdependence between faunal diversity in Malaysian mangrove forest and the stages in the harvesting and regrowth of mangroves are given by Sasekumar and Chong (1998), who found that although the epifaunal diversity is greatest in mature *Rhizophora apiculata* forests, and lowest in recently cleared forests, the opposite is shown by the infaunal taxonomic diversity. At the upper edges of the mangroves, near EHW, there are diverse, less salt-tolerant trees, including *Ficus* spp. (wild figs, banyans), palms, including coconut and thatch palm *Nypa*, and tropical shrubs with large colourful, nectar-producing flowers adapted for pollination by birds and large insects (Figure 10.22).

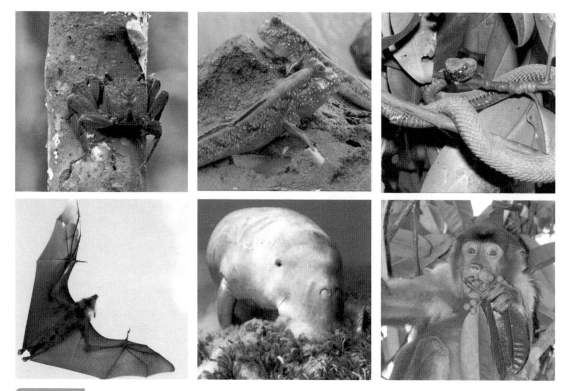

Figure 10.21 Common small animals of the Southeast Asian tropical mangal, excluding birds and insects. Top: left: mangrove crab (*Episesarma* sp.) (photo by blackbass, CC, Flickr); centre: *Periophthalmus* mudskipper (photo by Bjørn Christian Tørrissen, CC); Right: mangrove pit viper (*Trimeresurus purpureomaculatus*), Thailand (photo by Ria Tan, wildsingapore.com); Bottom: left: Flying fox (large fruit bat; *Pteropus*) (photo by jo3hug, CC, Flickr); centre: Dugong (*Dugong dugon*) (photo by Blue Carbon Working Group, AGEDI/ UNEP/GRID-Arendal); right: proboscis monkey (*Nasalis larvatus*), Malaysia (photo by Paulo Philippidis, CC).

10.4.4 East Asia: China, River of Sorrow and macrotidal mudflats

In the South China Sea between Northeastern Vietnam and Fujian (Southern China), mangroves continue to grow northwards along the tropical-subtropical coasts and shores of Hainan and Taiwan, from ~18 to 27° N. In this region there is a total of over 40 species in 21 families and 26 genera, most of which are thermophilic and environmentally wide-ranging. However, diversity declines northwards, decreasing to nine species at Fujian west of Taiwan (~23° N). Moreover, nearly two-thirds of China's mangroves have disappeared during the past 40 years because of reclamation for rice farming, aquaculture, and urban and industrial development (Mao *et al.*, 2006; 2012). Hainan Island has the largest remaining mangrove area and the most species in China (29 true mangrove species in 16 families, and 9 mangrove associate species in 6 families). The major mangroves are trees: *Avicennia marina*, *Ceriops tagal*, *Kandelia ovata* and species of *Bruguieria.*, *Rhizophora* and

Figure 10.22 High water areas of tropical mangrove forests in the equatorial region include a wide diversity of shrubs with colourful flowers and large trees. Left: buttress roots (up to 1 m high) of a fig tree, *Ficus macrophylla*; Right: red wild hibiscus with exerted stamens pollinated by birds (top) and the trumpet-shaped yellow flowers of sea hibiscus (*Hibiscus tileaceus*) (bottom) pollinated by butterflies (photos by P. J. Mudie). See also colour plates section.

Sonneratia; minor elements include horned mangrove, *Clerodendron inerme* (= wild jasmine), *Milletia* [*Pongamia*] *pinnata* (= Indian beech), *Scyphiphora hyrophyllacea* (chengam bush) and *Xylocarpus granatum*. The Hainan mangrove associates are yellow sea hibiscus and *Cerbera manghas* (sea mango or pink-eyed pong pong tree). The lowest diversity is on Kuyushu Island of southern Japan, ~27° N, where only *Kandelia obovata* is present today. Here, erosion from waves produced by recreational motorboats and high sediment infill around Okinawa are threats to the mangroves remaining after heavy over-harvesting and land clearance (Spalding *et al.*, 2010).

Further north, at ~27–40° N, the coasts of southern China, Taiwan and the Nansei Islands of Japan span the transition from subtropical to temperate climates and cooler ocean temperatures. There is also a change from mesotidal to macrotidal conditions. This temperate region includes some of Asia's largest rivers and extensive deltas, which have been settled and farmed on the margins since Neolithic times. Today, the northern deltas in China and Korea support only a few small remnants of the natural wetlands. In contrast, large areas of new tidal wetlands are being forming by accretion of sediment, particularly in the Yangtze Delta (Figure 10.23 B,D).

The largest Chinese deltas are in three estuaries: (1) the Yellow (= *Huang He*) and Liaohe Rivers east of Beijing, (2) the Yangtze River (= *Yangtze Jiang, Chang Jiang*) near Shanghai and (3) the Pearl River (*Zhu Jiang*) near Hong Kong – all of which are among the most

Figure 10.23 Relatively little remains of China's tidal original wetlands, but large areas of salt marsh still cover parts of the Yellow and Yangtze River deltas. A: meadows of native sedges fringed by taller cordgrass behind the heavily barricaded coast of Nanhui, near Shanghai, with fishing towers on the South Channel; B: rapidly aggrading mudflat east of Chongming Island, central delta, where cordgrass is colonizing the Dongtan Nature Reserve; industrial developments line the North Channel (top centre); C: cordgrass invading more diverse native marsh on Changxing mud island, South Channel; D: layers of peat and silt in the rapidly aggrading channel at Jiuduansha, a new island formed since 1950; note peat layers in rapidly aggrading reed marsh (photos courtesy of Li Yuang, E. China Normal University, Shanghai). See also colour plates section.

highly populated areas in the world. These deltas have long been modified by dam construction, canalized and converted to rice paddies and fish ponds, although extensive areas of wetlands remain around the Bohai Sea (Mitch and Gosselink, 2007).

The Yellow River is the second longest in Asia, arising on the high Tibetan Plateau and flowing eastwards through yellow, silty periglacial deposits (= loess), which give it the characteristic colour. The huge sediment load of this river (1 billion tonnes/yr) raises the river bed above the surrounding land, making it very prone to flooding, and hence it is called the 'River of Sorrow'. As early as 1270 CE, Marco Polo wrote of a wide, deep channel built by the Khan Emperor from the Yellow River to the Yangtze for use by large sailing vessels (today this Grand Canal is 1776 km long, the longest in the world). At that time, the Yellow River delta between Beijing (~39.9° N) and the ocean was the emperor's favourite hunting ground for shooting wild crane; the surrounding region also supplied wild game, including boar, roebuck, lions, tigers, bears, wild cattle and asses (Latham, 1958). However, archeological evidence shows that the upper Yellow River estuary (~37° N) was already under cultivation for foxtail millet and broomcorn millet as early as 8500–7500 yr BP. At that time, hunting, gathering and fishing were still the principal subsistence practices, but by 6300–4500 yr BP, rice and millet were cultivated and many species of livestock were domesticated (Guo et al., 2013). After 5000 yr BP, human activity extended seawards, the prehistoric cultures intensified their social stratification and the early city-states emerged. By that time, agriculture in the delta was widespread and a salt industry was established.

Between 540 BCE and 1964 CE, the Yellow River flooded 1593 times, drowning many people and triggering times of famine and spread of disease. The main tributary of the Yellow River has switched direction over ~40 km during the past millennium, the last time being in 1855, when the mouth moved from the East China Sea to the Bohai Sea. Subsequently, the new delta distributaries shifted 50 times up to 1976, so human habitation of the new delta only commenced around 1934. The tides in the Yellow River estuary, which presently empties into the Bohai Sea, come from both the Yellow Sea (between China and Korea) and the East China Sea, resulting in a tidal range of 4–7 m and a 1 m-high tidal bore with flood tide current velocities up to 4.5 m (Eisma, 1997; Liu et al., 2004). Erosion of the old estuarine sediment has built large tidal sand ridges offshore, which are periodically changed by typhoon storms. Overall, the coastline is rapidly retreating, with the mudflats being lowered up to 3 m per decade, but in some areas, the accretion rate is up to 50 m/yr. Offshore, a wedge of Yellow River silt 20–40 m thick wraps around the Shandong Peninsula and records the westward migration of the river mouth, which emptied into the East China Sea before about 9000 years ago (Liu et al., 2004).

The history of Shandong Yellow River Delta National Nature Reserve (a 152 000 ha Ramsar Wetlands Convention site) is given by Chen et al. (2007) (Box 10.3 Conservation statuses). The Shandong Nature Reserve protects mainly marine and coastal wetlands and the adjacent mudflats and marine waters. These wetlands support 393 plant species, including the threatened wild soybean (Glycine soja). Waterfowl include 48 shorebird species numbering up to a million/yr, 10 species of cranes, including the overwintering red-crowned crane (Grus japonensis), and 30 species of ducks, including a large wintering population of whooper swans. Thirty species of other waterbirds are also present, including breeding populations of Saunder's gulls. Unfortunately the area is prone to pollution from the large oil industry (65 oil fields; 515 drilled wells), which is a source of waste gas, contaminated wastewater and industrial residues. Eutrophication from sewage and agricultural runoff has also begun to trigger red tide outbreaks in the estuary.

Mitch and Gosselink (2007) summarize the successful bioecological engineering programme undertaken to stabilize China's estuarine mudflats by trial planting in 1963 and 1979 of several Spartina species, which could also be used for agriculture and reduction of soil salinity. However, the cordgrasses subsequently began to replace the native Phragmites australis in brackish wetlands and some are now considered aggressively invading species (Figure 10.23; see also Section 14.3). Using remote sensing imagery, Zuo et al. (2012) show that the area colonized by the less aggressive Spartina anglica has now diminished to <1% of its maximum extent and is replaced by the native reed and seepweeds (Suaeda spp).

Conservation statuses along East China coasts

Red-crowned crane (*Grus japonensis*)[a,b]
Great bustard (*Otis tarda dybowskii*)[a]
White-tailed eagle (*Haliaeetus albicilla*)[a,b]
Golden eagle (*Aquila chrysaetos daphanea*)[a]
White stork (*Ciconia ciconia boyciana*)[a,b]
Hooded crane (*Grus monacha*)[a,b]
Chinese merganser (*Mergus squamatus*)[a]
White-naped crane (*Grus vipio*)[b]
Little curlew (*Numenius minutus*)[b]
Spotted greenshank (*Tringa guttifer*)[b]

[a] Class 1 Priority protection in the Shandong Reserve (Chen *et al.*, 2007)
[b] Convention on International Trade in Endangered Species (CITES)

In the Sino-Japanese Agreement on the Protection of Migratory Birds and Their Habitats, the Yellow River Delta Nature Reserve has 152 species, constituting 67% of the total (227 species). For the Sino-Australia Agreement on the Protection of Migratory Birds and Their Habitats, 51 species have been found in the reserve, accounting for 63% of the total 81 species.

Several marine mammals (near-extinct Yangtze white dolphins *Lipotes vexillifer*), reptiles and fishes (Chinese sturgeon *Acipenser sinensis* and Reeves shad *Tenualosa reevesii*) have also been listed in CITES appendices or national priorities.

However, *Spartina alterniflora* continues to expand in both the Yellow River and Yangtze River estuaries, as well as in the high marsh of mangroves in Fujian Province between Shanghai and Hong Kong.

The Yangtze River is the longest in Asia and transports the world's fourth-largest sediment load to its fertile delta, which supports about one-fifth of China's economy. The headwaters are on the Qinhai–Tibet Plateau 6300 km inland and the river cascades 4 km down through mountain gorges to the coastal plain about 2000 km from the East China Sea (Luo *et al.*, 2012). The Yangtze Delta developed after the postglacial sea level transgressive maximum was reached about 8000 yr BP, and during most of the Holocene, delta development was dominated by a large fluvial sediment supply that filled the eroded valley. Six subdeltas have prograded >200 km seawards in a southward-migrating trend to the present-day megatide-dominated estuary north of Shanghai (Hori *et al.*, 2002). Today the Yangtze Valley is home to almost 0.5 billion people and one of the most heavily impacted rivers in the world. To provide freshwater for domestic and agricultural use, furnish hydroelectric power and minimize flooding, more than 10 000 dams (some over 100 m high) were built on the Yangtze after 1950, including the world's largest dam, Three Gorges Reservoir (TGR), which was completed in 2003 (Yang *et al.*, 2011). Within five years and 50 000 dams later, there was a 70% decrease in sediment supply to the Yangtze Delta, and the delta front devolved from a strongly aggrading system to a complex of eroding shorelines and channels with shoaling mudflats forming islands in the South Channel and along the delta front (Figure 10.23B). Despite the reduced post-TGR fluvial sediment load, however, the Yangtze subaerial delta continued to grow from 1974 to 2010, mainly in eastern Chongming Island and Nanhui bank, where the coast extended seaward about 8 and 6 km (Chu *et al.*, 2013).

Salt marshes of the megatidal Yangtze Delta are now largely invaded by smooth cordgrass that was introduced in 1995 to stabilize the rapidly forming mudflats. Tian *et al.* (2008) show that the low marshes (2.0−2.9 m asl) containing native sedges (*Scirpus mariqueter* and *S. triequiter*) and bare mudflats and tidal creeks below 2.0 m are the most important habitats for waterfowl in the Dongtan Nature Reserve, which is one of the largest Ramsar sites in East Asia. The high marsh above 2.9 m is fringed by cordgrass interspersed with *Phragmites* reeds and small areas of *Imperata cylindrica*, *Suaeda glauca*, *Juncus setchuensis* and *Carex scabrifolia*. This high marsh provides less valuable food habitat although it offers good cover protection (Table 10.5). Chen *et al.* (2012) investigate the stabilization benefits versus problems of continued expansion of cordgrass in the Dongtan Reserve, which causes lower soil microbial respiration and less carbon storage relative to that of the native *Phragmites*

Table 10.5 Habitat suitability of wetland land cover types for the different groups of birds (Tian *et al.*, 2008) and associated macrobenthos food types, as both biomass and approximate density (adapted from Jing *et al.*, 2007).

Habitat	*Scirpus*	*Phragmites*	*Spartina*	Bare mudflat	Shallow water	Deep water
Bird group: suitability per habitat						
Charadriidae (plovers, lapwings)	Very good	Poor	Poor	Very good	Fair	Poor
Ardeidae (herons, egrets)	Very good	Fair	Poor	Good	Very good	Poor
Laridae (gulls)	Fair	Poor	Poor	Very good	Very good	Good
Anatidae (ducks, geese, etc)	Very good	Fair	Poor	Poor	Good	Poor

Habitat	*Scirpus* = middle + inner *Scirpus*	*Phragmites*	*Spartina* = outer *Scirpus*	Mudflat	Sandflat	N/A
Invertebrates:	Categories (biomass: g/m²) VG = >100, G = 40–100, F = 10–40, P = 0–10 as individuals/m²					
Gastropods	Good; 11 000	Good; <1000	Good; 4000	Fair; 4000	Fair; 500	
Bivalves	Poor; <150	Poor; <20	Fair; 450	Good; 500	Very good; 500	
Polychaetes	Fair; 150	Poor; 100	Poor; 75	Poor; 150	Poor; ~100	
Crustaceans	Fair; 300	Poor; 75	Poor; >200	Poor; 75	Poor; 50	
Insects	Poor; 150	Poor; 20	Poor; 150	Poor; 10	Poor; <5	

(although mowing of cordgrass can enhance carbon sequestration). Jing *et al.* (2007) show that the relative abundance of epifaunal macrobenthos in the salt marshes is much higher than on the bare flats, but the opposite is true for the infaunal macrobenthos. This difference is reflected by waterfowl use: pause-travel shorebirds mainly feed on salt marsh fringes, while residents rely heavily on the bare flats, eating either bivalves (tactile feeders) or crustaceans (visual feeders). The steep-sided marsh creeks (Figure 10.23D) shelter both freshwater and brackish water fish, such as *Mugil cephalus*, *Liza carinatus*, *Odontamblyopus rubicundus*, *Synechogobius ommaturus* and *Boleophthalmus pectinirostris*, which are common in tidal creeks and shallow waters. Overall, the Yangtze River supplies 50% of China's fisheries, including the now-endangered Chinese sturgeon and Reeves shad (Box 10.3), but only 18% of these are dependent on the estuary. Macrophytobenthos, mainly crustaceans, gastropods, bivalves and polychaetes, are abundant in the intertidal zones (Table 10.5), and the dominant species include crab *Ilyoplax deschampsin*, Asian clam *Corbicula fluminea* and Korean mud snail *Bullacta exarata*.

Recently, the Yangtze Delta has become polluted by industrial and agricultural runoff, by siltation, gravel extraction and by loss of wetland and lakes which are natural buffers against seasonal flooding. According to Syvitski *et al.* (2009), both the Yangtze and Yellow River Deltas are now in dire peril of flooding by future sea level rise, while the about half of the Pearl Delta is below sea level and highly susceptible to typhoon floods despite extensive shoreline barricades. The Holocene evolution of the Pearl Delta and marshes is strongly intertwined with three driving factors: (1) sea level change, which controlled accommodation space, (2) the history of the Asian monsoons, which weakened and reduced flood volumes after 6800 yr BP and (3) recently intensified land use, which accelerated progradation after 2000 yr BP (Zong *et al.*, 2009). The intensity of monsoon rains is this region is amplified by interannual and interdecadal variations of tropical cyclone activity in the South China Sea, which are, in turn, linked to El Niño oscillations that appear to be on the rise (Wang *et al.*, 2007). However, the traditional role of the gigantic Tibetan Plateau 'heat pump' as the fundamental driver of the South Asian monsoons remains a subject of hot debate (Qui, 2013) that must be resolved before accurate forecast can be made for future flood trends.

11 Australasia: wetlands of Australia and New Zealand

Key points

Australia and New Zealand are in the Oceania biogeographic realm; regional wetland variations reflect different bedrock types, and tidal and wave regimes; Australia is tectonically stable, with a wide spectrum of climates, many estuaries and unique inland 'billabongs'; mangroves are similar to the Indonesian islands that were connected by Ice-Age land bridges; lack of Australian icesheets means the history of sea level change is different; there are four coastal wetlands types: Barrier Reef mangroves, northern mudflat mangroves, giant stromatolite carbonate bioherms and sandy headlands with the world's southernmost mangroves; New Zealand is a tectonically active part of the 'Pacific Ring of Fire', with volcanoes and frequent earthquakes; it has cool-temperate salt marshes and shrubby mangroves; one of the last Pacific regions to be colonized by humans, New Zealand has already lost 90% of its wetlands.

In this chapter, we review the salient features of the coastal wetlands in the continent of Australia and the islands of New Zealand (Figure 11.1), both of which are in the Ramsar Oceania biogeographic region (Figure 6.1). Australia is the smallest of the world's continents, but it covers a large latitudinal range, stretching from 11° S of the Equator to the edge of the cool temperate climate region at ~44° S; its coastline includes many estuarine areas (Figure 11.1), although it has no large deltas in serious danger of destruction by rising sea level. New Zealand is a much smaller country, extending from ~35 to 48° S; it is one of the last of the Pacific regions to be colonized by humans, but already has lost 90% of its wetlands (Mitsch and Gosselink, 2007; Silliman *et al.*, 2009). Southern Australia and New Zealand North Island have the southernmost of the world's mangroves, but these countries differ greatly in climate and tectonic setting. Australia has the world's oldest rocks, which largely have been tectonically stable for >65 million years, whereas New Zealand is part of the Polynesian active 'Ring of Fire'. The two Australasian subregions are therefore discussed separately in this book.

11.1 Australia: tropical mangroves to warm temperate salt marshes

Once a jolly swagman camped by a billabong, under the shade of a coolibah tree. . .

(From 'Waltzing Matilda' by Banjo Paterson, 1895)

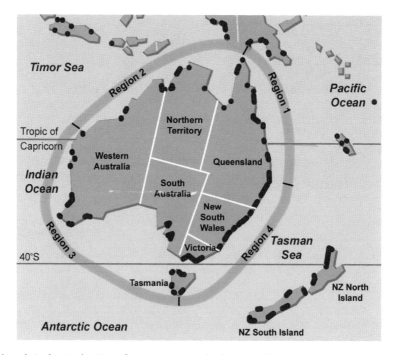

Figure 11.1   Map of Australasia showing locations of major estuaries and subregions of the Australian coastline (adapted from Ramsar, 2007 and Australian Government, 2009).

In the same way that Australian folksong is epitomized by the ballard of 'Waltzing Matilda', Mitsch and Gosselink (2007) characterize the Australian wetlands by their distinct seasons of dryness because of high evaporation and low rainfall over much of the desert-covered continent. These authors also highlight the wetlands of the distinctive inland *billabongs* in eastern Australia, which are semi-permanent pools periodically detached from overflowing river channels, often marked by small groves of *coolibah* trees (= *Eucalyptus* spp.). Thompsen *et al.* (2009) focus on the anthropogenic threats to Australia's coastal salt marshes, which include urban coastal development, eutrophication, oil and heavy metal pollution, fire, mosquito control, agriculture and aquaculture. In fact, the climate and coastline of Australia is regionally very variable, with a wide range of estuaries and associated wetland plant communities, as described by Jennings and Bird (1967) and in a book edited by Neil Saintilan (2009). The Australian coastline is 35 000 km long and together with the island state of Tasmania, contains 57 Ramsar coastal wetland sites. The coast can be subdivided into four regions (Figure 11.1) that correspond to different dominant bedrock types, tidal ranges and wave energy. *Region 1* in northern Queensland is low-lying and sheltered by the Great Barrier Reef and its islands. *Region 2*, the muddy north, has megatides and is prone to cyclone damage. *Region 3* in the west and south has carbonate rocks and small tides, but high wave and wind energy. *Region 4* in the east has rocky headlands with quartz sands, small tides and moderate wave energy (Australian Government, 2009).

Tidal salt marshes generally occur together with mangroves in all parts of Australia (Figure 11.1), covering about 991 000 ha and comprising 6.6% of the total global area of mangroves (Spalding *et al.*, 2010). In the wetter North and Northeast, there is a large region of almost exclusively tropical mangroves extending southwards from Indonesia (see Chapman, 1977 for details), and eastwards through the Micronesian Islands (Hosokawa *et al.*, 1977). The diversity of Australian mangroves is essentially the same (40 spp. in 14 plant families) as in Malaysia and the neighbouring island of New Guinea (30 spp.); however, progressively fewer species extend southward to Victoria (Australia) and New Zealand where only *Avicennia marina* var. *australasica* is present. The maximum number of 'true' (i.e. fully intertidal) mangroves listed for any Australian site is seven, at latitude ~26° S, which coincides with the southern limit of the mangrove fern *Acrostichum aureum* (Morrissey *et al.*, 2010). The mangrove tree species *Bruguiera gymnorrhiza* and *Rhizophora stylosa* have their southernmost limits at 29−30° S.

During the height of the last Ice Age about 20 000 years ago, global sea level was about 125 m lower than now which means that Australia was connected to both Tasmania and New Guinea, and it was separated by only short stretches of ocean from Sulawesi (Borneo) and Malaysia. At this time, the North and South Islands of New Zealand were also connected by dry land. The lower sea level means that the most cold-tolerant mangroves could extend their geographical ranges throughout the region during the late glacial and early postglacial time. Indeed, palynology records from offshore regions confirm that mangroves in northwestern Australia were more extensive during the middle Holocene, around 7000–6500 years ago (Woodroffe *et al.*, 1985). Then, as the sea level rose (Figure 11.2), the land bridges between the islands and Australia were disconnected until the mangrove populations became geographically and genetically isolated, as found today. With the development of simple techniques for measuring the affinity of plants by DNA, there is a lot of interest in trying to back-track the history of cross-breeding and genetic affiliation among the Australasian, Micronesian and Malaysian mangroves, in the hope of finding strains with high resistance to salty soils and/or pest infestations (e.g. Su *et al.*, 2006; Inomata *et al.*, 2009). So far, however, results are puzzling, pointing towards a high degree of genetic difference among some geographically isolated mangroves, e.g. *Lumnitzera* (Asian black mangrove) populations, and relatively low differentiation in others, such as *Rhizophora*.

The coastline of the tectonically stable Australian continent, with no mountains higher than 2.2 km, differs from most other continental and island shores that have not yet reached a relatively stable sea level position after the end of icesheet melting since the last glacial maximum (Australian Government, 2009). In previous chapters, we explain how most glaciated countries are still subject to postglacial rebound because of isostatic adjustment of the land and continental shelves on removal of the weight of ice after deglaciation. In other regions, coasts may be either subsiding due to excessive sediment loading and subsurface water extraction, or uplifted where tectonic plates collide. In contrast, although the postglacial sea level change in Australia was initially the same as other parts of the world (Figure 11.2), the Australian relative sea level curve reached a 'stillstand' of little or no change about 7000−6000 years ago, when coasts of the Northwest Atlantic were still rising steadily (Figure 11.2).

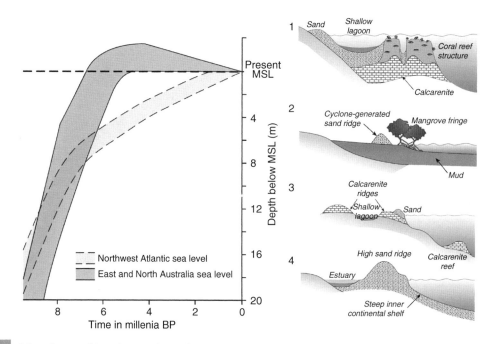

Figure 11.2 Salient features of Australia's coastline. Left: postglacial rise in sea level for the early Holocene period and the subsequent 'stillstand' around the Australian Coast compared to the more continuous gradual sea level rise along the Atlantic coast of the USA; Right: profiles of shorelines in the coastal regions 1–4 (from Australian Government, 2009).

11.1.1 Tropical Northeastern Australia: behind the Barrier Reef

Region 1 is the tropical coast of Northern Queensland west of the Great Barrier Reef, which consists of >3600 reefs and islands extending for 2300 km over the shallow continental shelf. This reef is relatively young, having mostly formed on top of older reefs in the last 6000 years since the sea level stabilized (Figure 11.2, top right). The mainland coast is low-lying with several small deltas that are periodically inundated during cyclonic floods. Storm surges and high winds also cause much economic damage to a coast long impacted by agriculture, tourism, fishing and other industry.

The city of Cairns is located in the centre of Region 1, at 16.5° S, 146.7° E, in a warm tropical climate zone, with monsoon rains from November to May and a relatively dry season from June to October. The mean annual rainfall is 2.016 m and the neighbouring township of Babinda is one of Australia's wettest towns, with an annual rainfall of over 4.2 m. The summers are hot and humid, but winters are milder, with mean maximum temperatures of 25.7 °C in July and 31.4 °C in January. Cairns is on the southeastern coast of Cape York Peninsula, between the Coral Sea and the Great Dividing Range mountains that rim the East Coast of Australia. The city is on Trinity Bay and the shores of Trinity Inlet at the bay head. Tides in this region are semi-diurnal, with a mean spring tide range of 2.1 m at Cairns, and storm surges up to 8 m. However, the Great Barrier Reef

Figure 11.3 Daintree National Park, Cairns Harbour, Region 1, Queensland, Northeast Australia. Foreground with estuarine mangrove; submerged seagrass beds and corals (darker colour) offshore (photo by Dean Jewell, www. deanjewellphotography.com.au).

offshore protects the coast from the South Pacific swell, so they are subject mainly to waves driven by coastal winds (Jennings and Bird, 1967).

Cairns city was founded in 1876 to serve miners heading for goldfields at Hodgkinson River, and later it became a major port for export of sugar cane, minerals and agricultural products from the surrounding region. In the 1870s, Cairns was an area of mangrove swamps and sand ridges that were gradually cleared and filled in with dried mud, sawdust, quarried rocks and railroad construction debris. The railway opened up land for farming on the lowlands after 1886, when sugar cane, corn, rice, fruit and dairy products were raised on the Tableland. The city became a major port and then an important World War II Airforce base. After an international airport opened in 1984, the city became a major tourist destination, with a population of about 150 920 in 2010.

Today, there are remnants of dense mangrove forests with tall trees throughout Northern Australia, but they are most diverse and best developed in Northern Queensland, particularly in the Daintree River estuary north of Cairns (Figure 11.3). Saenger *et al.* (1977) show profiles of the mangrove vegetation in this region and the zonation of the mangrove species, which corresponds to the tidal water levels. In general, there is a progression of 'true mangroves' landward from pioneer *Rhizophora stylosa* shrubs and *Avicennia marina* through a diversity of tall mangroves, including *Bruguiera gymnorrhiza*, *Xylocarpus australasicum*, *Lumnitzera* spp. and *Ceriops australis* (= *C. tagal* var. *australis*) to a fringe of *Exoecaria* and *Aegialitis* scrub with *Melaleuca*, *Eucalyptus* or *Casuarina* woodland at the landwood edge. Boyd's forest dragon is an endemic reptile occasionally found in these mangroves. The salt marshes include *Batis argillicola*, the globally widespread Mediterranean–subtropical species *Cressa cretica* and *Sesuvium portulacastrum*, and the Australian endemic *Tecticornia australasica* (Saintilan, 2009). However, there are many

Figure 11.4 Mangroves and animal habitats in Northern Australia. A: fish eagle on *Rhizophora*; B: Crocodile Island in the Great Barrier Reef near Townsville south of Cairns; C: prop roots of mangroves forest at Cairns; D: tidal channel in a *Rhizophora* swamp, inhabited by saltwater crocodiles; E: upper tidal creek and salt marsh, with mangroves at +5 m a.s.l. (photos by D. B. Scott).

variations on this theme, depending on substrate type, aerial extent and wave fetch. In Trinity Bay, there are also extensive intertidal and subtidal meadows of diverse tropical seagrasses, including *Zostera capricorni*, *Halodule uninervis*, *Cymodocea serrulata*, *Cymodocea rotundata*, *Halophila decipiens* and *Halophila ovalis* (Campbell *et al.*, 2001).

The seagrass meadows provide critical nursery habitat for regional prawn and finfish fisheries and are feeding habitats for dugong (*Dugong dugon*), green sea turtle (*Chelonia mydas*) and wading bird populations (Campbell *et al.*, 2001). The seagrass abundance is primarily associated with climatic conditions, but is also affected by land clearance, crop fertilization and high sediment load associated with clearing of stream bank vegetation. The sediments reduce light available for seagrass photosynthesis and physically smother seagrasses. Herbicides used in sugar farming are also toxic to seagrass, inhibiting germination and photosynthesis. At present, commercial fishers market 38 species of fish and crabs caught within the greater Trinity Inlet region, and recreational fishing targets barramundi, mangrove jack, fingermark, estuary cod, grunter, bream, trevally, queenfish, threadfin, barracuda and sickle fish. The inlet is also a significant cultural and economic resource for the Gunggandji, Yidinji and Yirranydji indigenous people, who have eleven primary fishing sites and four crabbing sites in the area.

11.1.2 Tropical Northern Australia: saltwater crocodiles and flying foxes

The Region 2 muddy coasts of tropical Northern Australia are subject to high tides and the impacts of tropical cyclones. Although the Gulf of Carpenteria has only diurnal tides and a lower tidal range than Cairns Bay, most of the northwest coast has a macrotidal regime with tidal ranges up to 8 m. Cyclone waves build low sand and shell ridges (cheniers) on the muddy shorelines, which are fringed by mangroves and fronted by extensive muddy tidal flats. Rainfall is high, especially in summer when rivers transport large amounts of sediment to the estuaries, resulting in complex interactions between river discharge, tidal flow and sediment deposition.

Essentially the same mangrove forests are found in the northern estuaries and westwards to the Timor Sea coast as are found around Cairns (Saenger *et al.*, 1977); however, the Northern Territory also includes an endemic mangrove, *Avicennia integra*. In the wetlands of the Adelaide River estuary, the mangrove trees *Avicennia integra*, *Rhizophora apiculata*, and *R. lamarkii* and the shrub *Acanthus ebracteatus* are present. Ball (1998) finds that the plant species richness along a soil salinity gradient from 7 to 180 psu is lowest where there is prolonged exposure to either freshwater or hypersaline conditions, regardless of whether the extreme is caused by tidal inundation or soil moisture cycles. In contrast, species diversity is highest is areas of moderate salinity and high soil moisture during the late dry season.

The Region 2 tropical mangrove fauna commonly includes the saltwater crocodile *Crocodylus porosus* and flying fox (= a black fruit bat, *Pteropus alecto*), and the marsupial wallaby (*Wallabia agilis*). The huge fruit bat has a wing-span up to 1 m and while it feeds mainly on *Eucalyptus* (gum tree) flowers, it may steal cultivated mangos and apples. Spalding *et al.* (2010) report there are 128 bird species and 3000 invertebrates; some of these birds are largely restricted to the mangroves – yellow white-eye, chestnut rail, white-breasted whistler and mangrove golden whistler. The Richardson's mangrove snake is an endemic reptile. The Australian giant saltwater crocodile is the world's largest reptile and biggest riparian predator; the crocodile evolved from the archosaurs, which are also the ancestors of dinosaurs and pterosaurs. However, unlike the dinosaurs that became extinct about 65 million years ago, the crocodile has survived to modern times (Wicander and Monroe, 2004).

Figure 11.5 Stromatolites in Western Australia. A: Large concentration of living stromatolites in the intertidal zone at Shark Bay (photo by D. B. Scott); B: a slice through one stromatolite shows layers of organic cyanobacteria and other microorganisms alternating with calcite mineral deposits (photo from University of Wisconsin, Board of Regents); C: another large assemblage of living stromatolites is found in Hamelin Pool at the southeast end of Shark Bay; the carbonate rock 'stools' look dried out here, but when covered by tides, they will expand and become slimy (photo by D. B. Scott, 2002). See also colour plates section.

11.1.3 Southwestern Australia: Shark Bay stromatolites and Leschenault Inlet Estuary

Region 3 on the semi-desert, limestone coast of Western Australia south of the Tropic of Capricorn has few modern-day wetlands. However, fossil intertidal mudflats at Shark Bay (~23° S) and Hamelin Pool ~25° S (Figure 11.5) are of great interest to botanists and geologists because they suggest what the Earth may have looked like about 3.5 billion years ago. Here, there are stromatolites like those found in the world's most ancient tidal mudflats: they are the oldest and largest 'living fossils' on Earth. Stromatolites may represent a time in the Earth's early evolutionary history when primitive cyanobacteria grew in marine shallows in many parts of the world and used the ancient nitrogen-rich air to produce the oxygen needed for larger plants to evolve and migrate onto the land. Today, the stromatolites mostly grow only in extremely saline lakes of Australia, South America and Mexico (see also Section 7.4.4 on salt marshes in northern Baja California). The cyanobacteria grow successfully at Hamelin Pool because here the seawater becomes twice as saline as normal seawater when it is trapped by a sand bar across the bay entrance and evaporates rapidly in the near-desert climate of this region.

The cyanobacteria forming the stromatolites at Shark Bay and at San Quintin lagoon in Mexico are transitional between bacterial and filamentous blue-green algae. The Hamelin Pond stromatolites up to a metre high are probably hundreds to thousands of years old because the algal formations grow very slowly, at a maximum of rate 0.3 mm/yr (Figure 11.5B). At Shark Bay, there are several kinds of stromatolitic microbes (Jahnert and Collin, 2013): (1) laminated forms (tufted, smooth and colloform/prismatic), (2) non-laminated thrombolitic forms (pustular) and (3) cryptomicrobial non-laminated types forming microbial pavement. The layers within the stromatolite 'stool' are calcium carbonate films precipitated over a growing mat of bacterial filaments when the bacterial photosynthesis depletes the carbon dioxide in the surrounding water. The minerals and other sediments precipitated from the water are trapped in the sticky layer of slimy mucilage that surrounds the bacterial colonies. The bacteria then continue to grow upwards through the sediment to form a new layer and multidecker 'sandwiches' of sediment alternating with bacterial matter. The layer thickness, shape, oxygen and carbon isotope composition of the stromatolite colonies record the history of Holocene sea level variations and indicate an interval of lower sea level from 1040 to 940 yr BP.

South of Shark Bay, near the southernmost tip of Western Australia, there are salt marshes and small mangroves at Leschenault Inlet Estuary near 34° S, about 180 km south of the state capital at Perth. This inlet off the Indian Ocean is an elongated coastal bay behind a beach barrier, and wetlands are present on the shoreline around the estuaries of the Collie, Brunswick and Preston Rivers (Mckenna, 2007). The climate here is warm-temperate, with a mean annual temperature range of 14–25 °C; most of the rainfall is during the austral winter, giving it a climate similar to that of the Mediterranean. This estuary is the southernmost limit of the mangrove (*Avicennia marina*) in Western Australia, where it survives in hyperarid soils at the edge of salt pans (Spalding *et al.*, 2010). The nearest location of other mangroves is 500 km further north on the Houtman Abrolhos islands offshore of Geraldton, but the red mangrove does not occur south of Port Hedland at 20.31° S, 15 degrees of latitude further north (Saenger *et al.*, 1977). Tides in the Leschenault Estuary are diurnal and microtidal, with a mean spring range of 0.5 m and a maximum range of 0.9 m. The estuarine waters are well mixed during the summer dry period, but stratified, with a low-salinity surface layer during periods of high runoff in winter. Salinity normally ranges from 33–47 psu, sometimes increasing to 57 psu in summer.

The subtidal area at Leschenault Inlet supports three species of seagrass – *Halophila ovalis*, *Ruppia megacarpa* and *Heterozostera tasmanica*. Salt marshes are dominated by *Arthrocnemon* spp., *Sarcocornia quinqueflora* and *Frankenia pauciflolia*, with wetter areas being covered by salt-tolerant sedges (*Schoenoplectus validus*), rushes (*Juncus kraussii*) and other herbs. The less-saline upland fringe typically has forests of the small saltwater sheoak (*Casuarina obesa*), saltwater paperbark (*Melaleuca cuticularis*), paperbark (*Melaleuca viminea*) and swamp paperbark (*Melaleuca raphiophylla*). In sandy soils of high coastal sand dunes and lower beach dunes, there may be scrub with *Jacksonia furcellata*, *Eucalyptus rudis* or *Acacia saligna*. There is little information on the fauna of these tidal wetlands, which were altered by farming during ~1830−1850, and subsequently, by river damming, siltation, dredging and channelization. For example, the natural tidal entrance to the barrier lagoon was blocked in order to develop the port of Bunbury.

11.1.4 Temperate Southeastern Australia: southern mangroves and paleotsunamis

Region 4 includes 11 significant coastal wetlands in estuaries and bays cut into rocky headlands of the southeastern coast, which is marked by smaller tides and more moderate wave energy than the other Australian coastlines. Saenger *et al.* (1977) describe the mangrove and salt marshes of this region where there is a transition from the high-diversity tropical forests of Northern Queensland in Region 1 to the Region 4 low-diversity mangroves at Townsville (~20° S, with four mangrove species) and at Careel Bay (~33.5° S, near Sydney) where *Avicennia marina* and *Aegiceras corniculatum* are the only mangroves. The temperate-region mangrove fauna of Southeastern Australia, particularly that around Sydney (33° S), is diverse (see Box 11.1 Opening a can of worms) and characterized by deposit-feeding oligochaete and polychaete annelid worms (e.g., capitellids and spionids); gastropods; small crustaceans (cumaceans, tanaids, isopods and amphipods) and nine species of crabs, including ocypodid, grapsid and sesarmid types (Morrisey *et al.*, 2010). The macrofauna in core samples from an urbanized mangrove forest in Sydney Harbour includes amphipods, insect larvae, oligochaetes, crabs, capitellid, nereid, sabellid and spionid polychaetes and gastropods. Six species of crabs (three ocypodids, two grapsids and a sesarmid), an alpheid prawn and five species of gastropods (including the pulmonate slug *Onchidium damelii*) dominate the mangrove habitats of Pittwater near Sydney. The commercially harvested portunid crab (*Scylla serrata*) occurs in the temperate mangroves as far south as the Bega River. Distinctive and often abundant mangrove gastropods include the potamidids *Pyrazus ebeninus* and *Velacumantus australis* and the amphibious pulmonates *Salinator* spp. Less-abundant taxa include large vermiform sipunculids and echiurids, acarids (mites) and dipteran (fly) larvae.

Newer research themes focus on two aspects of the wetlands in New South Wales, Southeast Australia: (1) the past history and future management of the mangroves at their southern limit and (2) evidence of tsunamis in a Pacific region that today is supposedly far removed from zones of high tsunami risk. The common thread for the two themes is concern for future management of coastal zone wetlands in light of rising global sea level and increased storminess associated with warming ocean surface water. The tsunami history research is explained in Chapter 12; here we focus on the mangrove history.

At Minnemurra River estuary ~34.5° S, about 85 km south of Sydney, the mangroves *Avicennia marina* subsp. *australasica* and *Aegiceras corniculatum* occur at the seaward edge of tidal wetlands where large salt marshes are dominated by dropseed grass and the succulent perennial *Sarcocornia quinqueflora*, mixed with areas of *Triglochin striata*, *Suaeda australis*, and *Juncus krausii* (Oliver *et al.*, 2012). Reeds or swamp-oak, *Casuarina glauca*, grow around EHW. Historical records of 1938–2011 reveal the gradual

Box 11.1	Opening a can of worms

Just as confusions with appropriate scientific and common plant names make comparing floral diversity challenging (see Box 8.1 What's in a name?), descriptions of fauna, especially invertebrates, can also involve extensive new terminology that is at first overwhelming to zoologists and non-zoologists alike. Worms can mean anything from true annelids like earthworms and clam worms, to parasitic flatworms to nematodes, let alone all obscure phyla with vermiform body shapes. Sometimes 'common names' really are not that common, so the readers are encouraged to reference current invertebrate biology textbooks. Even descriptor words are challenging, unless you are familiar with Latin or Greek languages and then they make immediate sense. The words just seems strange to use because they are handed down from early days when the animals were first described by scientists who used Latin and Greek as the common languages for communicating – just as we use mostly English today. The following list shows a short example of terms mentioned above in Section 11.1.4:

Pulmonate = air-breathing gastropods (modified 'lung')
Portunid = swimming crabs (paddle-like fifth walking leg)
Ocypodid = i.e. ghost crab
Grapsid = i.e. marsh crabs
Sesarmid = i.e. many mangrove crabs, often terrestrial species
Alpheid = snapping or pistol shrimp with asymmetrical claws
Potamidid = common name horn snail or mudwhelk amphibious snails
Capitellid = fragile polychaete worms that resemble terrestrial earthworms
Spionid = family of tube-making polychaete worms

invasion of *Avicennia marina* into the salt marsh. Models of sediment dynamics show that long-term surface accretion trends (0.61 and 0.26 mm/yr for mangrove and salt marsh, respectively) are much lower than the average local sea level rise over this time interval. Consequently, in the future there is likely to be more frequent flooding of salt marsh communities and continued landward encroachment of the mangroves.

At Hawkesbury River estuary ~33° S, *Avicennia marina* and *Aegiceras corniculatum* have also moved landwards into the upper intertidal salt marsh environments over the past four decades (Saintilan and Hashimoto, 1999). A study was made of peat layers in cores from the inner salt marsh plains to look for evidence of previous mangrove growth at these sites. On the prograding bayhead deltas, mangrove roots were found about 30 cm beneath the present-day salt marsh surface. These buried roots have radiocarbon ages of 500 to 1700 yr BP, indicating that there was an earlier normal progression from mangrove scrub to salt marshes as sediment accumulated on the mudflat and raised its elevation. The current landward transgression of mangroves is reversing the longer-term successional trend in these environments.

Near the border of New South Wales (NSW) and Queensland, the tropical mangroves *Rhizophora stylosa*, *Bruguiera gymnorhiza* and *Excoecaria agallocha* have their southern limits in sheltered estuaries along this high wave-energy coast. In a study of the relationship between mangrove diversity, biomass and latitude, Wilson and Saintilan (2012) examined the role of minimum temperatures on mangrove growth versus other factors such as less sunlight and slower growth rates during intervals of colder temperatures. It is well known that *Rhizophora* is frost-intolerant, but since the north coast of NSW is essentially frost-free, some other limiting factor is implied to explain the southern limit of the species distributions. The studies of *R. stylosa* along a ~300 km latitudinal gradient from Queensland to NSW (~28.5−30.88 °S) show that although the area of mangroves here is increasing and there is a recent rise in mean temperatures, measurements of leaf gain and growth rates in three estuaries do not reveal any trends corresponding to latitudinal gradients. However, leaf longevity and the number per shoot appear to be higher in the temperate NSW region than in the warmer latitudes of Queensland, possibly reflecting an adaptation to lower levels of photosynthesis in the cooler, less sunny temperate region. Overall, the results suggest that the red mangrove *R. stylosa* is not close to its physiological limits at the southern end of its geographical range and they negate the long-held idea that the southern limits of the mangroves represent absolute thermal boundaries. These results, although of a preliminary nature, have very important implications for understanding the true limits on growth of various mangrove species and they indicate the need for caution in interpreting past climates using mangrove pollen and peat records as proxies for winter temperatures, as discussed in Section 12.2.

11.2 New Zealand coastal wetlands: the outer edge of the colonized Pacific World

New Zealand consists of two large islands of continental crust broken off from Australia about 83 million years ago and now forming the exposed portion of a submerged micro-continent straddling the boundary between the Indo-Australian and Pacific tectonic plates. To the north, the Pacific plate is sliding under the Australian plate at a rate of about 6 cm/yr, and a line of surface and submerged volcanoes extends from Wellington, North Island, northeastwards. To the south, the western edge of South Island is sliding northwards on the Australian plate at a rate of 3 cm/yr, along the Alpine Fault zone of the Southern Alps. Consequently, the country experiences many small earthquakes in a zone stretching from the southwestern tip to the Bay of Plenty on the southeast side of North Island. This region is a segment of the 'Ring of Fire', the almost continuous belt of volcanoes and earthquakes that rims the Pacific Ocean (Figure 12.1).

In addition to the complex tectonic history of its shoreline, the New Zealand coastal marshes are particularly interesting for two reasons: this country is the last of the world's large islands to be inhabited by humans, and it is the southernmost limit of salt marshes in the Indo-Pacific biogeographic region. The timing of first colonization of New Zealand is known from radiocarbon dating of bones of the introduced Pacific rat (*Rattus exulans*) found at roosting sites of the now-extinct laughing owl (*Sceloglaux albifacies*) and from eggshells of the flightless moa bird found in Maori archeological burial sites (Wilmshurst *et al.*, 2008). New Zealand was uninhabited up until ~1280 CE, when both the North Island and South Island were first colonized by Polynesian settlers, who then formed the distinctive Maori culture. The early settlers appear to have extinguished 51 bird species, including the

3 m-high ostrich-like moa, and they cleared large amounts of lowland forest before the invasion and settlement of European farmers and traders around ~1840 CE. Today the total population of this temperate climate region (average 16 °C in the north, 10 °C in the south) is only around four million, but the coastal wetlands have been much altered by developments and they are heavily invaded (~50% in 2003) by non-native species of northern affinity, probably introduced mainly by the European settlers (Haacks and Thannheiser, 2003). Mitsch and Gosselink (2007) estimate that New Zealand has lost 90% of its total wetland area – amounting to >300 000 ha.

Most of the New Zealand tidal wetlands occur at the heads of estuaries where there is little wave action and where sedimentation is high. Representative marshes and mangroves are located in Awhitu Regional Park in Auckland (Figure 11.6) and the Manawatu Estuary near Wellington in the North Island (Figure 11.7), and in the Avon-Heathcote Estuary in Christchurch, South Island. In North Island, the grey mangrove *Avicennia marina* var. *resinifera* (= *A. marina* subsp. *australasica*) dominates the tidal wetlands from the northern tip at ~34° S to about 38° S on the east coast Bay of Plenty, and to ~100 km south of Auckland on the west coast (Mildenhall and Brown, 1987). Although most of these mangrove trees are relatively small (<8 m high), they form extensive shrubby thickets in many areas and also colonize volcanic boulders on the shores of North Island. Elsewhere in New Zealand there are salt marshes dominated by the native species *Sarcocornia quinque-flora* (beaded glasswort or chickens feet) or introduced *Puccinellia distans* (alkali grass) and

Figure 11.6 Miranda marsh is one of the most southerly mangrove marshes in the Southern Hemisphere, near Auckland in eastern North Island of New Zealand. A: Murray Gregory and David Scott sampling mangroves at the treeline; B: sampling in the *Sarcocornia* high marsh; C: Erica Hendy and friend on a small mangrove patch adjacent to the largest mangrove area in New Zealand ; D: a thicket of mangrove interspersed with *Sarcocornia* in open areas (photos by Bruce Hayward, Auckland, NZ).

Figure 11.7 Marshes at the southern tip of New Zealand North Island, near Wellington, south of the mangrove growth zone. A: overview of the *Sarcocornia* marsh and high marsh rush–grassland fringe adjacent to Pauatahunai Inlet, just north of Wellington; B: upper marsh dominated by *Sarcocornia* and *Juncus* spikerushes (photos by D. B. Scott).

Samolus repens (sea primrose) in the low marsh, together with the endemic herb *Selliera radicans* (bonking grass) and *Juncus kraussii australiensis* (sea rush) in the middle marsh (Saenger *et al.*, 1977). Where there is freshwater seepage, *Triglochin striata* var. *filiformis* and *Scirpus* sedges replace *Sarcocornia* as pioneer species.

Morrisey *et al.* (2010) list the benthic macrofauna and other animals in the mangrove marshes of the relatively pristine Whangateau Harbour (~36.3° S) near Auckland. Burrowing animals are rare except for the grapsid crab *Helice crassa* and the pulmonate mud snail *Amphibola crenata*, which characterizes New Zealand mangrove wetlands. Other common gastropods are *Diloma subrostrata*, *Zeacumantus lutulentus* and *Z. subcarinatus*. The main predatory species is the whelk *Cominella glandiformis*, which feeds on crabs. The high marsh benthic community at Whitford embayment east of Auckland is dominated by corophiid and paracalliopiid amphipods, oligochaetes and the crabs *Halicarcinus whitei* and *Helice crassa*. On suitable sites close to the intertidal sandflats, there are limpets (*Notoacmea helmsi*), several bivalves (*Paphies australis*, *Macomona liliana*, *Austrovenus stutchburyi*, *Nucula hartvigiana*) and the isopod *Exosphaeroma chilensis*. Burrowing deposit-feeders in the mangrove habitats are mainly polychaete worms (*Scoloplos cylindrifer*, *Heteromastus filiformis* and other capitellids), oligochaetes and the crab *Helice crassa*. On the east coast, the mangroves and seagrasses (*Zostera muelleri* and *Halophila ovalis*) in Matapouri Estuary (35.5° S) support different faunal assemblages, with seagrass patches having the highest abundance and species diversity. The mangrove fauna here includes the cockle *Austrovenus stutchburyi*, but there are few crabs, and no mud snails, shrimps or amphipods. However, *Helice crassa* is common in the subtidal seagrass beds. Saenger *et al.* (1977) report that important birds in the New Zealand marshes are the pied stilt (*Himantopus leucocephalus*) and oyster-catchers (Haematopodidae) that feed on oysters (*Saccostrea cucullata*) growing on the mangrove pneumatophores. The native black swan (*Cygnus atratus*) is a large grazer of the intertidal seagrass in estuaries of the Bay of Plenty, and recent studies suggest that overgrazing by these birds may be causing a decline of the seagrass biomass (Dos Santos *et al.*, 2012). The birds graze at high tide and although the reason for the overgrazing is uncertain, it might be related to overcrowding of the birds in areas protected from boats and other disturbances.

As in Australia, there have been studies to determine the factors that limit the latitudinal distribution of the grey mangrove in New Zealand (de Lange and de Lange, 1994). The New Zealand grey mangrove *Avicennia marina* var. *resinifera* is the same species as in Australia, and although it is often thought to have its southernmost natural growth limit at about 38° S on North Island of New Zealand, in fact it reaches its limit almost a degree of latitude further south at 38.913° S in Victoria, Southern Australia. However, it has been planted successfully farther south in New Zealand, and one tree survived many years as far as 41.22° S before it was killed by land infill that smothered its pneumatophores. At this location, coastal frosts exceed −2.2 °C and occur at least once every 5–10 years, which was believed to the limiting factor for flower production and long-term survival. In a review of all the evidence for frost sensitivity of the species, however, de Lange and de Lange conclude that the New Zealand mangrove distribution is limited not by climate, but the by distance of dispersal for its crypto-viviparous propagules (= germinated embryos that look like the first stage of sprouting in a broad bean seed; Figure 3.6a). The propagules only remain viable for 4–5 days in the coastal currents, and because the inshore water around North Island has a net movement of less than 175 km over five days, this is insufficient to transport the propagules to new inlets on the largely unsuitable rocky shoreline. It is intriguing, however, that pollen of this mangrove occurs in early Holocene sediment of cores from Poverty Bay ~38.7° S, far removed from its present natural limit (Mildenhall and Brown, 1987). The sediment was radiocarbon dated as 9840 yr BP, from wood fragments at a depth of ~30 m below present sea level, suggesting that the climate at that time was either warmer than now or, more likely, that there were more extensive coastal habitats available for colonization by the mangroves of North Island before sea level reached its present position ~6000 years ago.

Another study of the New Zealand tidal wetlands documents the distribution of modern-day foraminifera in the southernmost salt marsh of Australasia, at the mouth of the Catlins Estuary on the east coast of South Island, ~46.5° S, about 180 km south of Dunedin. This area is in a region of tectonic stability which was largely undisturbed before European settlement in 1861. The mean tidal range near the mouth of the estuary is ~1.5 m, and four zones of vegetation are found along a profile across the salt marsh (Southall *et al.*, 2006). The lowest zone above the unvegetated tidal sand flat is dominated by *Sarcocornia quinqueflora* and *Samolus repens*, above which is salt meadow with *Selliera* and *Samolus*; at the upper marsh limit, there is a zone of the jointed wire rush (*Leptocarpus similis*) or, in areas of freshwater inflow, a zone of shrubby ribbonwood (*Plagianthus divaricatus*) with minor occurrences of foxtail grass (*Schedonorus phoenix*), New

 Figure 11.8 Visible traces of earthquake activity in New Zealand South Island. Top: offset of a fence and crop rows by a 13 km-long trace (GNS Science, 2011); Bottom: 4 September 2010 Christchurch earthquake damage (photo by Malcolm Teasdale/Kiwi Rail, 2010).

Zealand flax (*Phormium tenax*) and ferns. Statistical analyses of the foraminiferal species abundances show that different species are significantly correlated with marsh elevation and can be used as a proxy for changes in marsh elevation to within ±5 cm or better. This study of the foraminifera in a stable part of South Island provides a baseline for future investigation of the history of vertical shoreline displacement in the tectonically active regions further north. Here the Australian Plate carrying North Island is rotating southwards over the Pacific Plate that carries most of South Island, and earthquakes are common. The devastating 7.1 and 6.3 magnitude earthquakes that shook the Christchurch and Canterbury regions of South Island in 2010 and 2011, respectively, are examples of the strong tectonic activity in the region (Figure 11.8). The variable nature of the New Zealand coastline and its tectonic activity make it particularly difficult to determine the likely impacts of future sea level rise along the coast of New Zealand (Beavan and Litchfield, 2012).

Applications in geological monitoring: paleoseismology and paleoclimatology

---

Key points

Coastal ecosystem knowledge is essential for understanding earthquake mechanics and forecasting catastrophic shoreline movement and flooding; multiple sources of fossil proxy-data – including microfossils, pollen and sediment – are best used to reconstruct patterns of earthquakes and tsunamis in time and space; multi-disciplinary studies are also needed to distinguish tsunami from tropical storm events; correct measurement of timing and speed of paleoseismic events depend on accurate dating methods, best provided by tree roots and salt marsh peat; foraminifera provide the most precise estimates for amounts of vertical shoreline change; pollen of mangroves and salt marsh plants provide best estimates of climate change; diatom and dinoflagellate paleotransfer functions are best for tracking the prehistoric sea-ice changes.

12.1 How wetland archives are used in paleoseismology and paleotempestology

---

The past is all we know about the future.

(Barbara Kingsolver, *The Lacuna*, 2009)

In Chapters 3 and 4, we explained how study of foraminifera (Box 4.1 Tidal wetland foraminifera) and pollen grains (Figure 3.3) in present-day tidal wetlands can be used to analyse and interpret geological archives of past changes in sea level, salinity and coastal vegetation. Barlow *et al.* (2013, p. 90) state that, 'Understanding late Holocene to present relative sea level changes at centennial or subcentennial scales requires geological records that dovetail with the instrumental era. Salt marsh sediments are one of the most reliable geological tide gauges.' Here we give additional examples of other micro-fossils and geochemical tracers that can be used as proxies in studies of coastal wetlands, and we describe various case histories for applications in paleoseismology, which is the study of prehistoric earthquakes and tsunamis – particularly their location in space and time. Paleotempestology is the related study of storms and hurricanes from a primarily geological perspective (Liu, 2004, 2007). Throughout, however, we recognize that all 'geological tide gauges' demand detailed chronologies, where it is certain that apparent relative sea level fluctuations are not simply a consequence of the age–depth model derived from a few radiocarbon ages (see Box 12.1 About time).

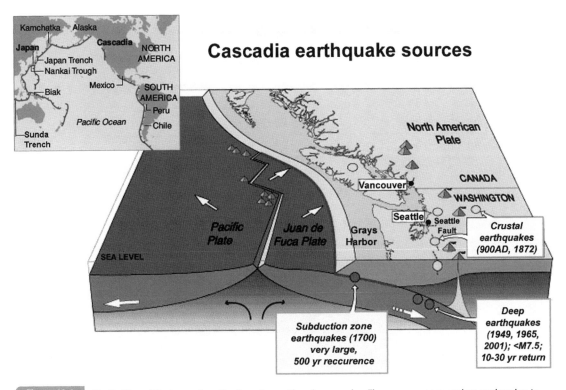

Figure 12.1 'Pacific Ring of Fire' tectonism: the Cascadian earthquake examples. Three source zones produce earthquakes in Cascadia, between Washington State and Alaska: deep events start below ~30 km, beneath the subduction zone; shallow earthquakes originate in the overlying North American Plate; the largest earthquakes are on the subduction zone (USGS). The inset shows the upper edges of the plate boundaries for the Pacific Ring of Fire; the teeth point towards the subduction (from Atwater *et al.*, 2005, USGS).

Earthquakes and the associated tsunamis (giant tidal waves) are a natural outcome of tectonic activity, either where mountains are forming on the surface of the Earth, like the Himalayas, Rockies, Andes and the Pontic mountain range in northern Turkey, or where submarine tectonic plates are sliding under continental plates (Figure 12.1), as off Sumatra, Japan and western America, on the borders of the 'Pacific Ring of Fire'. To try and develop better early warning systems for these catastrophic events so that fewer lives are lost, geologists search for better earthquake detection methods and improved understanding of tsunami records. It is important to know how these earth movements and tidal waves have impacted coastlines with different geomorphologies and sedimentary environments, whether eroding or aggrading shorelines, and how to interpret the sediment and fossil records left by receding waves (Kumar and Nistor, 2012). There is also a need to distinguish between earthquake-driven paleotsunami deposits and records of climate-driven large storms (hurricanes and cyclones), both of which cause marine overwash on land. In these situations, multidisciplinary studies that use a combination of proxy-tsunami tools, such as sedimentology, micropaleontology (foraminifera, thecamboebians, diatoms) and paleobotany (pollen and peat), allow the most reliable reconstructions of events. Paleoseismic studies also provide geologists with unique signals of 'silent' earthquakes and a better ability to assess the probable timing and severity of future earthquakes. In most countries, paleoseismic risk analysis is also a mandatory requirement for planning of nuclear power plants, dams, waste repositories and other critical structures. The text book by McAlpin (2009) provides a thorough discussion of most aspects of the science of paleoseismology.

Historically, foraminifera and thecamoebians (= testate amoebae) are the preferred proxies used for tracing the changes in sea level that accompany paleoearthquake and paleostorm events. These microfossils are amply illustrated by Scott *et al.* (2001) and scientific references therein. The full suite of proxies used for paleoseismology studies, however, now includes other microfauna, specifically ostracods (Ostracoda; seed shrimp), and microflora – mainly coccoliths (Coccolithophorida) and diatoms (Bacillariophyta) – as well as macrofossils such as shells (Mollusca), tree roots and peat beds, sediment texture and the geomagnetic properties of the sediment (cf. Chagué-Goff *et al.*, 2011). These other microfossil types are illustrated in Figure 12.2 and Box 12.1, and their strengths or weaknesses as proxies are summarized in Table 12.1. Essentially, the value of the microfossil varies with its potential for long-term preservation (= taphonomy), including its likelihood of being rearranged in the sediment. Many living foraminifera and all coccoliths and molluscs are encased in shell walls made of calcium carbonate that form calcareous fossils after death. These calcareous shell walls, like chalk or limestone, will slowly dissolve in highly acid soil. Diatoms have cell walls made of biosilica, which is resistant to acid corrosion, but may be fragile and shatter like very thin glass. The other microfossils, along with pollen grains and plant spores, have cell walls of organic material that is highly resistant to decay in the absence of oxygen, but is destroyed by long exposure to sunlight and air.

New geochemical paleoecological methods attempt to use diagnostic chemical tracers as proxies instead of the well-known and traditional microfossils. Examples include: (1) measurement of biosiliceous opal as a proxy for diatoms and the lipid IP_{25} (= Ice Proxy with 25 carbons) in place of sea-ice diatoms, (2) use of mangrove tannin alkaloids or triterpenol lipids (from leaf wax) in place of mangrove pollen or peat, and (3) employing the ratio of the stable carbon isotopes ^{13}C and ^{12}C as a proxy for salinity (marine shells have $\delta^{13}C$ values of around 0 to +6‰, while freshwater shells have values lower than –8‰). Quantitative geochemical measurements can often be applied faster than microfossil analyses, but their value is presently limited by the degradation of the biochemicals over time, and weak understanding of processes controlling $\delta^{13}C$ values, which vary among C_3 and C_4 plants (see Box 3.1 Comparison of plant photosynthesis and metabolism, for descriptions of these plants). Water-leachable chloride measurement is another new geochemical tool being applied to paleotsunami deposit studies (e.g. Goto *et al.*, 2011), but can be ambiguous to interpret where saline groundwater is near the surface, as in the Mackenzie Delta studies by Solomon *et al.* (2002).

12.2 Paleoearthquakes and earthquake prediction

12.2.1 Alaskan case histories

A good example of a paleoearthquake study is provided by the proxy data from Girdwood Flats marsh in the Cook Inlet area, Alaska (see Section 7.2.1, Figure 7.4).

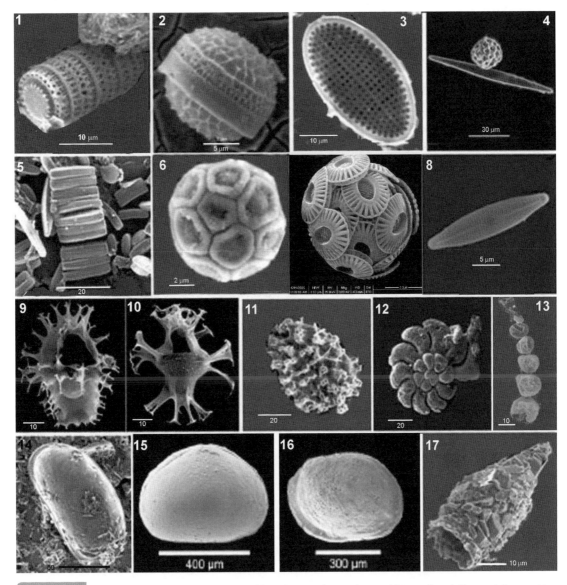

Figure 12.2 Scanning electron images of proxies used for paleoseismology studies in addition to foraminiferan and thecamoebian microfossils, from work by P. Mudie and from Scott *et al.*, 2011 (salt marsh ostracods). Macrofloral proxies are illustrated in Section 12.3. 1–4: cylindric and pennate diatoms from Alaskan tidal marshes (4 includes a round *Sarcocornia* pollen-grain); 5 and 8: cylindric and pennate diatoms from an inland saltflat; 6 and 7: Coccolithophores from shallow and deep water, respectively (7 is *Emiliania huxleyi*, photo by Alicia Kahn); 9–11: organic walled dinoflagellate cysts from the Black Sea; 12 and 13: organic linings of calcareous and arenaceous foraminifera, respectively, in Saanich Inlet, BC; 14–16: Ostracods, 14 is from a saline pond in The Yukon; 15 and 16 are from San Diego County coastal wetlands; 17 is a thecamoebian from an Alaskan tidal marsh, with broken diatoms for its wall cover.

Box 12.1   About time: dating of sediments for paleoenvironmental studies

The value of predictive paleoseismology and paleoclimatic studies usually requires measuring the age of the salt marsh or mangrove swamp sediments in order to calculate when a change in relative sea level (RSL) or climate took place, how fast the catastrophe was and the time for recovery after the event. Radiocarbon dating (^{14}C or C-14 dating) of organic (carbon-containing) material is the most common method used for measuring the age of sediment older than 1950 CE, and this method can be applied fairly reliably to sediments as old as ~35 000 yr BP. The radiocarbon dating method is based on measured records of changes in the ratio of radioactive (^{14}C) to stable carbon (^{12}C) in the atmosphere. Coastal wetland soils usually provide excellent sources of wood or leaf material for dating because the trees and herbs that produced the dated plant materials absorbed their carbon from the atmosphere while alive. Plant macrofossils (seeds, leaves, cones) from salt marsh peat or wood from 'ghost trees' drowned by seawater (Box Figure 12.1) therefore give more accurate ^{14}C ages than shells (molluscs, ostracods, or foraminifera) that obtain their carbon from seawater, which includes variable amounts of fossil carbon from recycled 'old' marine or river-transported land carbon. However, the release into the atmosphere of excess radioactive isotopes during the twentieth century atomic bomb tests caused erratic changes in the ratio of ^{14}C to ^{12}C, making the radiocarbon method unreliable for marsh sediments younger than 1950. To compensate for this 'bomb-spike' effect, other radioactive dating methods are used to measure the post-1950 sediment ages, primarily ^{210}Pb for the past 125 years and ^{137}Cs for sediment younger than ~1963 (Marshall et al., 2007). These short-lived radioactive isotopes are further used to correct errors in the ages obtained from ^{14}C dating of older samples. However, there are several sources for errors in the radioisotope methods that are best evaluated using independent dating sources such as marker pollen of exotic species (see Figure 3.7 and Section 12.2), spheroidal carbon from oil furnaces, dating by tree rings (Roberts, 1998; Atwater et al., 2001, 2005) and the counting of tree rings in 'ghost forests' or finely layered estuarine sediments called rhythmites or varves (Mudie et al., 2002). More specialized dating methods include optically stimulated luminescence (OSL) and thermoluminescence (TL) for sand and amino-acid racemization for bivalve shells.

Ghost forest at Yakutak, Alaska (P. J. Mudie).   Stumps of spruce trees drowned by rising sea level, Price Edward Island, Canada (R. Taylor, GSCA).

Box Figure 12.1

Table 12.1 Proxies used in selected paleoseismology and paleoclimate studies, indicating their strengths and weaknesses at the present state of research

Proxy source	Environment	Example	Strengths (+) versus limits (−)	Reference
Microfauna Foraminifera	Marine to brackish water	Alaska – Cascadia earthquakes and tsunamis; Pacific Island tsunamis	+ Sensitive to water depth and marsh elevation + Taxonomy and ecology well known − Shells dissolve in acid − Arenaceous species fragile	Nelson *et al.*, 1996; Scott *et al.*, 2001; Goff *et al.*, 2011
Thecamoebians	Freshwater to brackish water	Abrupt postglacial sea level adjustments in New Brunswick and Nova Scotia	+ Mark marine–freshwater transitions + Tracers of landslides − Fragile and blow around − Fossils are easily oxidized	Honig and Scott, 1987; Scott *et al.*, 2001; Nixon *et al.*, 2009
Ostracods	Marine to freshwater	Lake Manyas: Anatolian Fault Paleotsunami; 2004 Sumatran tsunami in Chennai, India	+ Quantifiable salinity + Relatively large, not easily moved by water currents + Datable with radiocarbon − Taxonomy complicated − Carbonate shell dissolution	Leroy *et al.*, 2002; Ruiz *et al.*, 2010
Microflora Diatoms	Marine to freshwater; sea ice	1964 Alaskan earthquake	+ Very widespread habitats, including sea ice + Geochemistry: opal, IP_{25} − Small, easily moved by currents or blown by winds − Thin walls break easily	Atwater *et al.*, 2001; Belt and Müller, 2013
Coccoliths	Exclusively marine	2004 Sumatran tsunami	+ Very small, transported far inland by tsunami waves − Very small (clay size)	Paris *et al.*, 2007
Dinoflagellate cysts; algal spores	Marine to brackish; some freshwater	'Noah's Flood' wetlands drowning ~7600 yrs BP	+ Organic wall resists decay + Salinity and temperature quantifiable − Low diversity in salt marshes	Mudie *et al.*, 2013; Chmura *et al.*, 2006

			Table 12.1 (cont.)	
Proxy source	Environment	Example	Strengths (+) versus limits (−)	Reference
Macroflora				
Tree stumps	Above storm tide level	1700 AD Orphan Tsunami	+ Not easily moved by waves + Datable by ^{14}C or tree rings − Absent in polar and desert regions	Atwater *et al.*, 2005
Mangrove roots and leaves	Intertidal zone	Campo Saldago, N. Brazil sea level	+ Roots not easily moved by waves; datable by ^{14}C + Geochemical biomarkers − Rare/absent outside tropics − Biomarkers degrade in time	Koch *et al.*, 2011

Here, offshore drilling rigs are in operation and the seafloor is crossed by a network of now-aging pipelines carrying gas and oil onshore. Megaearthquakes in this tectonically instable region clearly represent severe danger to humans and long-term environmental damage from pollutants released by oil pipeline rupture and drill-rig or tanker accidents (see Section 5.7). The 1964 Great Alaskan Earthquake was a megathrust earthquake of 9.2 M_w, also called the Good Friday Earthquake because it began on Good Friday, 27 March, 1964. This megaquake lasted about four minutes and is the most powerful recorded in North American history, and the second most powerful ever measured by modern seismographs. The next day, there were also 11 aftershocks of ~6 M_w and thousands of smaller aftershocks were recorded during the year. Along the south-central Alaskan coast, ground fissures, collapsed buildings and tsunamis resulting from the earthquake killed 143 people of the sparsely populated subarctic region. By comparison, the devastating 2011 Tōhoku-oki or Great East Japan Earthquake that triggered the Fukushima nuclear plant disaster of 11 March, 2011 was smaller (9.0 M_w), but it killed far more people (~16 000) along the densely populated, industrialized coast of northeast Japan (USGS NEIC record posted 14 March, 2011).

The 1964 Alaskan earthquake epicentre was positioned underwater about 120 km east of Anchorage. Onshore, this earth0quake caused soil liquefaction and other ground failures that resulted in major structural damage and landslides, including the salt marshes around Girdwood (Plafker, 1969; Shennan and Hamilton, 2006). Some areas near Kodiak ~280 km to the southwest were permanently raised by 9.1 m, while southeast of Anchorage, shorelines near Girdwood and Portage dropped down by 2.4 m. Near the epicentre, an 8.2 m-high tsunami wave destroyed a village and in Valdez Inlet, the combined tsunami

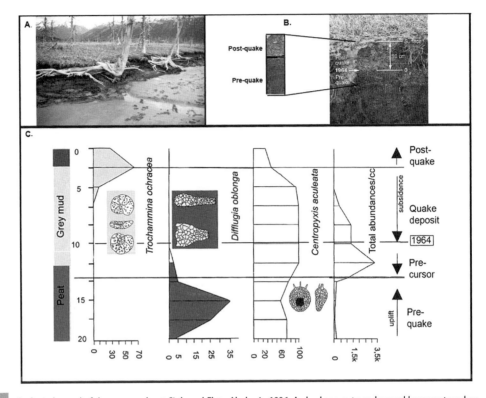

Figure 12.3 Geological record of the megaquake at Girdwood Flats, Alaska, in 1996. A: dead spruce trees drowned by seawater when the earth subsided after the 1964 quake; B: sediment section from the marsh showing an abrupt transition from pre-quake forest peat to tidal mudflat before 1964; grey sediment is mud deposited when the coast subsided before and during the earthquake; C: simplified diagram to show the microfossil distributions in the section just before 1964, where the freshwater thecamoebian (*Difflugia*) and forest peat disappear and are replaced by abundant brackish water *Centropyxis*. The earthquake is marked by the grey mud with fewer *Centropyxis*, and after the 1964 earthquake, the salt marsh foraminiferan *Trochammina* appears (from Scott *et al.*, 2001). See also colour plates section.

wave and local seiche generated by underwater mudslides flooded the shore up to 70 m. Post-quake tsunamis severely affected many other Alaskan communities, as well as people and property in British Columbia, Oregon, and California. The tsunami even caused damage in Hawaii and Japan, and some evidence of motion directly related to the earthquake was reported from all over the Earth (Atwater *et al.*, 2005).

Figure 12.3 shows the landform change and the microfossil traces of the pre- and post-Alaskan earthquake interval, as indicated by paleoseismology studies in 1997 (Shennan *et al.*, 1999) and later (Hawkes *et al.*, 2005). Thirty-five years after the 1964 earthquake, spruce forest was growing over former intertidal salt marshes deposited after the earthquake, but below the salt marsh and tidal mudflat sediment was an earthquake 'precursor' that occurred 1−15 years *before* the Great Alaskan Earthquake, as estimated from radioisotope dating using ^{210}Pb and ^{137}Cs (Box 12.1 About time). The sediment core from Girdwood Flat near Anchorage (Figure 12.3 B, C) demonstrates this preseismic earthquake precursor as an

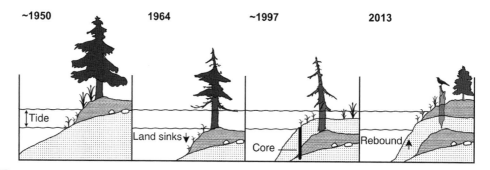

Figure 12.4 Sketch of the events associated with the 1964 megathrust earthquake as uncovered by application of regional salt marsh transect studies to cores from Girdwood. The sequence of events that shook much of the Northern Hemisphere could not be measured by conventional geophysical seismographic studies.

abrupt change of the fossil microfauna from a forest assemblage to brackish water assemblages. Diatoms and spores of peat moss (*Sphagnum*) from the same samples as the microfauna showed the same abrupt switch from freshwater indicators to more salt-tolerant species. This paleoseismological signal marks a subtle downward shift of the shoreline about 15 years before the megathrust earthquake that shook much of the Northern Hemisphere, yet this preseismic event was not recorded at any seismic station across the globe. This type of seismic event recorded by the microfossils, but not seen by geophysical methods is called a 'silent' or 'slow' earthquake and cannot be detected on seismographs, but only by continuous measurements with expensive geographical positioning systems in strategically located areas of the anticipated earth movement. The reconstructed sequence of megaquake events at Girdwood is summarized in Figure 12.4.

The 1997 discovery of a preseismic tracer at Girdwood marsh led to a wider search and uncovering of the same 1964 earthquake 'precursor' event in five other sediment cores from the same region (Hawkes *et al.*, 2005). Additional studies based in diatoms showed the presence of similar microfossil precursor events before older megaquakes as far back as ~3050 yr BP, as dated by radiocarbon ages for tree roots and other plant macrofossils imbedded in the peat layers (Shennan and Hamilton, 2006). Further microfossil studies have investigated the application of the same techniques in other areas of the tectonically active West Coast of North America – including the states of Washington, Oregon and California, where similar megaquakes were found to have occurred over the last several thousand years (Shennan *et al.*, 1999; Hawkes *et al.*, 2005). Most recently, Kemp *et al.* (2013) have begun studies aimed at reconstructing earthquake histories and sea levels from islands on the Alaska–Aleutian subduction zone (Figure 12.1), using foraminiferal assemblages. Here, intertidal mudflats and higher marsh zones are characterized by different communities of foraminifera, which mark the MSL tidal datum to within ±9% of the tidal range known for the offshore islands. Notably, the marker species in the Aleutian Islands differ from those in the Girdwood marshes, which emphasizes the need to obtain local data from transect studies like those in Chapters 3 and 4 before sediment core studies are made. This requirement for transect 'data training sets' is particularly emphasized for diatoms (Barlow *et al.*, 2013; Watcham *et al.*, 2013), which are smaller and lighter than foraminifera, hence more easily displaced by strong waves or winds. In general, however, there is a need for tight statistical calibration between the transect microfossil data and local environments before using salt marshes as accurate Holocene tidal gauges. The pros/cons of some statistical methods for calibrating and tuning modern data sets are summarized in Box 12.2 Paleoecological transfer functions.

Box 12.2   **Paleoecological transfer functions: qualitative versus quantitative methods**

Traditional methods of assigning paleoecological values to fossil proxy data were developed by Xenophanes in 540 BCE and they use simple visual comparison rather than statistical analysis of data sets. This traditional Taxonomic Uniformitarian Method matches species distributions in a compiled table (matrix) of many surface samples or core tops (X_i), with a corresponding set of selected modern-day features such as height above relative sea level (RSL), salinity and temperature (Y_i), and then visually compares these 'training data' graphs (XY_{ct}) to a set of fossil samples from a dated core (X_{core}). For each core sample, the section of the graphs where the modern data (XY_{ct}) and microfossil (X_{core}) samples from the core overlap is then interpreted as having the equivalent present-day environment, the 'paleo-Y' values shown here.

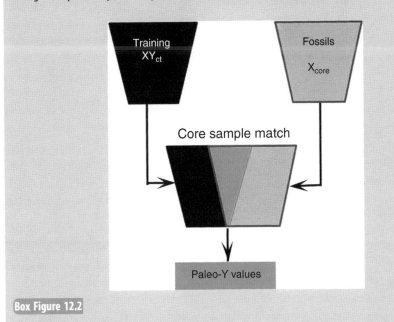

Box Figure 12.2

This traditional method relies heavily on the presence or absence of high-fidelity environmental indicator species, but is surprisingly reliable for temperature and sea-ice estimates of dinoflagellate cysts (Mudie, 1992) and is a standard method for foraminifera, where living specimens are used for the calibration of surface samples (Scott et al., 2001; see also Section 7.3). Other methods use ratios of selected species with very well-known correlations to modern-day conditions such as sea surface temperature. Quantitative methods, e.g. Q-mode factor analysis (QFA) and principle components matching analogue (MAT) methods rely on multivariate statistical analysis of the training data sets of surface samples and their cross-correlation with the environmental conditions. This method was not successful for RSL studies when first applied to subarctic benthic foraminifera (Mudie *et al.*, 1984), but now is validated for salt marsh foraminifera and diatoms (Barlow *et al.*, 2013), and is highly successful when used for dinoflagellate cyst temperature and sea-ice reconstructions. Details of the methods are given in the book *Proxies in Late Cenozoic Paleoceanography* by Hillaire-Marcel and de Vernal (2007). The continued need for salt marsh records as proxies for late Holocene tide gauges is the outcome of the wide variability in instrumental tide gauge records for the North Atlantic region, where sea level trends are systematically larger than the long-term sea level trends. Measured values of past temperature from sediment archives are also required to constrain computer-generated models. An unresolved problem for high-resolution 'tuning' of the training data for RSL proxy data, however, is uncertainty about fluctuating local variations in salt marsh water levels related to recently accelerated melting of the world's glaciers (Gardner *et al.*, 2013).

12.2.2 Cascadia and the 'Orphan Tsunami' case history

After the initial finding of preseismic traces at Girdwood, the same microfossil techniques were applied to other paleoseismology studies of the North American West Coast – particularly the Cascadia region from northwestern USA to central British Columbia (Figure 12.1). Here, some crustal earthquakes are known from historical records and others are total surprises, e.g. the 28 October 2012 Haida Gwaii superquake (M7.8) occurred offshore in a region well beyond the known subduction zone and although the tsunami wave was forecast as merely 50 cm, in fact, waves up to 7 m tossed fish and seaweed high into tree branches (Szeliga, 2013). Cores from Willapa Bay in Washington State reveal repeated earthquakes over several thousand years (Atwater, 1987).

One of most the detailed paleoseismology case histories involves the 'Orphan Tsunami' of Japan (Atwater *et al.*, 2005) – a case in which Japan's largest ever tsunami on 26 January 1700 was described in early Japanese books, but its parent source remained unknown for almost 300 years until ultra-high-resolution dating of coastal wetlands fossil data traced it to the Cascadia coast >7500 km away. Figure 12.5 shows how earthquake and tsunami paleoseismic proxy data in Willapa Bay marsh record the 1700 CE Cascadia earthquake event. The accurate dating of this earthquake source on the opposite side of the North Pacific comes from fossil wood and tree ring 'climatic barcodes' which pinpoint the event to between 1695 and 1720 CE – the exact time window indicated in the Japanese books. The fossil salt marsh sequence shows that the Cascadian coast, once thought not to host earthquakes, was actually the source of a megaquake event of magnitude 8.7–9.2.

In Netarts Bay, Oregon, Hawkes *et al.* (2005) found several ecostratigraphic units that indicated seismic events in the last 300–3000 yr BP (Figure 12.6) and linked these with sequences in Alaska, ranging from the 1964 giant earthquake to older records dated to ~1450–1700 CE (Combellick and Reger, 1994). Further south, in the Humboldt Bay region of Northern California, cores collected by Li (1992) and by Scott *et al.* (unpublished) showed the onset of a M6.9 earthquake eight months before it happened (Figure 12.7), and a longer

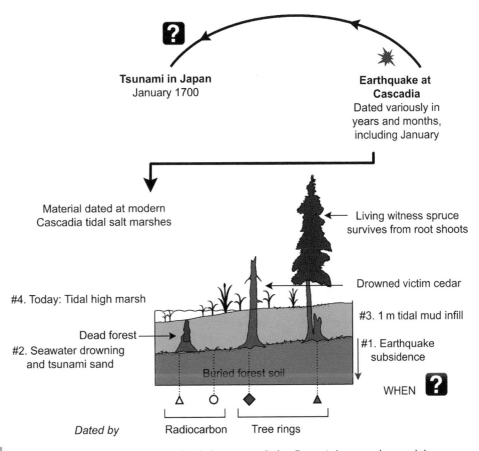

Figure 12.5 Cartoon summarizing the events associated with the Japanese Orphan Tsunami. A megaquake caused the subsidence of the old tidal salt marsh and adjacent spruce forest (#1), rapidly followed by a tsunami wave that smothered the plants with sand, probably several times in quick succession (#2), after which tidal mud flats and a new salt marsh aggraded over the peat layers (#3), and 'living witness' spruce trees (#4) regenerated from buried stump shoots (from data in Atwater *et al.*, 2005, USGS).

sediment record from vibracores showed at least four quake events in the last 2000 years. However, it is probable that this area has received far more earthquakes per 2000 years because in the last 50 years alone, there has been an average of two large earthquakes every five years. Until more sensitive geophysical methods for continuous monitoring are developed, ongoing monitoring of marsh sediments in this area could be used for megaquake forecasting up to 4–5 years before the event because the microfossil indicator method is more sensitive for local events than either regional seismographs or sedimentology methods that use traces of landslide mud flows (turbidites) in submarine cores. For example, a study of a 12 000-year record obtained by deep ocean drilling in Saanich Inlet, eastern Vancouver Island (Blais-Steven and Clague, 2001), suggested that earthquake-triggered underwater mudflows occurred at roughly 150 year recurrence intervals over the past ~3000 yr BP, but more frequent historical earthquake events were not recorded in the varved sediments.

Netart's bay

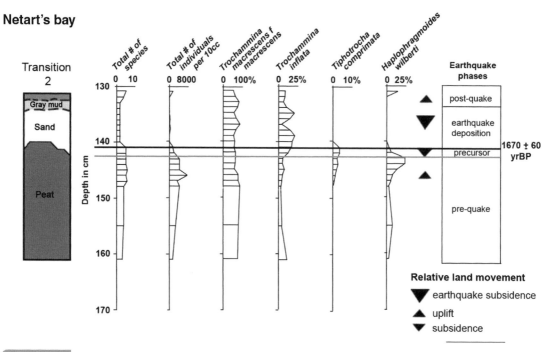

Figure 12.6 Marsh sediments at Netarts Bay, Oregon show an earthquake sequence that was dated at ~1670 yr BP (see Hawkes *et al.*, 2005, Geological Society of America). The precursor event here is marked by disappearance of the high marsh species *Halophragmoides* just below a probable tsunami sand deposit.

12.2.3 The Peruvian earthquake and Hawaiian tsunami

The 1960 Chile earthquake, the 'Gran terremoto de Valdivia', was larger than the 1964 Alaskan earthquake, and is the strongest quake ever recorded, measuring 9.5 M_W (see Section 8.4). The earthquake epicentre was ~1000 km north of Valdivia, which was the most severely impacted city. This megathrust earthquake, generated by the subduction of the Nazca plate below the South American plate, was a succession of three quakes between 21 and 22 May. The first major shock of 8.3 M_W lasted only 35 seconds, but destroyed 33% of the buildings in the city of Concepción. The main earthquake created local tsunami waves up to 25 m high along the Chilean coast and waves also swept across the Pacific as far as New Zealand in the south and the Aleutian Islands in the north. The main tsunami devastated the downtown area of Hilo, the largest city in the Hawaiian Islands and waves up to 10.7 m were recorded 10 000 km from the Chilean epicentre, including Japan and the Philippines. The 1960 Chilean earthquake caused subsidence of ~1.5–2.7 m along about ~1000 km of the Chilean coastline. Study of microfossils in a core collected in Valparaiso Estuary in 1989 (Figure 12.8) showed that, just as in Alaska and Cascadia, the Chilean megaquake was preceded by a preseismic subsidence event detected from matching of foraminifera in Valparaiso marshes with those in the sediment core. However, the magnitude of the subsidence event in the Valdivia estuary cannot be used to estimate the amount of abrupt

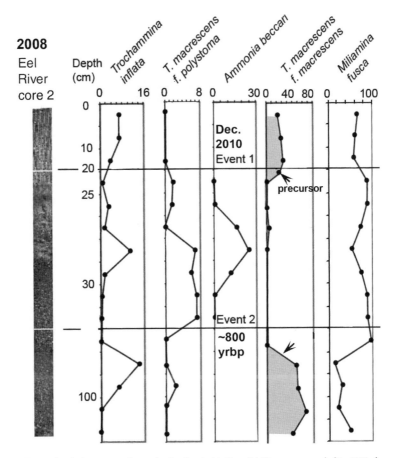

Figure 12.7 Sediments and microfossils from a marsh core in the Humboldt Bay–Eel River area sampled in 2008 show a seismic sequence predicting an earthquake in the near future — which actually occurred in January 2010 when a quake of 6.9 magnitude caused extensive infrastructural damage, but fortunately no loss of life.

sea level rise because the core contains only grouped low marsh–mudflat assemblages, which are indistinguishable from one another. The absence of microfossils that clearly define ranges of the low marsh/mudflat and upland zones limits the precision of estimates of relative sea level change in this estuary. In contrast, at Bahia Quillaipe, Fernandez and Zapata (2010) could distinguish low marsh and mudflat foraminifera, so higher-precision paleoseismic studies might be made there in the future.

12.2.4 Australia: tsunamis or not?

Another geological study on the wave-exposed New South Wales coast of Southern Australia focusses on the question of the shoreline susceptibility to damage by tsunamis. This tectonically stable region is supposedly far removed from zones of high tsunami risk (Bryant *et al.*, 1992) and Australia historically does not record any tsunamis. However, geological evidence of Holocene-age tsunamis along the New South Wales coast include: (1) anomalous boulder

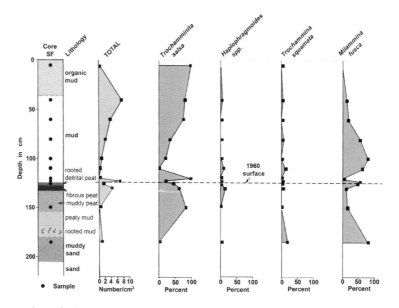

Figure 12.8 Sedimentary and microfossil events associated with the 1960 Valdivia earthquake. The quake precursor is indicated by a peak of *Milammina fusca* marking possible subsidence of ~10 cm about 5–10 years before the 9.5 magnitude megaquake (Jennings *et al.*, 1995).

masses, either chaotically tossed onto rock platforms or jammed into crevices, (2) highly bimodal mixtures of sand and boulders and (3) dumps of well-sorted coarse debris. Many coastal aboriginal middens were also disrupted by these kinds of deposits – apparently being abruptly abandoned. Within estuaries, tsunami evidence consists of shelly ridges that were previously misinterpreted as chenier beach ridges associated with low-energy mudflats. Dating of such deposits using both radiocarbon ages and thermoluminescence methods shows that several catastrophic overwash events have affected this coastline since 3000 yr BP. The tsunami paleorecords can be distinguished from storm wave records by their chaotic sorting and mixing of sediments from either different coastal environments or disparate sediment sizes. It seems that Pacific Ocean tsunamis in Australia cannot be ruled out as a possibility on a millennial timescale. Possible regional trigger sources for tsunamis in New South Wales include seamount formation or submarine faulting in either the Tasman Sea or off western New Zealand. Tsunamis can also be triggered by submarine slumps like the 1929 mudslide at the mouth of the Gulf of St. Lawrence in Canada that broke undersea telegraph cables and caused major damage and human deaths along the coast of Newfoundland, and minor damage >500 km south in Nova Scotia (Tuttle *et al.*, 2004).

12.3 Paleoclimate and paleo-sea levels

In Chapters 3 and 4, we have given several other examples of the combined use of foraminifera and pollen for monitoring of past changes of climate and sea level, which often go hand-in-hand (with the climate warming or cooling being one step ahead of the rise or fall in sea level). Where the changes in proxy indicators are not synchronous, then the sea level shift is likely driven either by paleoseismic events or sedimentary changes in mud deposition and/or erosion. The ability to discriminate between climatic and sedimentological or tectonic forces is an important aspect of refining our present idea that during the last interglacial period (LIG) ~125 000 yr BP, when mean global temperatures were comparable to those forecast for continued modern global warming (Figure 5.2), RSL was 4–9.4 m higher than today because of greater melting from the Greenland and Antarctic Icesheets (Dutton and Lambeck, 2012). The global mean temperature rise that created this greater LIG polar icesheet melting is estimated to be approximately that of the forecasted +1.8 °C by 2090 (IPCC, 2007). In the next sections, we show some applications in which microfossil and sedimentological studies alone are insufficient to reveal short-lived millennial-scale climate changes that occurred near the boundaries of subtropical and warm temperate regions in prehistoric times, and may be happening today in Southeastern Australasia (see Chapter 11).

12.3.1 Los Peñasquitos Lagoon case history: mangroves and immigrations

In a geological timeframe, all lagoons are ephemeral because the vertical growth of marshlands and channel infill eventually leads to their conversion to dry land. The rate of this infill process, however, depends largely on the amount of sediment and peat accumulation in the lagoon. Studies of pollen from three sediment cores in the salt marsh in Los Peñasquitos Lagoon (see Section 7.3) show several crucial features that were not revealed by the foraminifera and ostracods alone (Figure 12.9). This Southern California lagoon is located in a non-glaciated, aseismic sector of the Pacific coast, so RSL changes here reflect climate-induced and anthropogenic impacts. The combined pollen, microfossil, sedimentological and radiocarbon age data show how the wetland environments changed with time during the past ~8000 years, as part of a natural geological aging process that gradually fills an estuary unless sea level continually rises faster than sediment accretion (Scott *et al.*, 2011). The microfossils and sediments show that the lagoon was fully open to the ocean between ~6000 and 5000 yr BP, but they do not indicate any climate change. On the other hand, the appearance of, first, white mangrove pollen (*Laguncularia racemosa*), then also red mangrove tree pollen (*Rhizophora mangle*) means that in this mid-Holocene period, the California winters were 2−3 °C warmer than now, and summers were probably wetter, like those of the subtropical climate in Baja California Sur where these mangroves grow today (see Section 7.4 for present-day distributions). A parallel pollen-microfossil study of mangrove development at Kino Bay in NW Mexico, near today's northern limit of red mangroves, shows a remarkable coincidence of timing (6600–6300 yr BP); here the wetlands growth is also ascribed to warmer, wetter climate conditions (Caballero *et al.*, 2005).

At Los Peñasquitos, the pollen of non-native plants introduced by European settlers provide a means of dating the youngest sediments and calibrating 'bomb spike' effects in the ^{210}Pb measurements used to measure the change in sediment infill rates associated with land clearance around San Diego (Mudie and Byrne, 1980). A series of exotic pollen

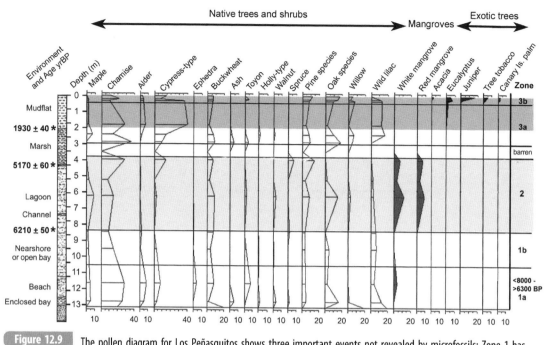

Los Peñasquitos Trees (% Pollen)

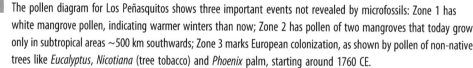

Figure 12.9 The pollen diagram for Los Peñasquitos shows three important events not revealed by microfossils: Zone 1 has white mangrove pollen, indicating warmer winters than now; Zone 2 has pollen of two mangroves that today grow only in subtropical areas ~500 km southwards; Zone 3 marks European colonization, as shown by pollen of non-native trees like *Eucalyptus*, *Nicotiana* (tree tobacco) and *Phoenix* palm, starting around 1760 CE.

markers of known introduction times show that cattle farming had little impact on watershed sediment erosion compared to urbanization of the watersheds around San Diego; in contrast, in unpopulated rural areas like Bolinas Lagoon of Marin County north of San Francisco, the combination of forest clearance and ranching increased sedimentation rate about fivefold (Figure 12.10).

12.3.2 Australasian and Southeast Asian examples

Pollen in radiocarbon-dated cores from the South Alligator River in Arnhem Land, Northern Australia, show that, as in Southern California, mangroves in Northwestern Australia were more extensive during the middle Holocene, around 7000–6500 years ago (Woodroffe *et al.*, 1985). Like Los Peñasquitos, this period of Australian mangrove expansion lasted about 1000 years. However, this time was not interpreted as a warmer period, but as reflecting changes in RSL and sedimentation that were followed by development of a flood plain with tidal channels in areas that today are mid-plain environments. It is predicted that increasing global sea level will restore the inland tidal conditions here in the future. In Sections 11.1 and 11.2, we outlined other applications of mangrove peat and pollen studies to understanding

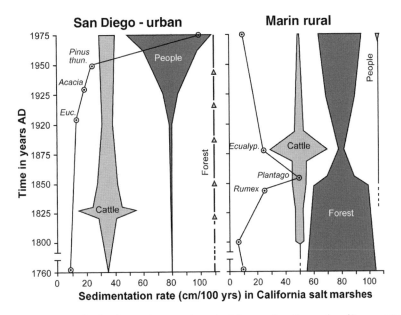

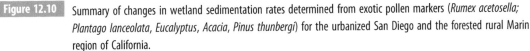

Figure 12.10 Summary of changes in wetland sedimentation rates determined from exotic pollen markers (*Rumex acetosella; Plantago lanceolata, Eucalyptus, Acacia, Pinus thunbergi*) for the urbanized San Diego and the forested rural Marin region of California.

mid-Holocene and recent changes in coastal wetland distributions in Southeastern Australia and Northern New Zealand, particularly as demonstrated by evidence of about 1000 years of expansion from ~6000 to 5000 yr BP.

The wide diversity of mangrove species in Southeast Asia (see Figure 10.20) is also reflected by a large range of mangrove pollen species found in the wetland sediments, which leave a record of the changing floral composition of the mangrove vegetation over thousands to millions of years (Blasco, 1984; Engelhart *et al.*, 2007). Figure 12.21 shows some of the beautiful mangrove flowers and their distinctive pollen grains; Mao *et al.* (2012) illustrate a wider range of pollen types for the mangrove flora of South China.

The distribution of mangrove pollen in Sulawesi (Celebes) was related to tidal inundation ranges in a test of a possible application for tracing past sea level changes (Engelhart *et al.*, 2007). Multivariate canonical component analysis demonstrates that at three sites, elevation is a significant control of pollen assemblages in surface samples; therefore mangrove pollen can be used in a regional transfer function to relocate past sea levels with an error of ±0.22 m. In India, this principle was applied to pollen profiles from Hadi, in Southwestern India near Mumbai (Figure 12.12). The Hadi palynological data show that ideal conditions for mangrove development prevailed during the mid-Holocene between 7220 and 4770 yr BP, which coincides with the Holocene Climate Optimum, when southern Asia had much higher monsoon rainfalls (Limaye and Kumaran, 2012). The mangroves responded to environmental changes due to relative abundance of freshwater, leading to change in species composition. During the late Holocene, there was another distinct change, a *Rhizophora–Sonneratia* transition, before the mangroves declined after 3500 yr BP. Further degradation

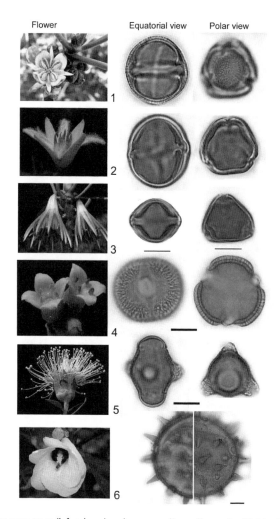

Figure 12.11 Flowers of tropical mangrove trees (left column) and corresponding pollen grains (right column). 1. *Rhizophora apiculata*, 2. *Rhizophora mucronata*, 3. *Bruguiera sexangula*, 4. *Avicennia marina*, 5. *Sonneratia caseolaris*, 6. *Hibiscus tiliaceus*. Scale bar for pollen: 1–4: 10 microns; 5, 6: 20 microns (data from Mao Limi, adapted from Mao *et al.*, 2012). See also colour plate section.

has been attributed to the prevailing arid climate and weakening trends of the monsoon until 1500 yr BP, but there seems to be a positive trend in emergence of mangroves in the least disturbed areas that is attributed to strengthening of summer monsoon in the recent past. Other applied studies use tree rings of mangrove associates to better understand the ocean–atmosphere processes that govern the spatiotemporal complexity of the Monsoon Asia Drought – a system that affects the lives of over half of the world's inhabitants (Cook *et al.*, 2012). New research is also examining possible use of mangrove tree rings as monsoon archives in Gazi Bay, Kenya (Robert *et al.*, 2011), but growth in six species (*Sonneratia alba*, *Heritiera littoralis*, *Ceriops tagal*, *Bruguiera gymnorrhiza*, *Xylocarpus*

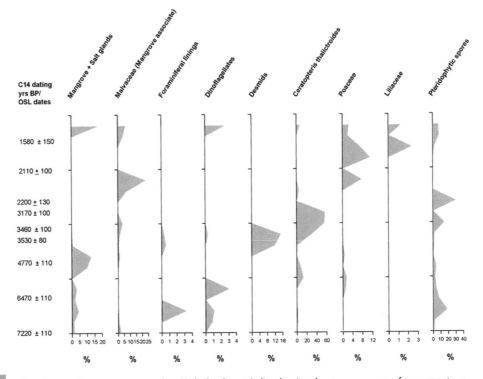

Figure 12.12 Pollen diagram from mangrove marsh at Hadi, Southwest India, showing the strong presence of mangrove trees, dinoflagellate cysts and organic remains of foraminifera until ~4000 yr BP, followed by a shift towards transitional marsh–upland conditions with ferns (pteridophytes) and grasses (Poaceae), before a return to mangrove conditions after ~1580 yr BP (from Limaye and Kumaran, 2012, with permission from Elsevier).

granatum and *Lumnitzera racemosa*) responds to local salinity and flooding in addition to climate, thus requiring evaluation of records on a case-by-case basis. Other mangroves, such as *Diospyros littorea*, however, do record precipitation cycles in multiyear tree-ring widths.

12.3.3 Caveat: the present limits of climate applications

Overall, the study of mangrove pollen in radiocarbon-dated cores from widely separated regions of the world, from Southern California, Amazonia (Section 8.3), Western India, Northern Australia and New Zealand (Section 11.2), reveals that the mangroves became more extensive during the mid-Holocene, around 7000–6500 years ago, regardless of the tectonic setting and ocean circulation. Nonetheless, it is clear that more detailed knowledge of coastal wetland ecology is required before exact temperature ranges can be assigned to fossil mangrove deposits (Ellison, 2009). Refinement of the paleoclimatic reconstructions by further study of mangrove pollen production and selective transport studies (e.g. Kholeif, 2007) and use of quantitative 'training data sets' from modern wetlands studies will play an important role in future evaluation of sea level change dynamics – separating climatic from geophysical drivers of change. Other refinements will come from better understanding of causes for periodic die-offs on stable coastlines which are associated with rising water levels that exceed the peat accretion rates of 1 mm/yr (low islands) to 1.5 mm/yr (mainland shores); it is important that this driver be separated from die-off caused by excess salinity and/or pencil-root burial (see Section 10.3).

13 Applications in conservation of plant biodiversity and agriculture

Key points

Land plants evolved from marine algal ancestors, but only halophytes retained their ancestral salt tolerance; humans have used salt marsh biological resources to their advantage since at least Neanderthal times; coastal wetlands support a wide range of plants useful for food, fibre, oil and medicine; because modern agriculture began in highland areas, early crops were not selected for salt tolerance; crop loss from soil salinization has led to collapse of societies and/or warfare from Neolithic to modern times; agricultural salinization problems have increased as human population growth drives increased irrigation and depletion of groundwater. Coastal wetlands are storehouses of genetic salt specialization, which offer solutions for a new Green Revolution; novel foods and biofuels from wild halophytes allow desert agriculture, aiding the need to conserve freshwater, which is considered the greatest challenge to human survival.

13.1 Salt of the Earth

In Chapter 6 we discussed the role of long-term changes in the distribution of Earth's continents and seas, which has shaped the diversity of coastal wetland floras over the past >66 Ma. In fact, environmental salt is a fundamental legacy of Earth's evolution for more than 550 million years. As far back as Pangean time when there was only one supercontinent, there have been major sea level changes like those described in Chapter 2, and these have required ongoing adaption to alternating terrestrial and tidal regimes for coastal plants. Both aquatic algae and land plants (excluding mosses) responded to these shoreline changes during their evolution by developing salt-tolerance mechanisms across a wide range of taxonomic groups (Plants in Action, 1999). This convergent evolution in various mangrove trees, intertidal salt marsh and salt desert halophytes is manifest as similar mechanisms for salt exclusion by roots, salt excretion by leaves and as compartmentation of salt within fleshy leaf or stem tissue across widely separated taxonomic groups (Chapter 3). Other less well-developed mechanisms for salt exclusion and cellular compartmentation are also found in some non-halophytic wild plants and their cultivated varieties. This genetic variation in salt tolerance can be exploited for genetic improvement of commercially significant species or cultivars, especially when the mechanisms and inheritance of the salt tolerance is well understood.

In many regions of the world, secondary salinization of soils, in addition to direct marine influence, follows intensive land clearance, overcropping or grazing of drylands, particularly in areas of low summer rainfall and where salt remains from former marine seaways. For example, the Caspian Sea, the site of the first Ramsar Wetlands Convention (Box 1.1 The Ramsar Convention) in Iran, Central Asia, is a relict ocean left over from the end of the Tethys Seaway ~40 Ma, but it still has a water salinity of 13 psu. In Late Cretaceous time, there was also a seaway stretching from the Gulf of Mexico to the tar-sands of Canada (Figure 6.6); hence, parts of the American Prairies also have relict saline soils. In addition to these natural salt sources, irrigation agriculture frequently results in soil degradation because of salts in water pumped from rivers that have crossed salt-bearing bedrock. Salt also accumulates in soils where drainage is inadequate, preventing leaching of soil salts (FAO/UNESCO, 1973). Salt-tolerant cultivars and salt-excluding rootstocks allow cropping under those circumstances, and are of enormous economic significance throughout the world, not just in deltas and lowlands, which are occasionally inundated by storm tides.

With the expansion of cropping areas worldwide, the availability of easily accessible arable land has diminished so that multiuse farming practices and reclamation of salt-affected land have become expedient or essential. Soil amelioration is technically feasible, but it is costly and takes decades to centuries to achieve. Salt-tolerant plants are a valuable asset to either of these farming processes, and their inherent halophytic usefulness, which was once confined to natural coastal wetlands, has now extended to various intensively managed landscapes.

13.2 Biodiversity of salt marshes and mangrove swamps

13.2.1 Traditional uses of coastal wetlands

Archeological records of large shell middens (= food refuse heaps) on the edges of salt marshes and mangrove swamps (Figure 13.1) in both tropical and temperate regions of the world testify to their traditional use as a source of food, including shellfish, fish, birds and sea mammals (Mudie and Lelièvre, 2013). Shell middens are present from Arctic to tropical regions worldwide and they have been dated to as far back as 140 000 yr BP – long before the first anatomically modern humans (*Homo sapiens*). These shell middens are of great interest to archeologists because they are archives of past sea levels, human dispersal, trading and specialization in coastal resource management (Álvarez *et al.*, 2010) and they can be archives of drought cycles and human diets (Culleton *et al.*, 2009). For example, in Chapter 10 (Box 10.1 Ancient Troy), we showed the geoarcheological history of salt marsh and shellfish usage in the once-powerful city of Troy (Troia) from ~4500 to 2000 years ago, when trade with Greece, Georgia, Lebanon and/or Egypt was vigorous; today, this location is just a small, isolated farm village far inland. It is also notable that Western agriculture is based on grain and pulse crops first cultivated in the highlands of Iran from ~12 000 to 9800 years ago (Riehl *et al.*, 2013), which spread from there to the warmer, drier Mediterranean coasts and the Mesopotamian Valley. Early droughts and soil salinity required the development of hydraulic irrigation systems in these drier regions (Roberts, 1998), and some of these irrigation methods spread far eastwards along the Silk Road to deserts of Northwest China,

Figure 13.1   Top: a modern shell midden in the shore of the Gambia River estuary is over 2 m high; note the women's dug-out canoes used to navigate the mangrove swamps seen in the background (photo by P. J. Mudie); Bottom: canoes in Belize are used for travel in red mangrove forest where propagules are collected and cooked for food; in Mayan time, canoes transported salt from the coast to inland cities (photo by Helen Pease).

where today, grapes are still grown alongside underground tiled aquifers of traditional Persian design. Some of our most salt-resistant crops (e.g., barley and tall wheatgrass) have their ancestors in salt marshes and salt deserts of these Middle East regions. In the Nile valley, irrigation by shaduf and canal dates back over four millenia (see Section 9.4). In contrast to these 'hydraulic civilizations' of West Asian origin, which culminated in the vast diked lands of the Netherlands and France, the East and South Asian 'aquatic civilizations' were founded largely on the impoundment of freshwater, e.g. as rice paddies and terraced hillslopes. As a consequence, the least salt-tolerant crops are found among the Asian staples of rice and soybeans. Less is known about the history of salt in crop development of the Mayan civilizations of Central America, but it is certain that coastal salt production and its transport inland by canoe was an important industry during the Late Classic era, 600−900 CE (McKillop, 2005).

13.2.2 Historical uses of coastal wetland plants

In Chapters 7 through 10 we give examples of the direct uses of coastal wetlands and mangroves made by people today – from the common European tradition of grazing goats,

Figure 13.2 Traditional uses of wetlands by New and Old World bovines. Cows and cattle introduced by Spanish ranchers graze on mangroves at Puerto Chale (top, photo by P. J. Mudie) and at Laguna Mormona (centre, photo by Hester Bell) in Northwest Mexico; in Borneo, feral water buffalo live in fringe mangroves; wild buffalo are probably extinct or inbred (photo by P. W. Mudie).

sheep and horses in high marshes of coastal lagoons (Chapter 7) to the Amazon crab-hunters (Chapter 8), the oyster women of West Africa (Chapter 9), riverside homes on the Danube and Tigris–Euphrates reed-swamps to the floating markets of Southeast Asia (Chapter 10). Additional examples and illustrations of 'wetlander' human cultures are given for North America and for freshwater wetlands by Mitsch and Gosselink (2007) and Keddy *et al.* (2009). Interesting modifications of these traditions are found in Baja California, where the non-native cow (introduced by Spanish settlers) grazes on both mangroves and salt marshes (Figure 13.2), while in Borneo, domesticated water buffalo escape and hide in wetland fringe forests, the home of their now extinct ancestors.

Historical uses of mangroves as sources of food, medicine and construction materials, including the keels of prehistoric ships in the Persian Gulf area, are reported by Gerald Walsh (1977), and Paul Keddy (2010) gives a concise account of the ways that coastal salt marshes have been used and altered by humans since Roman times (~2300 yr BP). However, less is known about the traditional uses of salt marsh plants and mangroves as direct food sources of vegetables and plant protein (Mudie *et al.*, 2005b). Recently, the traditional knowledge of North American aboriginals reveals that the succulent halophytes *Salicornia* and *Sarcocornia* – the dominant plants in most intertidal salt marshes from the subarctic to the tropics – are a favoured plant food among many aboriginal people, sometimes being exchanged as special gifts in addition to their regular use as a salad or a salty pickle (hence, the common name 'pickleweed'). In a unique forensic archeopalynology study of pollen grains in the stomach and intestines of a prehistoric frozen man, it has been shown that *Sarcocornia perennis* (locally called beach asparagus) was used as a travel food, source of vitamin C and/or a medicine on journeys from the coast of Alaska across the glaciated coastal mountains of the southwest Yukon, Canada, long before European colonization of the region in the Gold Rush era (Mudie *et al.*, 2013).

In Britain, the annual *Salicornia europaea* and related species are called 'samphire' – from the French words, *herbe de Saint Pierre*. Marsh samphire is a still prized salad delicacy that is harvested seasonally from salt marshes in traditional mudflat commons along the coast of Norfolk. These juicy, slightly salty greens were served to the British Royal Family at the brunch following the famous wedding of Prince Charles and Princess Diana in June 1981. Collection of samphire can be dangerous in areas of tidal bores, such as the Severn and Humber Rivers, eastern England, where untrained immigrant labourers are sometimes trapped by the tidal bore. However, it is important not to confuse the obligate halophytic samphire of the genus *Salicornia* (Family Chenopodiaceae) with another coastal plant *Crithmum maritimum* (now called *Cachrys maritima*), which is also called samphire (or rock samphire). *Crithmum* belongs to the carrot/fennel family *Apiaceae* and its salty leaves have a slight flavour of anise. *Crithmum* is a facultative halophyte, meaning that it can grow optimally without any sea salt (sodium chloride) – in contrast to the salt-dependent *Salicornia* samphire which cannot grow fully without salt. Rock samphire is actually very easy to distinguish from salt marsh samphire because *Crithmum maritimum* has attractive sprays of yellow flowers that become conspicuous anise-shaped 'seeds' in the fall, whereas the flowers of annual *Salicornia* species are barely visible and the seeds are buried within the fleshy stems (Figure 13.3).

13.3 Agriculture and soil salinity: past and future problems

Salinity, like drought, remains one of the world's oldest and most serious environmental problems. Mistakes made by the Sumerians in the Tigris and Euphrates basin of Mesopotamia over 4000 years ago are being repeated today...

(McWilliam, 1986)

Although geologic timescales of marine and terrestrial salinization were slow enough to allow the natural evolution of salt-tolerant varieties of algae, ferns and flowering plants, human population growth has greatly accelerated soil salinization during the past two centuries. As a result, bioengineering of salt-tolerant plants is now a matter of urgency for both food production and coastal wetland reclamation (Plants in Action, 1999). Today,

Figure 13.3 Comparison of flowers in *Salicornia europaea* (left), and *Crithmum maritimum* (right). The *Salicornia* flowers have no petals and are buried inside the fleshy stem and leaves. The *Crithmum* flowers are bright yellow and easy to see (photos by P. J. Mudie). See also colour plate section.

Figure 13.4 Salt accumulation (white soil) in fields near the Salton Sea after 15 years of irrigation, and FAO 2012 data (graph) showing worldwide rise in irrigated farmland over the last century (photo by P. J. Mudie).

almost one billion hectares of land (>6% of the Earth's land surface) is salt-affected, along with about 20% of the irrigated land that produces a third to half of the world's food supply (Munns and Tester, 2008; FAO, 2012a; Figure 13.4). Most of the potentially arable soil still unfarmed in the world is saline to some degree and the water available for its potential irrigation is also salty. Historically, early agricultural civilizations in the Middle East, India and China ended when the groundwater became too saline for irrigation (Roberts, 1998), even in the presence of vast rivers like the Tigris, Euphrates and Indus. In the USA, irrigation in the Colorado Desert and San Joaquin valleys led to soil salinization on ~25% of the farmland within 25 years, because each year of irrigation transferred the equivalent of 57 railroad car-loads of salt to the hot desert region (Letey, 2000). Overall, since the 1960s, the presence of excessive amounts of salt in irrigation or soil water has become a major threat to the permanence of irrigation agriculture throughout the world (Mudie, 1974; Munns and Tester, 2008).

Despite these historical problems, the expansion of irrigation agriculture in the arid and semi-arid regions of the world is still considered to be one of the most important possibilities for increasing food production in the near future (Rains and Epstein 1967; Flowers, 2004). Furthermore, as the human population approaches 8 billion – twice as much as the 4 billion in 1974 when the spotlight was first turned on halophytes as food plants of the future – agricultural scientists now realize that another 'green revolution' is needed for crop yields to meet food demands of the future (von Caemmerer *et al.*, 2012). The feasibility of using slightly saline urban and agricultural waste water, large underground reservoirs of brackish water, or even seawater, for cultivation of economic plants also presents challenging and important opportunities for conserving dwindling freshwater supplies – without which humankind cannot survive (Ramsar 2013: Year of the Wetlands and Water).

The ability of salt marsh and mangrove halophytes to grow well in highly saline soils provides opportunities for combating salinity problems and for expanding agriculture into regions of limited freshwater. Consequently, in the last few decades there has been a major effort to inventory the plant biodiversity of formerly spurned 'dismal, mosquito-ridden coastal wastelands' in the search for obligate and facultative halophytes that might provide direct food sources in the future – or that might be sources of genetically determined salt tolerance. In the simplest terms, obligate halophytes are salt marsh plants that must have a salinity of ~15 psu or more for optimum growth and which cannot survive without sodium chloride (NaCl, the main mineral element in seawater), while facultative halophytes can grow well at lesser salinities, with or without NaCl (see Waisel, 1972; Flowers and Colmer, 2008; Aslam *et al.*, 2011 for details of this complex subject). Lists of halophytes and their known uses as food, medicine, domestic animal forage and other values to people are given by Mudie (1974) and Chapman (1974) in global surveys, by Jaradat (2003) for sustainable biosaline farming in the Middle East, by Giesen *et al.* (2007) for Southeast Asia, and for Australasia, by the Australian and New Zealand Plant Society programme 'Plants in Action' (1999). The concept of sustainable agriculture with so-called 'cash-crop halophytes' irrigated with saline water (up to seawater salinity) is being explored at the same time that halophytes are being used as model plants for the development of salt-resistant crops from current salt-intolerant conventional crops (e.g. Kingsbury *et al.*, 1976; Glenn *et al.*, 1999; Aslam *et al.*, 2011).

Commercial farming of halophytes has now been tested for vegetable production, forage and oilseed crops in agronomic field trials. The most productive species yield 10 to 20 tonne/ha of biomass on seawater irrigation, equivalent to that of many conventional crops. The oilseed halophyte, *Salicornia bigelovii*, yields 2 tonne/ha of seed containing 28% oil and 31% protein, similar to soybean yield and seed quality, while other species of *Salicornia* are commercially grown in Egypt and India as sources of medicines, including treatment of tuberculosis. In the tropics, the mangrove *Barringtonia racemosa* has a long tradition of use as an anti-inflammatory and anticancer drug; it is currently being investigated as a source of the antioxidant lycopene that protects animals against damage from free radicals and suppresses the growth of cancer cells in laboratory studies (Behbahani *et al.*, 2007). The most novel development is a project in which aerospace engineers (Boeing and Etihad airlines) have teamed up with United Oil Products Honeywell to use effluent from shrimp farms to grow *Salicornia* for biofuel production in the coastal desert of the United Arab Emirates (AEM, 2010).

13.4 Salt marsh biodiversity: emerging studies of halophyte genomes and ionomes

Mining of the biodiversity of plants is currently 'a revolution in the making' because only a small fraction of the diversity of plant metabolism has been explored for production of new medicines and foods important to human wellbeing (De Luca *et al.*, 2012). Halophytes are of particular interest because of their ability to tolerate salt concentrations that kill 99% of other plant species (Flowers and Colmer, 2008; Munns and Tester, 2008; Conn and Gilliham, 2010). Earlier experiments aimed to understand the genetics of salt tolerance in halophytes in the hope of being able to isolate and capture salt-tolerance genomes from wild varieties of domesticated beets, barley, wheat and tomatoes that grow on the edges of salt marshes or within the wave splash zone of ocean cliffs (Mudie, 1974; Kingsbury *et al.*, 1976; Glenn *et al.*, 1999). For example, much effort has been made to increase salt tolerance in rice, which is the staple food for over half the world's population. Archeological and genetic studies of rice indicate that the lower Yangtze River area of China was the evolutionary centre of rice cultivation (Zong *et al.*, 2007). Recent paleo-ecological studies of pollen, algal and fungal spores in sediment cores near the coast of Hangzhou Bay now show that the earliest agricultural experiments on rice cultivation were made in brackish water marshes ~7700 yr BP, including bunding (making low dikes) to prevent flooding by brackish or seawater. Thus domestic rice (*Oryza sativa*) has been subject to more than 7000 years of selection against salt tolerance, but the remnant salt marshes of the Lower Yangtze or other Southeast Asian estuaries may support salt-resistant wild-types which could be used in cross-breeding for increased salt tolerance (Figure 13.5). Likewise, there is interest in the development of salt-tolerant soybeans from the wild-types remaining in the Yangtze River delta. Another good example of the potential usefulness of wild facultative halophytes comes from programmes to increase salt tolerance in tomatoes (Läuchi and Epstein, 1984). The cultivated tomato (*Lycopersicon esculentum*) is very salt sensitive, but it has a wild relative (*Lycopersicon cheesemanii*) that grows in the salt spray zone of sea cliffs on the Galapagos Islands. Cross-breeding of the two species produces a 'cherry' tomato which has a significantly higher tolerance to high levels of sea salt (Figure 13.5).

Figure 13.5 Genetic experiments with halophytic plants. Top left: salt-tolerant Galapagos Island tomatoes with tiny yellow fruits have been hybridized with the commercial tomato species to produce an intermediate-sized 'cherry' tomato (Läuchli and Epstein, 1984); Top right: a selection of salt-tolerant cultivars available today (photo by Shawna Henderson); Bottom left: field trial of salt-resistant hybrid rice (photo by IRRI); Bottom right: ancestors of the new salt-resistant forage crop, tall wheatgrass (*Thinopyrum ponticum*), come from the remnant salt marshes on the Turkish shores of the Black and Marmara Sea (photo by P. J. Mudie). See also colour plate section.

The earlier gene transfer and hybridization work had limited success, however, because of the lack of understanding of the diversity of metabolic pathways in halophytes (C_3, C_4 and CAM pathways), and limited knowledge about the physiology of halophytes and the genetic basis for their adaptation to variable salinity. However, development of new ultra-sensitive miniprobes for measuring sodium concentrations and pathways of transport within cells are revolutionizing the understanding of halophyte physiology. Furthermore, a gene for sodium resistance (*Nax2*) in the germplasm of ancestral wheat varieties (*Triticum monococcum* vars) has now been successfully transferred to the salt-sensitive commercial durum wheat (*Triticum turgidum* ssp. *durum* var. *Tamaroi*) by cross-breeding; this genetic experiment resulted in a 25% increase of durum wheat yields in field trials with saline soils (Munns *et al.*, 2012). The search for increased salt tolerance in conventional grain crops like wheat and rice, which are mostly C_3 plants with mechanisms to exclude salt uptake, must be weighed up against advantages to be gained from exploiting more efficient, heavy-yielding C_4 grasses like *Spartina* and *Sorghum* (millet) – and ongoing research aims to identify the genes that would be required to transfer C_4 photosynthesis to rice (von Caemmerer *et al.*,

2012). Unfortunately, the challenges involved in capturing and transferring the genes controlling the CAM metabolism of obligate halophytes like *Salicornia* are much greater, and will probably limit further development of the economic usefulness for these extremely salt-tolerant plants.

Development of inexpensive, high-throughput methods for sequencing of plant proteins, however, has led to the fast unravelling of previously unknown enzymes and metabolic pathways in wild plant populations (de Luca *et al.*, 2012). This accelerated technology and DNA sequencing of halophytes are creating huge libraries of biologically active products and their genetic affiliations that can be screened for potential new drug applications. For example, studies of DNA sequences were recently made in the yellow mangrove *Ceriops tagal*, which is a valuable source of durable hardwood and a substitute for quinine in the treatment of malaria. The DNA sequences of both chloroplasts and nuclei across the geographical range of the yellow mangrove show a low level of variation compared to the genetic diversity within a region (Huang *et al.*, 2012). This information is important in planning to protect the genetic diversity of this inner marsh mangrove, which is threatened by human populations encroaching on the margins of tropical coastal wetlands.

Focus on the elemental composition of plant tissues, called the ionome, and the way the so-called 'ionome' of halophytes changes with shifts in soil salinity may lead to better understanding of how soil chemistry can shape plant genetics (Munns and Tester, 2008; Conn and Gilliham, 2010; Milo and Last, 2012). This understanding is predicted to provide the solutions to agricultural challenges like salt intrusion in coastal regions and salt build-up in irrigated soils far removed from the ocean, such as the Canadian Prairies and Russian Steppes. These novel approaches to solving the growing demand for greater food production are very important, given that together, seawater and brackish water make up 98% of the world's water supply.

Using mesocosms as a way to study coastal wetlands

Key points

Anthropogenic impacts have destroyed many salt marshes and mangroves; we are now trying to rebuild them for the ecosystem services they provide; although individual studies and experiments may not answer all questions, they provide valuable insights to effective restoration means; appropriate experiments lead to best methods for achieving high success rates; mesocosms can provide information on effects of future impacts from sea level rise, pollution and biological invasions; more global collaboration with experimental efforts is required to reduce wasted time, energy and finances on overlapping 'trial-and-error' experiments and evaluators of success; examples are given of various coastal wetland restoration and construction projects worldwide; there is also strong need for individual research teams to search current subject literature and collaborate with multidisciplinary teams to achieve the best outcomes efficiently and economically.

14.1 Why make experimental studies in coastal wetlands?

There are many ways to study coastal wetlands, for just as many purposes. Experimentally, microcosms, mesocosms, whole-system studies (i.e. *in situ*) and even mathematical models (defined in Table 14.1) can give detailed information on the ecology, sedimentology and hydrology of a salt marsh or mangrove system, answering specific research questions that might be missed in basic observational field studies. The purposes of experimental work include pollution impact and remediation, creating and restoring salt marshes, impacts of biological invaders, to modelling effects of sea level rise. This chapter introduces the principles of mesocosm studies and some of the experimental work done in coastal wetlands, giving various global examples.

Mesocosms (from m^2 to ha) have been used in ecological research for decades and accommodate more ecological complexity than microcosms, while still being a controlled setting. They are the mid-point between the oversimplified microcosms and the whole-ecosystem studies. Eugene Odum clearly writes: 'The mesocosm provides an environment where parts (populations) and wholes (ecosystems) can be investigated simultaneously by a team of researchers' (Odum, 1984, p. 558). Mesocosms are relatively easy to study in a short time period and for little cost, whereas whole-ecosystem studies often require teams of researchers, long time periods and a permanent, often undisturbed study location (Ahn and Mitsch, 2002). Some small mesocosm studies include those of Erik Kristensen and colleagues

Table 14.1 Main types of ecosystem experimental studies		
Term	Definition	Example/Reference
Microcosm	A small (centimeter-scale) artificial and very simple replicate of a natural system. Can be either opened or closed to external conditions. Often testing small parameters under greater control than mesocosms.	David Latimer's 'garden in a bottle' (de Bruxelles, 2013); small cultures in petri dishes or aquaria.
Mesocosm	A relatively small (m^2 to ha) replication of a natural setting under controlled conditions; often made replicable for statistical experimentation. Can occur in both the laboratory and in natural settings in enclosed areas (*in situ*).	Gribsholt and Kristensen, 2002; Frail-Gauthier *et al.*, 2007.
Whole-system studies	Similar to an outdoor *in situ* mesocosm, on a much larger, 'entire ecosystem' scale and includes most, if not all, natural processes.	Studies done in entire lakes, salt marshes, islands, etc. See Mitsch and Day, 2004; Zedler and West, 2008
Mathematical modelling	Using mathematical equations to predict scenarios and carry out simulations of the natural environment on a theoretical scale.	Weigert's models of Sapelo Island, Georgia; Chen and Twilley (1998) model of Florida mangroves basal area (see Mitsch *et al.*, 2009 for more reading).

studying the effects of biota on sediment biogeochemistry over a few months, including mangroves (Kristensen and Alongi, 2006) and salt marshes (Gribsholt and Kristensen, 2002). Larger, more permanent studies include those by Joy Zedler and colleagues for over 30 years in the Tijuana Estuary of Southern California (see Section 7.3.1.6; Zedler and West, 2008). Both small mesocosm and large ecosystem studies have been carried out at the same location by William Mitsch and colleagues at the Olentangy River Wetland Research Park at Ohio State University, USA (Ahn and Mitsch, 2002; Mitsch *et al.*, 2009).

Conflict between proponents of mesocosms and large-scale ecosystems have led to a simultaneous mesocosm study comparing small, $1\,m^2$ wetlands to a larger experimental marsh in a $10\,000\,m^2$ large research park (Ahn and Mitsch, 2002; Figure 14.1). Although there are many similarities between small mesocosms and large wetlands, the conclusions drawn from mesocosms cannot be extrapolated to whole ecosystems without appropriate scaling considerations (Ahn and Mitsch, 2002). Overall, whole-ecosystem studies yield better results, as they do not purposely simplify the system, but mesocosms can provide useful information on various effects and processes, especially when the predicted outcomes are unknown and could potentially lead to harm of the natural system. If the limitations of mesocosms are taken into account, in that they cannot always be applied to entire ecosystems, they still provide valuable information on processes and functions within the coastal

Figure 14.1 A: large-scale wetland ecosystem (10 000 m²) and B: small-scale mesocosms (1 m²) from Olentangy River Wetland Research Park at Ohio State University, Ohio, USA, to study the scaling effects between small and large experimental systems (from Ahn and Mitsch, 2002, with permission from Elsevier).

wetland (Mitsch *et al.*, 2009). Mesocosms are particularly important in studies of impact assessments, understanding natural ecosystems, and for the testing of various parameters such as changes in temperature and salinity (Kangas and Adey, 1996). The advantages of using mesocosms as model systems include being able to replicate and reproduce multiple scenarios and to control environmental variables (Lawton, 1995), although it is necessary to strike a balance between nature and theory. As long as the limits of mesocosms are known, they can be extremely useful in studying numerous questions in ecological settings, such as those in salt marshes.

The distinctive characters and spatial distributions of coastal salt marshes and mangroves make them highly suitable ecosystems for scientific research and they have been used for many hydrological studies, chemical analyses, energy production and flow, geology and paleoclimatology, conservation and biological studies (species compositions and zonations, ecology and food webs, etc.). Salt marshes are relatively well understood and well constrained in regards to species diversity and nutrient dynamics, making them fairly simple ecosystems to study and manipulate (Wiegert and Freeman, 1990). As such, salt marshes have been used as model systems for decades, no matter what the discipline of research. Most studies of salt marshes, however, are restricted to field studies and manipulations because of the difficulty in simulating a low-energy tidal system in the laboratory. Field studies of temperate and polar marshes are even more challenging because they are frozen over for parts of the year (see Box 14.1 The value of indoor salt marsh mesocosms). Because of the intertidal location and soft mud substrates flooded daily, many marshes are also

Box 14.1   **The value of indoor salt marsh mesocosms: example from Halifax, Nova Scotia, Canada**

In northern climates, such as Nova Scotia, Canada, the marshes freeze over in the winter, covered in snow and ice, although they recover quickly after the spring melt. Therefore, year-round and long-term studies of the biota in the salt marsh are restricted, leaving large seasonal gaps in data. To solve this problem, an indoor salt marsh system was designed by Frail-Gauthier *et al.* (2007) in a saltwater facility at Dalhousie University, Halifax, Nova Scotia, Canada. This unique facility pumps and filters seawater from the local Halifax Harbour of the Atlantic Ocean (few universities and intuitions can provide continuous flow of ambient seawater into a laboratory setting). As a result, this 'temperate' indoor marsh can be sampled anytime, regardless of the season.

The primary aim of the Dalhousie laboratory salt marsh is to create and simulate the natural marsh system from a cool temperate, well-studied area like Chezzetcook Inlet (Box Figure 14.1) so that frequent and regular sediment samples can be analysed year-round. Initial designs determined missing parameters of the system, until almost all environmental parameters were controlled (six-hour diurnal tide cycles, continuous freshwater input, 8–12 h UV light cycles, and ambient lab temperatures). The only thing not controlled is the temperature of the freshwater and seawater entering the system, which ranges from 23 °C in summer to slightly less than 10 °C in winter, therefore never freezing. Mudflats, low marsh, middle marsh and high marsh are scaled within a 2.5 m long × 0.5 m wide tank, with the heights of the marsh segments (40 cm tall for high marsh to 5 cm tall for mudflat) reflecting their elevations above sea level and approximate exposure time during tidal cycles. These marsh segments were taken from two previously studied transects at Chezzetcook Inlet, based on *Spartina* zonations. Both a high-salinity outer marsh transect (closer to the mouth of the inlet) and a low-salinity inner transect (near the head of the inlet) are replicated in the laboratory (Box Figure 14.2). The low-salinity transect receives a continuous freshwater input to the high and middle marsh subsegments. No freshwater is received in the high-salinity transect. Salinities range from 23 to 29 psu in the high-salinity mesocosm, and from 20 psu (mudflat) to approximately 4 psu (high marsh) in the low-salinity mesocosm. Despite their laboratory location and environmental control, the marshes still replicate seasonal cycles of vegetation. The *Spartina* grasses die off during the winter months (November to March) and return as shoot and leaf growth during the spring and summer. Faunal changes also follow these seasonal trends; after two years, seasonal foraminiferal and flora assemblages resemble the seasonal data collected from the natural marsh at Chezzetcook Inlet. Extremely detailed data have been collected on foraminiferal assemblages and their associated meiofauna over continuous biweekly sampling, which is otherwise impossible to do in this cool temperate (boreal) environment. This laboratory mesocosm marsh also supplies living meiofauna and foraminifera used for biological studies and experiments in microcosms to aid in detailed paleoenvironmental interpretations of these microfossils in sedimentary archives (see Chapter 12; also Frail-Gauthier *et al.*, 2011).

Although this Dalhousie University laboratory marsh primarily is used to study the time series assemblages of foraminifera and other associated meiofauna, in principle, a laboratory mesocosm marsh like this can be applied in a wide variety of disciplines (examples given throughout this chapter). Variables that cannot be completely controlled in a natural marsh can be avoided in such systems, including, but not limited to, migrations and dispersal, storms, intense weather (rain or drought), invasions and anthropogenic pollutants. Once in a controlled environment, key parameters can be manipulated and their exact cause and effect can be determined. Pollution and remediation studies can be done without damaging the natural environment. Studies of invasive species can be carried out without introducing foreign invaders to the native marsh itself. Physical parameters, such as daylight, salinity and tidal heights, can easily be changed and their resulting effects monitored.

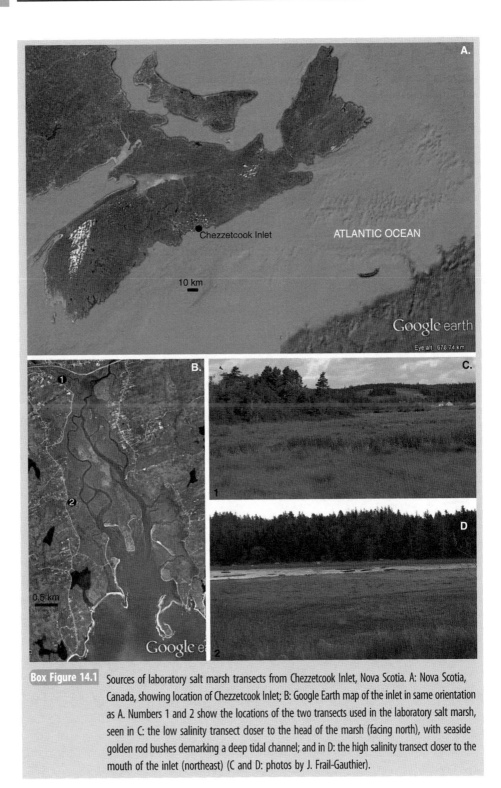

Box Figure 14.1 Sources of laboratory salt marsh transects from Chezzetcook Inlet, Nova Scotia. A: Nova Scotia, Canada, showing location of Chezzetcook Inlet; B: Google Earth map of the inlet in same orientation as A. Numbers 1 and 2 show the locations of the two transects used in the laboratory salt marsh, seen in C: the low salinity transect closer to the head of the marsh (facing north), with seaside golden rod bushes demarking a deep tidal channel; and in D: the high salinity transect closer to the mouth of the inlet (northeast) (C and D: photos by J. Frail-Gauthier).

Box Figure 14.2 Laboratory mesocosms of the Chezzetcook Inlet marsh. A: side-view of the two transects, with high marsh to the right. B: system at low tide, C: system at mid-high tide. Both show the low marsh in the foreground (photos by J. Frail-Gauthier).

challenging environments for obtaining frequent, and most importantly, regular sampling required for quantitative analyses, and this limitation makes cohesive, comparable studies difficult. Whether *in situ* or in the laboratory, mesocosm experiments provide valuable information for examining physical parameters for restoration or sea level rise, effects of changes within the sediment, pollution studies and biological interaction studies. While whole-ecosystem studies may yield the most accurate results, they are the most challenging in terms of research team, facilities and funding.

Despite the aforementioned drawbacks and limitations to small mesocosm experiments, their results yield much potential to studies of coastal wetland restoration and creation, mitigation of anthropogenic impacts and combatting sea level rise, especially in times of intense coastal development and coastal wetland loss.

14.2 Examples of coastal mesocosms: introduction

As early as the 1970s, salt marsh mesocosms were developed in both indoor and outdoor settings to answer various research questions. After 40 years, there are enough studies to fill an entire book, so this chapter aims just to introduce the reader to the method, using examples providing a global perspective versus the highly emphasized studies in the USA (i.e. see Mitsch *et al.*, 2009; Silliman *et al.*, 2009). Many of the experiments aim to study vegetation recruitment and growth of *Spartina* as one of the main environmental engineers of a salt marsh and highly important to restoration studies, which dominate ecoengineering experiments in coastal wetlands (for example, see Section 10.4.4). Macrophytes, particularly tall grasses (*Spartina*) and pioneer mangroves (*Rhizophora*) are often the focus in experimental work for many reasons, as they are one of the most visible determinants of coastal zonation and also visible indicators of coastal wetland health. As ecosystem engineers, once the plants colonize, their presence leads to increased sedimentation and further tidal wetland accretion for spread of mangroves and salt marsh. The physical and biological thresholds of many coastal wetland ecosystem engineer plants are well known, so monitoring their growth and assemblage changes is a good way to track overall changes in the ecosystem: from success/failure of restoration studies, to physical changes such as sea level rise. For example, rudimentary 'wave' studies in sloping sand troughs by Gleason *et al.* (1979) examine sediment retention by *Spartina alterniflora*, emphasizing the crucial role of these grasses in creating and maintaining early salt marsh colonization and expansion. In the natural setting of Southern California, Joy Zedler and colleagues have created marshes in the Tijuana Estuary at the Pacific Estuarine Research Laboratory (PERL), to better understand management and restoration of a valuable estuarine preserve (Zedler *et al.*, 1992; Zedler and West, 2008). Nonetheless, the importance of further use of salt marsh mesocosms in restoration projects is emphasized and recommendations for future studies are explained in detail in Callaway *et al.* (1997), as mentioned later in this chapter (Section 14.3).

Another example of an indoor marsh creation project is work by Padgett and Brown (1999) using a very similar design to the Chezzetcook laboratory marsh (Box 14.1 The value of indoor salt marsh mesocosms), including continuous seawater inflow and a timed drain to simulate tidal cycles. They wanted to determine the effect of soil drainage and the amount of organic content on the growth of *Spartina alterniflora*. Instead of using segments from a local salt marsh, they started their study with clean sand and peat moss in varying percentages to manipulate substrates for *Spartina* seedlings. In these manipulations, they studied the growth of the vegetation and recruitment of invertebrate infauna to gain insight to marsh creation and restoration. They concluded that soil organic content is crucial for the recruitment and growth of a new salt marsh (Gleason *et al.*, 1979).

Studies carried out in marsh mesocosms, whether laboratory or *in situ*, are also important for contemporary environmental monitoring and development of an ecosystem that has been greatly degraded over the last few hundred years. Whether the mesocosms focus specifically on one research question, or they are used in integrated observational studies similar to the Chezzetcook marsh work, they reduce the impacts imposed by field monitoring a natural system, while answering questions about them. Studying the effects of various changes in a laboratory setting can help predict and mitigate contemporary problems and also better interpret past (historical and geological) changes to these vital, yet sensitive, coastal ecosystems. Here, some major themes of experimental work in marshes are introduced: (1) coastal wetland restoration and construction, including vegetation experiments, (2) measurement of various physical effects, e.g. sea level and salinity, (3) biological effects of invasions and (4) various other studies, including effects of pollution and concerns for human health. However, these four themes are often studied in combination. For example, when considering restoration and marsh construction projects, the biogeochemical and physiochemical properties of the sediment play a large role in determining if plants can grow and the restoration project will be a success (Lara *et al.*, 2009).

14.3 Restoration and construction

Much of the experimental and mesocosm work in coastal wetlands is done with a purpose for construction or restoration. Mitsch (2000) gives many examples of large coastal wetlands projects that have been underway in the USA for the last 20 years, including the Delaware Estuary, the Florida Everglades, the Louisiana coast and in the Gulf of Mexico. It is public knowledge that there is urgent need to restore and maintain valuable coastal wetlands that provide us with invaluable ecosystem services, from flood and erosion protection, to carbon retention, to high amounts of biological productivity (see Chapter 5). Wetlands can be restored or created in order to create agricultural land, as a way to stabilize dredged materials for benefit of waterfowl, to prevent or reduce shoreline erosion and to accumulate sediment for defence against rising global sea level (Broome and Craft, 2009). A good example of the ecoengineering benefits of marsh construction and agriculture is given by the Chinese Yangtze River salt marsh project outlined in Section 10.4.4.

There is a clear differentiation between created wetlands and restored wetlands. Wetland creation changes the landscape so that a coastal wetland develops where it did not exist in the recent past. In restored wetlands, one 'simply' attempts to re-establish the functions of a wetland that existed in the location in the past (Broome and Craft, 2009). The salient question behind many restoration and creation studies is what makes it successful? Is success just physically making a new wetland, or is it successful when it has achieved the same functions (physical, chemical and biological) as natural, undisturbed wetlands? The length of time is also important, as some short-term evaluations may study all aspects of the restoration (e.g. hydrology, soil biogeochemistry, biotic interactions), but for only a few years post-restoration, whereas some long-term (20+ years) studies only specifically monitor some of the functioning aspects of the coastal wetland (Craft *et al.*, 2003). The lack of unified protocols for wetland restoration success has been questioned for years, including many important tools for measuring and evaluating success that are not being used. This is because each project has its own goal and purpose, and definitions of success itself vary from project to project.

Experimental work at Tijuana Estuary also shows promise for restoration of salt marshes, through implantation of halophytes into bare mudflat areas for colonization and creation of new productive salt marshes (Zedler *et al.*, 1992). The experiments reveal many restrictions to successful implantation of ecological engineer halophytes, including the choice of suitable sediment conditions and drainage, length of inundation, salinity and shoreline slope. The project goal was to determine hydrologic effects of tidal flushing on marsh vegetation for better understanding the processes of wetland restoration in lagoons with channels previously closed by man-made changes (Zedler *et al.*, 1992). Unfortunately, these outdoor '*in situ*' mesocosms did not show any clear effects of hydrologic treatments on establishment of marsh vegetation (Callaway *et al.*, 1997). Zedler's experiments showed that force of wave action is one of the main reasons for implantation failure. Even if marshes existed previously in the same area, when nearby coastal and inland anthropogenic developments altered the natural coastline, the hydrology may be changed enough that the marsh plants will not grow successfully in the former area.

In addition to hydrology, correct elevation and substrate are two other important physical barriers to effective salt marsh restoration. Establishing the correct elevation above sea level, and the proper sediments for plant growth and propagation are the most costly aspects of restoration projects, especially those on smaller spatial scales (Sparks *et al.*, 2013). Interestingly, Willis *et al.* (2005) used sediment from processed drill cuttings in excavated oil and gas drill holes for restoring marshes along the Louisiana coast, USA. They tested the processed drill cuttings (i.e. removed drilling fluid additives from muds) as sediment for a restored marsh using 200-litre containers in a temperature-controlled greenhouse mesocosm system, and examined the sediments' suitability for growth of various coastal wetland plants. With further refinement and processing, these processed drill cuttings not only increase the elevation of the restored marsh area and provide a viable substrate for plant growth, but also help minimize the wastes in landfills from old drill cuttings and balance out costs of wetland creation to drill-cutting waste disposal. Because of the possible nature of toxic sediments, one can see how laboratory mesocosms prove useful in various restoration and creation experiments in temperate-subtropical regions. In contrast, similar mesocosm experiments in the Arctic tundra of Mackenzie Delta (Section 7.1) have not been successful and result in thermakarst pond formations and soil salinization.

Determining the success of created wetlands is as variable as that of successful restoration, including how long it takes for the new wetland to become the same as nearby natural

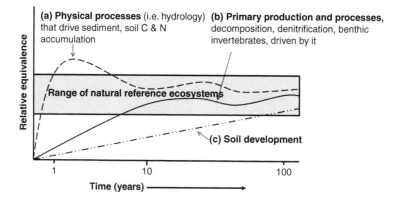

Figure 14.2 Various trajectories showing the changes in physical, environmental and biological processes in a created marsh compared to a reference natural marsh, over time. Trajectory 'a' shows physical functions and services from hydrological processes (i.e. sediment accumulation, plant density), 'b' shows biological processes (i.e. primary production, faunal feeding activity) and 'c' shows wetland soil development (i.e. soil organic carbon and nitrogen). Soil organic carbon and nitrogen take much longer to reach equivalence to a natural marsh (from Craft *et al.*, 2003).

wetlands, and once it is similar, do the services and functions of the new wetland persist (Craft *et al.*, 2003). One effective study method is the *chronosequence approach*, where a natural comparison site is selected that is very similar to the created site, i.e. the created wetland is paired to a natural marsh with very similar hydrology, substrate and biota. This allows the researcher to 'compress time' and monitor a created site over a shorter time frame, including hydrological processes, biological processes and wetland soil development, averaged over multiple comparisons to rate relevant equivalence to natural marshes (Figure 14.2; Craft *et al.*, 2003). This method has successfully shown that constructed *Spartina alterniflora* marshes in North Carolina are similar to natural marshes, but tend to have less carbon and nitrogen organic content in the soil. If the soil organic content is high enough, plants can become well established and the wetland creation can be deemed a success. Therefore, testing soil organic carbon content may be the most effective way to determine the success of wetland creation, and even restoration.

On the other hand, many restoration projects are immensely successful with little cost or effort. The highly invasive *Spartina alterniflora* can completely take over a healthy salt marsh, out-competing natural species and decreasing diversity and overall marsh health (Grosholz *et al.*, 2009; Strong and Ayres, 2009). In other places, such as China, *Spartina* has been purposely cultivated and implanted to restore damaged marshes and regain ecosystem services pertaining to protection from sea level rise (Chung, 2006; An *et al.*, 2007). In 1963, *Spartina* seeds were first brought to the east coast of China (Beijing) and the plants have since covered over 112 000 ha and vertically accreted the coastal wetland over 78 cm in just 6 years. The success of *Spartina alterniflora* implantation and growth along the east coast of China for protection from sea level rise and tsunamis is apparently so great that many other restoration projects in Jiangsu and Zhejiang provinces have been cancelled, saving millions of yuan. On the other hand, there remain concerns about the aggressively invasive properties of some of the non-native *Spartina* species and the fact that bird populations favour the areas of natural vegetation (see Section 10.4.4).

In addition to experimental restoration projects described above, mesocosm restoration projects often begin within small spatial scales, metre by metre plots in the study area, or they may occur within a greenhouse or other laboratory setting. The following examples describe a few of the smaller-scaled restoration experimental studies of seagrasses and mangroves.

Seagrass beds tend to be the first 'step' in coastal wetland creation, where pioneering grass species (i.e. *Zostera*) establish and lead to further sedimentation and vertical accretion of the marsh or mangrove. A good review of seagrass restoration work can be read in Paling *et al.* (2009). Many of the seagrass restoration studies have taken place in developed countries; however, much of the damage to intertidal seagrass beds is occurring in countries without legislation for protection of these sensitive areas (Rodríguez-Salinas *et al.*, 2010). Even fairly 'simple' soda-bottle transplant experiments can show that small-scale restoration is possible and can mitigate possible damages from coastal development (Figure 14.3; Rodríguez-Salinas *et al.*, 2010).

Many of the experimental studies done in mangroves focus on restoration, as they are being destroyed at an alarming rate: so much so, that to achieve a no-net-loss of mangroves worldwide, we would need successful restoration at 150 000 ha per year, and without any more loss (Lewis, 2009). Unfortunately, mangrove creation and restoration has not been as successful as those projects in salt marshes, as the trees have more restricted hydrological regimes, requiring a raised sedimentary platform above mean sea level where they are inundated less than 30% of the time, and some are more susceptible to storms and tsunamis in comparison to salt marshes (Lewis, 2005; Spalding *et al.*, 2010). Despite the lower success than salt marsh restoration, mangrove replanting is highly worth any potential costs

 Figure 14.3   A cost-effective and simple method of transplanting *Zostera marina* from a natural seagrass bed to an intertidal mesocosm study area for restoration in Baja California, Mexico (from Rodríguez-Salinas *et al.*, 2010, with permission from Elsevier).

involved in replanting experiments (see values in Walton *et al.*, 2006). Much of the experimental work is occurring in Africa (*Rhizophora* plantings in the Niger Delta; see Section 9.2), India and Southeast Asia, and ironically, most success has happened in areas where timber production and monotypic communities were the goal of planting, not restoration of natural mangrove habitats.

In Mexico, small reforestation projects of black mangrove (*Avicennia germinans*) had the goal to create simple and inexpensive nursery methods for producing seedlings from propagules in Laguna de Balandra, Baja California, Mexico. Small mesocosms (900 ml biodegradable plastic bags) containing black mangrove propagules were reared in a greenhouse and transplanted to an outdoor nursery only after roots were established. Once transplanted to the cleared reforestation site, there was almost a 75% survival rate of the planted black mangrove. Compared to the much slower growth rate of naturally germinating seedlings, nursery-reared black mangroves may be the only answer for effective restoration in these arid coastal lagoons (Toledo *et al.*, 2001).

Similar studies of seedling establishment for restoration experiments have been done in Singapore, using laboratory mesocosms and field experiments to monitor physical processes on initial seedling establishment and growth of *Avicennia alba* (Balke *et al.*, 2011). Propagules from natural mangrove forests in Singapore were cultured in a laboratory setting using individual PVC flumes, simulating a natural tidal flat. Balke and colleagues monitored inundation, hydrodynamic stress and sediment disturbance, which all can cause seedling dislodgement, noting that three thresholds of germination need to be reached before a propagule can successfully be established as a mature seedling. Knowing the conditions of these sensitivity-limiting thresholds is crucial for effective reforestation projects, as the planting areas must be essentially free of harsh physical processes during germination and radical growth. If the mangroves never grew in an area before, it is likely that the hydrological and sediment conditions are not supportive of propagule germination and establishment (Lewis, 2005). This same principle applies to all coastal wetland restoration and construction experiments, except where changing climate might permit growth of formerly temperature-limited species (see Section 12.3.1 for *Rhizophora mangle* in San Diego).

14.4 Experiments testing physical parameters

Coastal wetland restoration and construction projects have become of high importance in the last 20 years, but they cannot be studied in isolation. For example, physical processes such as sea level rise and tectonic activities must be taken into account in planning a successful restoration project (Stevenson *et al.*, 2000; Adam, 2009); it would waste thousands of dollars of research money to create a marsh that is unable to accrete vertically in pace with a rapid sea level rise. Lateral accretion is also important. Marshes have been able to accommodate sea level rise over the last ~6000 years, but with human shoreline reshaping, sediment dredging and coastal developments such as infilling, damming and diking, marshes may not have the available space to accrete landward (Adam, 2009) – a phenomenon called *coastal squeeze*. Mathematical models by Craft *et al.* (2009) tested the response of coastal marsh area and resultant ecosystem services (plant biomass and nitrogen levels) to varying heights of projected sea level rise on the Georgia coast (USA). A 20% reduction in salt marsh areas is predicted from a 52 cm increase in sea level to 2100, whereas an 82 cm increase in sea level will result in a 45% reduction of salt marshes by 2100. Although there are limitations to mathematical models because they do not take into account feedback processes, the emphasis on marsh loss as a result of sea level rise alone is astounding and we need to find a way to protect and mitigate these coastal wetland losses.

Many of the experiments and mesocosms testing physical parameters also focus on responses of the coastal plants. Recent studies test the growth capability of resident coastal halophytes exposed to various stressors, such as flooding and salinity. This is because sea level rise is associated with increased tidal inundation and salt intrusion further up the estuary. Using indoor mesocosms, Woo and Takekawa (2012) grew the dominant halophyte pickleweed (*Sarcocornia pacifica*) under various conditions of salinity and inundation. Their results show that saltwater inundations negatively affected seedling growth and adult height, which would lower their competitive ability during projected sea level rise. However, they also noted that sediment type greatly affects how these plants grow, and most marsh plants require highly organic sediments in which to become established enough to allow for successful restoration projects. A similar study by Sharpe and Baldwin (2012) used greenhouse mesocosms to determine the effects of increasing salinity and inundation on salt marsh grass species compositions and the susceptibility of freshwater versus oligohaline communities to increased inundation and salinity. Using greenhouse mesocosms allows for replicated experimentation and controlled changes in salinity and inundation that cannot easily happen in a natural field setting. After experimentation, they suggest that increases in sea level, which increases inundation and salinity further up the marsh, will significantly reduce the richness and biomass of plant communities in the freshwater to slightly saline zones.

In addition to plant community changes, studies of effects below-ground have shown the projected changes from sea level rise and increased inundation in salt marshes. Recently, Kirwan *et al*. (2013) used outdoor experimental mesocosms in Chesapeake Bay (USA) that implicated sea level rise by manipulating mesocosm heights within the tidal creek (Figure 14.4a). They determined that in order to effectively survive sea level rise, salt marsh plants need to increase root production in order to increase organic matter accumulation for the necessary vertical accretion. Although these mesocosms were studied *in situ*, they could not account for horizontal water flow in the soil, which may have affected the results. A similar *in situ* mesocosm design by Langley *et al*. (2013; Figure 14.4b) tested the effects of increased sea level (along with other anthropogenic impacts) on marsh plant survival in the upper marsh (i.e. *Spartina patens* zone) in enclosed 'marsh organs' (Morris, 2007). In this experiment, sea level (and not carbon dioxide or nitrogen levels) was the dominant factor affecting the plant growth, but emphasized the connectedness between all three factors, such as increased nutrient competition between species (Langley *et al*., 2013).

Some other smaller mesocosm studies using glass or greenhouses also test various physical effects. In Europe, small tanks with *Salicornia* and *Aster* plant species tested growth rates under varying degrees of tidal inundation and sediment addition to determine interactions between vertical accretion and submersion related to sea level rise. These small

Figure 14.4 A: a small outdoor mesocosm experiment testing effects of manipulating sea level rise (increased tidal inundations by changing individual elevations) on organic matter decay in Chesapeake Bay, USA (from Kirwan *et al.*, 2013); B: open-top chamber mesocosms in an *in situ* tidal creek, which can monitor and manipulate gases, nitrogen levels and water inundation in a controlled manner (from Edgewater, Maryland, USA; Langley *et al.*, 2013, photo by T. Mozdzer, © 2013 Blackwell Publishing Ltd).

tanks were designed in a greenhouse which used reservoir tanks and timers to simulate varying tides. As long as sediment is continuously added, as is predicted along the coasts of Europe, sea level rise will probably not have any deleterious effects on these salt marsh plants (Boorman *et al.*, 2001). Small mesocosms also make it easy to test different effects of salinity on plant physiology (e.g. Redondo-Gómez *et al.*, 2010), and to test changes in temperature (e.g. outdoor chambers by Idaszkin and Bortolus, 2011).

14.5 Experiments testing biological parameters

Biological interactions play just as important a role as physical parameters in shaping and controlling marsh and mangrove growth and resilience to changes. From consumers causing diebacks of vast areas of marshes (e.g. snow geese in Arctic marshes, see Section 7.2.3), to success of biological invaders changing the dynamics of the resident salt marsh (e.g. *Spartina alterniflora* invasions; Chung, 2006; Silliman *et al.*, 2009), coastal wetlands are dynamic systems often subject to changes caused by biological forces. In many studies, it is important to not initiate irreversible damage in a natural coastal system by introducing biological change, so mesocosm work here plays an important role in testing effects and concluding appropriate mitigation techniques in areas where these biological interactions and invasions are already occurring.

Invasions of introduced species are a problem throughout the world and can lead to reductions in native biodiversity as they out-compete often multispecies zones for mono-specific expanses of new invasive species, which can lead to changes in food webs, energy flow and macrofaunal assemblages; e.g. *Spartina alterniflora* is disrupting healthy marshes in China (Chung, 2006). Other studies in *Spartina alterniflora* are covered in Silliman *et al.* (2009). It is not just the marsh plants that can be invasive. In the Black Sea, invasions of large molluscs, jellyfish (Ctenophora) and blue-green seaweeds have led to a trophic cascade of the natural marine ecosystem and an onset of creeping dead sea conditions (total anoxia) in some of the bays (Sorokin, 2002). Another example, the invasion of the European green crab *Carcinus maenus* along the northwestern Atlantic coast, is well documented throughout the literature, but in some cases, it may actually promote the recovery of salt marshes (*Spartina alterniflora* growth) that were once highly affected by die-off from overgrazing by herbivorous *Sesarma* crabs (Bertness and Coverdale, 2013). In using $30 \times 30 \times 30$ cm mesh mesocosms to keep crabs together, they tested the competitive ability of the invasive green crab to take over the destructive *Sesarma* crabs' burrows, effectively promoting regrowth of *Spartina* cordgrass.

Other experiments with invasive fauna were done using the acorn barnacle (*Balanus glandula*) in Patagonian salt marshes. This barnacle was introduced to the Argentina coast in the early 1970s and soon completely dominated the upper zones of rocky shores. The barnacle is now invading soft sediment marsh habitats, usually around the roots of resident plants. A study by Mendez *et al.* (2013) uses manipulative field mesocosms to test the colonization of the barnacle on different substrates within 100 cm^2 plots and determines that the barnacles can colonize a wide array of substrates in a salt marsh, emphasizing the capability of certain fauna to successfully invade very different habitats.

Species interaction experiments have been done using fauna, plants and a combination of both. In Patagonia, greenhouse mesocosm experiments and field reciprocal transplant experiments test the interspecific interactions between low and high marsh plant species to see what primarily affects the vertical zonation of salt marshes: physical factors or interspecific competition (Idaszkin *et al.*, 2010). In addition to zonation, it is important to understand how plants and animals influence the sediment biogeochemistry of coastal wetlands. In salt marshes, mesocosm experiments using the polychaete *Nereis diversicolor* and the plant *Spartina anglica* show the influence of plants and animals on the mineral processes within the sediment (Gribsholt and Kristensen, 2002). Using four separate mesocosms, they test the individual and combined effects of biota on salt marsh sediment processes. In mangroves, Kristensen and Alongi (2006) test the same interactions using *Uca vocans* fiddler crabs and *Avicennia marina* control on sediment processes, using outdoor container mesocosms with replicated tides. Independent mesocosms worked well for this mangrove study and show similar sediment characteristics to nearby natural mangroves, demonstrating the use of mesocosms for studying biological effects in coastal wetlands without disrupting the natural system.

14.6 Experiments and mesocosms testing other parameters

In addition to restoration, sea level rise and biological interactions, mesocosm experiments in coastal wetlands are used to examine effects of pollution and even for medical purposes (i.e. mosquito control against spread of infectious diseases; see Section 5.6).

Here, mesocosms play a crucial role in keeping natural systems free from damage. For example, in China, the invasion of *Spartina alterniflora* into healthy marshes has led to developments of possible eradication methods. Laboratory experiments have developed herbicides that can kill the entire *Spartina* plant in less than two months, although the effects on other plants are not yet known and the chemical may be potentially detrimental (Liu *et al.*, 2005). Further examples of pollution studies where laboratory mesocosms are used include those examining the effects of oil spills and remediation on coastal wetland communities. Oil spills are a common occurrence and have received much public attention, including that from the *Amoco Cadiz* 1978 spill off France, and more recently, the 2010 *Deepwater Horizon* oil spill in the Gulf of Mexico. Although the salt marshes affected by those events could not be protected, mesocosm and laboratory studies have tested various remediation treatments under controlled conditions. Wright and colleagues used greenhouse mesocosms to test the sediment and microbial recovery of a simulated crude oil spill in *Spartina* marsh containers under various conditions. They emphasized that in the natural marsh, oil would disperse too quickly and would not give replicable results for testing specific remediation techniques (Wright *et al.*, 1997). A multifactorial laboratory experiment evaluated consequences of various amounts of 'spilled' oil followed by multiple remediation techniques, such as soil aeration and nutrient addition, on growth of certain species of freshwater marsh plants (Dowty *et al.*, 2001). Different responses of various plant species determine which species would be most suitable for initial revegetation of oil-damaged sites. In the *Deepwater Horizon* spill, mesocosm studies were conducted along the Louisiana coast to determine recovery of *Spartina alterniflora* and *Juncus roemeranus*. Here, six oil treatments in various locations and amounts of oil coverage test growth in small containers with monospecific sods of either *Spartina* or *Juncus* (Lin and Mendelssohn, 2012). Field results show that *Juncus* is more negatively affected by the oiling, with *Spartina* recovering within weeks after initial toxicity from the oil spill. For both plants, however, the amount and severity of the oil spill is most detrimental to their long-term recovery.

Another interesting type of pollution is saline water effluent from water desalination treatments, which could have negative effects on the natural seagrass beds along the Mediterranean in Spain. Transplants of *Posidonia oceanica* seagrass were cultured and maintained in a controlled laboratory mesocosm with circulating filtered seawater (124 times per day) and tightly controlled light, pH and temperatures (Marín-Guirao *et al.*, 2011; Figure 14.5). Salinities were manipulated to test the effects on various saline effluents, from 37 to 43 psu. Results show that the growth and photosynthetic capabilities of *Posidonia* are most negatively affected at hypersaline conditions of 41–43 psu. The resulting decline in growth could reduce the area of Mediterranean seagrass beds near effluent sites and the coastal ecological productivity.

Coastal wetland mesocosms are not only needed for testing and monitoring processes that affect the entire wetland. Some experiments may have larger goals, including those for medical purposes. One main example is ongoing effort to control mosquitoes, the greatest killer of humans on Earth, infecting millions each year with vector-borne diseases, such as malaria. There is a balance between environmental concerns and those of human health (Rose, 2001; Dale and Knight, 2012). In many situations, salt marshes are completely

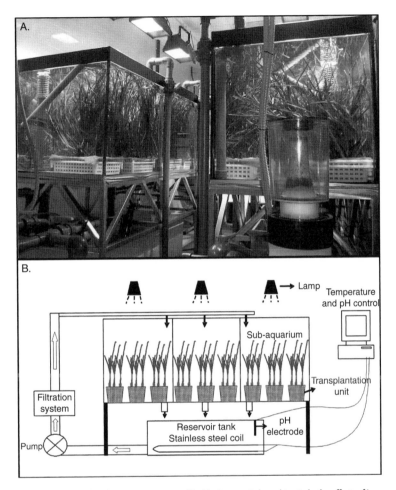

Figure 14.5   Laboratory mesocosm system of Mediterranean seagrass (*Posidonia oceanica*) used to study the effects of increased salinity to test response of effluent from desalination projects (from Marín-Guirao *et al.*, 2011, with permission from Elsevier).

ditched and subsequently destroyed to eliminate standing water which harbours the mosquito larvae (see Section 5.6). In the industrialized countries, costly insecticides are used to kill mosquitoes, which may save the wetlands, but often secondarily kills many other insects (Russell *et al.*, 2009). Many field studies are being done to test various eradication treatments; however, laboratory mesocosms are a good option, allowing for more replicable testing of controlled parameters. Additionally, genetic and other microbiological work can lead to breakthroughs in disrupting the intricate relationships between the parasite (i.e. *Plasmodium*) and the mosquito, and ultimately preventing transmission of malaria to humans (see Section 5.6 and Regev-Rudzki *et al.*, 2013).

As illustrated with malaria eradication, a single approach cannot solve the issues discussed in this chapter. A combination of field studies, genomic research (i.e. Chapter 13), experimental mesocosms and paleoecological work (Chapter 12) centred on coastal wetlands can provide valuable insights to effective restoration means, proper methods to study success rates, and can give us information on effects of future impacts, such as sea level rise, pollution, biological invasions and medical problems. Unfortunately, there is little global collaboration with experimental efforts, and much time, energy and resources are being wasted to repeated 'trial-and-error' experiments and evaluators of success (Adam, 2009; Lewis, 2009).

15   Conclusions and future directions

In this book, we have shown where and how tidal salt marshes form worldwide and the kinds of conditions required for coastal marsh formation. There are many different examples of the dangers to survival of the remaining coastal marsh formations in light of the sensitivity of these environments to anthropogenic and natural forces of climatic and oceanographic change. We discussed the different marsh vegetation types, and their zonations, in the world's main climatic regions. Also included were lists of the kinds of animals living in these environments, many of which are now endangered species because of worldwide coastal wetland destruction or pollution. Ecologically, we discussed main patterns of energy flow, feeding relationships and the extraordinarily high production in coastal wetlands. Study of both the current biology and ecology of salt marshes and mangroves is extremely important in fully understanding various global paleoenvironments, and therefore possible future scenarios within these vital systems. For example, we have outlined the impacts expected to follow the unprecedented recent changes in Arctic sea ice and wetlands and we have shown ways in which we can unravel the natural history and speeds of past coastal wetland changes using fossil pollen combined with microfossils that record the salinity and elevation of the marshlands. These geological tools can also be used to forecast the future of the wetlands, including the likelihood of megaquake occurrences, as indicated by 'silent earthquake' traces. Experiments with marsh mesocosms provide an important tool for further understanding the complex interactions between salt marsh plants, animals and their intertidal environment. Examples are also given of practical applications in agriculture and industry that have arisen from research on some unique products offered by salt marsh and mangrove plants – and the importance of preserving marshland biodiversity as a reservoir of salt-tolerant ionomes and genomes for the new 'Green Revolution' and to meet global water conservation targets on which human survival depends.

Five important themes emerge from this worldwide tour of salt marshes and mangroves:

(1) Tidal wetlands depend for their existence on space in coastal regions throughout the world (outside the Antarctic region) wherever sediment accumulation allows pioneer plants to take root and there is regular tidal flooding to supply the intertidal biota with minerals, nutrients and water. Once established, the vegetation forms a natural barrier against coastal erosion. Their natural formations buffer shorelines from floods, where the high runoff after storms is intercepted and absorbed within the wetland system, slowing the water (and contaminants) before reaching nearby coastal waters. This process is vital to the health of coastal water ecosystems. The high organic production in coastal wetlands is the main source of blue carbon in the world's oceans, and their high carbon production is the same source of future gas and oil resources as the supply that is being tapped for gas and oil in many regions of the world, including the Mississippi Delta, Beaufort Sea and deltas of West

Africa, the Nile and East Asia. Mangrove trees and the flowering plants of salt marshes have existed and persisted for many millions of years, before which marine blue-green algae and possibly ferns may have dominated the intertidal areas. As long as coastal conditions were right (as outlined in the first three chapters), tidal wetlands flourished, accreted, migrated, disappeared and reappeared as continents collided, climates changed, and sea level rose and fell with glaciations. Yet, in the twenty-first century, we are seeing threats that these dynamic systems can no longer handle and keep up with. Since they depend on low-energy tidal flow, and high rates of sedimentation, anything that disrupts the flow and/or removes their migration space can drastically affect the longevity of the marshes, making them unable to adapt and change as they have for millions of years.

(2) The coastal wetlands support a vast variety of plants and animals that are highly adapted to the dynamic shoreline environment. These special morphological and physiological adaptations are repeated throughout the climatic regions of the world, giving the superficial appearance that most salt marshes and mangrove forests contain the same species. In fact, there is a huge diversity of tidal wetland species on a global basis, and the specialized biota is a storehouse of genetic diversity that is now being used to combat accelerating problems of salinity in both coastal and inland farming regions throughout the world.

Despite their regional differences, all the coastal wetlands provide us with valuable ecosystem services due to their extremely high plant biomass production. This high production, sometimes reaching over 3000 g C/(m^{2}yr) (above- and below-ground), supports immense biological activity, including the role of nursery grounds for coastal fisheries. Wetlands are such important nurseries for coastal fishes that declining fish catches are correlated with diminishing wetland area. Another ecosystem service of coastal wetlands is their ability to be a carbon sink for excess carbon being burned through fossil fuels. Without coastal wetlands, the amount of carbon dioxide in the atmosphere might be dramatically higher, accelerating unnatural climate change. Also, as salt marsh plants are nitrogen-limited, and the abundant bacteria within salt marshes help with nitrogen processing, salt marshes provide valuable services for limiting the nutrient runoff from agriculture and industry in temperate regions, especially when inland wooded areas have been destroyed, removing the first line of defence from coastal eutrophication.

(3) Throughout most of the world, the old perception of tidal marshes as coastal 'wastelands' has since changed, and continues to change as values of these ecosystems are better understood, acknowledging that they are one of the most biologically productive habitats on Earth, rivalling tropical rainforests and providing us with a large list of ecosystem services from fossil fuel to food. The recognition of wetland importance worldwide was embodied in the 1971 Ramsar Wetlands Convention. Additionally, salt marshes are now protected by legislation in many countries to look after these ecologically important habitats. In the USA and Europe, they are now accorded a high level of protection by the Clean Water Act and the Habitats Directive, respectively. With the impacts of this habitat and its importance now realized, a growing interest in restoring salt marshes through managed retreat of anthropogenic disturbance or the reclamation of land has been established, but this interest must be maintained and supported by binding legislation. Without continued science and public education, there is a danger of marshes returning to our 'wasteland' view, becoming

destroyed to satisfy human population growth and coastal development. Presently, many Asian countries still need to ratify their recognition of the value of marshland protection on a national level, although there is an encouraging amount of local support. With their ever-growing populations and intense development along the coast, the value of salt marshes and mangroves tends to be over-ridden by the real or perceived need for more food and industrial-scale wealth, and the land continues to be reclaimed. In wealthy North America, however, infilling and draining marshes has been the most destructive, whether it is for expanded agriculture or coastal development. Habitat fragmentation from coastal development has caused stagnant freshwater marshes to form, allowing invasive *Phragmites* reed to colonize and take over, even in tidal marshes. These *Phragmites* invasions have often been augmented by increased agricultural runoff, where the nutrients allow them to grow and out-compete high marsh species. Not only has biological diversity been reduced, but the longevity of the marshes is lessened because the *Phragmites* roots raise the level of the marshland and lead to the sediment drying out. This one species therefore can cause loss of wetland area, disruption of natural biodiversity, and loss of marsh ecosystem services.

Like salt marshes, mangroves continue to be destroyed through human impacts; most importantly logging and the use of tree materials for building, coastal development (whether it is for residential or resort development), destruction for shrimp aquaculture ponds, or for palm-oil production. Another threat to the future of mangroves is their sensitivity to sea level rise and human-driven global warming. As sea levels rise, mangrove soils can become water-logged, anoxic and highly acidic. These negative effects are compounded if sedimentation is decreased by reduced river supply and/or coastal sand sources, and mangroves will not be able to grow and accrete to keep up with sea level changes. Just as for salt marshes, the loss of mangroves leads to loss of the ever-important local ecosystem services that many non-industrialized societies depend on in tropical regions. We consider that the traditional importance and potential demise of mangroves throughout the world is best illustrated by a selection of case histories from different continents, as shown in the central chapters of our book. It should also be noted that different threats face the sustainability of mangal communities in developing African and Asian nations than in highly developed industrialized countries, but the net results are likely to be the same: a >50% loss of mangrove forests worldwide from 1980 to 2100, and a corresponding decrease in export of organic carbon nutrients to the coastal regions. Some scientists predict that the mangroves will be completely eliminated before the global rise in sea level floods the tropical lowlands allows the inland spread of mangal.

Despite the large quantity of research pertaining to salt marshes and mangroves, there is still much to learn. The tropicals wetlands should become a main focus, as the least is known about them and they are in developing countries with high population growth pressures. In these areas, much of the destruction of wetlands occurs because of urban residential development, which is unlikely to receive any foreign aid funds to mitigate. In emerging countries especially, many environmental laws are poorly developed and are not strictly enforced, which will result in wetland destruction for coastal development before the ecological dangers are understood. This is a perfect example of the need to be proactive instead of reactive; however, 'we cannot fix what we do not understand' (Bertness, 2007; p. 373).

(4) The importance of geological archives from salt marshes and mangrove forests for forecasting megaquakes and tsunami impacts is of worldwide concern because the effects are not confined just to the exact region where the earthquake is triggered. A good example of the hemispheric impacts of tsunami are provided by the 1960 Chilean earthquake that took place in southern South America, but caused a landslide in Hawaii in the Northern Hemisphere and was felt in Japan about halfway around the world from its origin. Fine-tuned understanding of the relation between salt marsh ecology and proxy-indicators such as foraminifera and diatoms is also an essential component for monitoring 'silent earthquakes' and understanding better the dynamics of earth movements in areas bordering megaquake zones, such as the Cascadia margin. The geological records provided by mangrove pollen and plant roots in sediment cores are also of interest in unravelling the way climate has changed since the last ice age and their sedimentary archives indicate how the recent trends in global sea level rise and warmer climates may alter highly populated coastal wetlands in the future. These archival data are an essential component for setting limits for parameters in computer models designed to forecast sea level, global warming and storminess trends.

(5) The value of salt marshes encompasses more than what is addressed in our book. For example, on a local scale, it is possible to use marshes as natural sewage treatment systems. This is because their large area of low energy would slow the runoff, increase sedimentation, and therefore remove particulate wastes, where the bacteria and decomposers, in addition to chemical processes, would help to remove organic as well as inorganic wastes in the water, cleaning it before entering sheltered coastal waters susceptible to eutrophication. Additionally, the importance of coastal wetlands and their vegetation as storm and tsunami buffers could save lives and billions of dollars in damages, easily replacing any cost incurred from restoration or conservation projects. These types of tangible 'return on investment' projects help the public and government officials see the immediate (monetary) value of these vitally important ecosystems.

In order to safely implement possible remediation strategies, and to better understand specific negative impacts in salt marshes and mangroves, scientific studies are used (e.g. Chapter 14). Some work, such as understanding responses to pollution and species invasions, would unnecessarily cause damage to already fragile ecosystems. This is where mesocosm and laboratory studies become important, as manipulations can be made in an isolated and controlled manner, with appropriate baseline information and specific parameter controls. Laboratory salt marsh mesocosms can provide the scientific community with a valuable tool and practically unlimited uses and potentials for salt marsh studies (remediation, implantation, invasive species, salinity fluctuations, etc.). Experiments on the restoration of salt marshes through implantation of flora into mudflat areas can create new salt marshes and help to grow back some of what we have destroyed. There is ongoing need, however, to be wary of aggressive implanted species destroying native salt marsh plants of potential importance as gene-stock for the new Green Revolution to combat soil salinization.

The adage that 'The sea knows no boundaries' is nowhere else so true than for people living on the edges of the world's largest wetlands – from Arctic to Equator. We have shown that Arctic coastlines are particularly vulnerable and fast-changing – with the consequences being felt as far south as New York. Melting glaciers are impacting coastal wetlands both directly as rising river volumes and sediment loads (e.g. the Amazon and Southeast Asian megadeltas) and indirectly, as surges reshaping fiords, and as estuaries and rising temperatures that melt the permafrost on Arctic shorelines, releasing more greenhouse gases into the already CO_2-enriched global atmosphere. Megaearthquakes that abruptly lower salt marshes on one continental margin are inevitably registered as tsunami surges on the opposite shores of the Pacific and Indian Oceans. The geological archives provided by wetland sediments, combined with sound understanding of the relationship between modern tidal wetland plants and animals, also provides us with the most powerful tools for understanding how our oceans and coasts have responded to El Niño, monsoons and tropical storms in the past, and this baseline information is the foundation for predicting the future.

References

AbdAllah, A. M., El-Gindy, A. A. and Debes, E. A. (2006). Sea level changes at Rosetta Promontory, Egypt. *Egyptian Journal of Aquatic Research*, **32** (1), 34–47.

Adam, P. (1995). *Saltmarsh Ecology*. Cambridge: Cambridge University Press.

Adam, P. (2009). Salt marsh restoration. In *Coastal Wetlands: An Integrated Ecosystems Approach*, ed. G. M. E. Perillo, E. Wolanski, D. R. Cahoon and M. M. Brison. Amsterdam: Elsevier BV, pp. 737–762.

Adams, J. B., Colloty, B. M. and Bate, G. C. (2004). The distribution and state of mangroves along the coast of Transkei, Eastern Cape Province, South Africa. *Wetlands Ecology and Management*, **12**, 531–541.

Adams, J., Rajkaran, A. and Pike, S. (2011). Die-back of mangroves in the Kobonqaba Estuary. *SANCOR Newsletter* #192, 11–12.

Adebayo, O. F, Orijemie A. E. and Aturamu, A. O. (2012). Palynology of Bog-1 Well, Southeastern Niger Delta Basin, Nigeria. *International Journal of Science and Technology*, **2** (4), 214–222.

AEM (2010). Progress on aerospace biofuels gaining momentum. *Aerospace Engineering Magazine*, February 2010. Online at http://www.sae.org/mags/aem/power/7696 (accessed October 2013).

Ahn, C. and Mitsch, W. J. (2002). Scaling considerations of mesocosm wetlands in simulating large created freshwater marshes. *Ecological Engineering*, **18** (3), 327–342.

Alberti, J., Escapa, M., Iribarne, O., Silliman, B. and Bertness, M. (2008). Crab herbivory regulates plant facultative and competitive processes in Argentinean marshes. *Ecology*, **89** (1), 155–164.

Allan, J. A. (1998). Mangroves as alien species: the case of Hawaii. *Global Ecology and Biogeography Letters* **7**: 61−71.

Allanson, B. R. and Baird D. (editors) (1999). *Estuaries of South Africa*. Cambridge: Cambridge University Press.

Alongi, D. M. (2009). Paradigm shifts in mangrove biology. In *Coastal Wetlands: An Integrated Ecosystems Approach*, ed. G. M. E. Perillo, E. Wolanski, D. R. Cahoon and M. M. Brison. Amsterdam: Elsevier BV, pp. 615–640.

Al-Otaibi, Y., Ait Belaid, M. and Abdu, A. (2006). Impact assessment of human activities on coastal zones of eastern Saudi Arabia using remote sensing and geographic information systems techniques. *The International Archives of the Photogrammetry, Remote Sensing and Spatial Information Sciences*, **34**, Part XXX.

Álvarez, M., Briz, I., Balbo, A. and Madella, M. (2010). Shell middens as archives of past environments, human dispersal and specialized resource management. *Quaternary International*, **239**, 1–7.

Al-Yamani, F. Y., Bishop, J. M., Al-Rifaie, K. and Ismail, W. (2007). The effects of the river diversion, Mesopotamian marsh drainage and restoration, and river damming on the marine environment of the northwestern Arabian Gulf. *Aquatic Ecosystem Health and Management*, **10** (3), 277–289.

Amaral, A. C. Z., Migotto, A. E., Turra, A. and Schaefer-Novelly, Y. (2010). Araçá: biodiversidade, impactos e ameaças. *Biota Neotropica*, **10** (1), 219–264.

An, S. Q., Gu, B. H., Zhou, C. F., *et al.* (2007). *Spartina* invasion in China: implications for invasive species management and future research. *Weed Research*, **47** (3), 183–191.

Anthony, E. J. (1996). Evolution of estuarine shoreline systems in Sierra Leone. In *Estuarine Shores: Evolution, Environments and Human Alterations*, ed. K. F. Nordstrom and C. T. Roman. New York: John Wiley and Sons Inc., pp. 39–61.

Anthony, E. J. and Gratiot, N. (2010). Coastal engineering and large-scale mangrove destruction in Guyana, South America: averting an environmental catastrophe in the making. *Ecological Engineering*, **47**, 268–273.

ASEAN Regional Centre for Biodiversity Conservation (2013). Information site, online at http://www.arcbc.org.ph/wetlands/vietnam/vnm_mekdel.htm (accessed October 2013).

Aslam, R., Bostan, N., Nabgha-e-Amen, Maria, M. and Safdar, W. (2011). A critical review on halophytes: Salt tolerant plants. *Journal of Medicinal Plants Research*, **5** (33), 7108–7118.

Atwater, B. F. (1987). Evidence for great Holocene earthquakes along the outer coast of Washington State. *Science* **236**, 942–944.

Atwater, B. F. and Hemphill-Haley, E. (1997). *Recurrence Intervals for Great Earthquakes of the Past 3,500 Years at Northeastern Willapa Bay, Washington*. US Geological Survey Professional Paper, 1576, 108 pp.

Atwater, B. F., Yamaguchi, D. K., Bondevik, S., *et al.* (2001). Rapid resetting of an estuarine recorder of the 1964 Alaska earthquake. *Geological Society America Bulletin*, **113** (9), 1193–1204.

Atwater, B. F., Tuttle, M., Schweig, E. S., *et al.* (2004). Earthquake recurrence inferred from paleoseismology. In *The Quaternary period in the United States, Developments in Quaternary Science*, ed. A. R. Gillespie, S. C. Porter and B. F. Atwater. New York: Elsevier, pp. 331–350.

Atwater, B. F., Musumi-Rokkaku, S., Sataki, K., Tsugi, Y., Ueda, K. and Yamaguchi, K. K. (2005). *The Orphan Tsunami of 1700: Japanese Clues to a Parent Earthquake in North America*. US Geological Survey Professional Paper, 1707, 133 pp.

Atwell, B. J., Kriedemann, P. E. and Turnbull, C. G. N. (1999). Chapter 17 – Salt: an environmental stress. In *Plants in Action*. Electronic edition, 2010. Melbourne: Macmillan Education Australia Pty Ltd. Online at http://plantsinaction.science.uq.edu.au/edition1 (accessed October 2013).

Australian Government (2009). Australia's coasts and climate change, Chapter 1. Online at www.climatechange.gov.au/climate-change/adapting-climate-change/australias-coasts-and-climate-change/coast (accessed May 2013)

Ayyad, S. M., Moore, P. D. and Zahran, M. A. (1992). Modern pollen rain studies of the Nile Delta, Egypt. *New Phytologist*, **121** (4), 663–675.

Baker, J. M., Leonardo, G. M., Bartlett, P. D., Little, D. I. and Wilson, C. M. (1993). Long-term fate and effects of untreated thick oil deposits on salt marshes. *International Oil Spill Conference Proceedings, March 1993* (1), 395–399.

Balke, T., Bouma, T. J., Horstman, E. M., *et al.* (2011). Windows of opportunity: thresholds to mangrove seedling establishment on tidal flats. *Marine Ecology Progress Series*, **440**, 1–9.

Ball, M. C. (1998). Mangrove species richness in relation to salinity and waterlogging: a case study along the Adelaide River floodplain, Northern Australia. *Global Ecology and Biogeography Letters*, **7** (1), 73–82.

Banister, K. E., Backiel, T. and Bishop, J.S (1994). *Fisheries*. Prepared for The Wetland Ecosystems Research Group, University of Exeter.

Barbier, E. B., Hacker, S. D., Kennedy, C., Koch, E. W., Stier, A. C. and Silliman, B. R. (2011). The value of estuarine and coastal ecosystem services. *Ecological Monographs*, **81** (2), 169–193.

Barbosa, C. F. (1995). Foraminifera e Arcellacea ('Thecamoebia') Recentes do Estuário de Guaratuba, Paraná, Brasil. *Anais da Academia Brasileira de Ciências*, **67**, 465–492.

Barbosa, C. F., Scott, D. B., Seoane, J. C. S. and Turcq, B. J. (2005). Foraminiferal zonation as base lines for Quaternary sea level fluctuations in south-southeast Brazilian mangroves and marshes. *Journal of Foraminiferal Research*, **35** (1), 22–43.

Barker, R. (1998). *And the Waters Turned to Blood: the Ultimate Biological Threat*. New York: Touchstone.

Barlow, N. L. M, Shennan, I., Long, A. J., *et al.* (2013). Salt marshes as late Holocene tide gauges. *Global and Planetary Change*, **106**, 90–110.

Barnett, L. K. and Emms, C. (2005). *Common Reptiles of The Gambia*. Halsham, UK: Rare Repro, 24 pp.

Barnhardt, W. A. (editor) (2009). Coastal change along the shore of northeastern South Carolina: the South Carolina coastal erosion study. *US Geological Survey Circular* **1339**, 77 pp.

Baroni, C. and Orombelli, G. (1991). Holocene raised beaches at Terra Nova Bay, Victoria Land, Antarctica. *Quaternary Research*, **36**, 157–177.

Baye, P. (2007). *Selected Tidal Marsh Plant Species of the San Francisco Estuary: A Field Identification Guide*. Berkeley: San Francisco Estuary Invasive *Spartina* Project, California State Coastal Conservancy. Online at www.spartina.org/project_documents/field_guide_tide_plants_low-res_200703.pdf (accessed October 2013).

Beavan, R. J. and Litchfield, N. J. (2012). *Vertical land movement around the New Zealand coastline: implications for sea level rise*. GNS Science report 2012/29, September 2012. Lower Hutt, NZ: GNS Science, 41 pp.

Beeftink, W. G. (1977). The coastal salt marshes of western and northern Europe: an ecological and phytosociological approach. In *Ecosystems of the World 1. Wet Coastal Ecosystems*, ed. V. J. Chapman. Amsterdam-Oxford-New York: Elsevier Scientific Publishing Company, pp. 109–155.

Behbahani, M., Manaf, A., Muse, R. and Mohd, N. B. (2007). Anti-oxidant and anti-inflammatory activities of leaves of *Barringtonia racemosa*. *Journal of Medicinal Plants Research*, **1** (5), 95–102.

Bell, H. L. (2010). A new species of *Distichlis* (Poaceae, Chloridoideae) from Baja California, Mexico. *Madroño*, **57** (1), 54–63.

Belsare, D. K. (1994). Inventory and status of vanishing wetland wildlife of southeastern Asia and an operational management plan for their conservation. In *Global Wetlands: Old World and New*, ed. W. J. Mitsch. Amsterdam: Elsevier BV, pp. 841–856.

Belt, S. T. and Müller, J. (2013). The Arctic sea ice biomarker IP25: a review of current understanding, recommendations for future research and applications in palaeo sea ice reconstructions. *Quaternary Science Reviews*, **59**, 1–17.

Berczik, A., Dinka, M. and Kiss, A. (editors) (2012). *Proceedings of 39th IAD Conference: Living Danube*. Göd/Vácrátót, Hungary: Danube Research Institute Centre for Ecological Research, Hungarian Academy of Sciences, and General Secretary of International Association for Danube Research, Sciences.

Berger, W. H. (2008). Sea level in the late Quaternary: patterns of variation and implications. *International Journal of Earth Science*, **97**, 1143–1150.

Bertness, M. D. (2007). *Atlantic Shorelines: Natural History and Ecology*. Princeton, NJ: Princeton University Press.

Bertness, M. D. and Coverdale, T. C. (2013). An invasive species facilitates the recovery of salt marsh ecosystems on Cape Cod. *Ecology*, **94**, 1937–1943.

Bianciotto, O. A., Pinedo, L. B., San Roman, N. A., Blessio, A. Y. and Collantes, M. B. (2003). The effect of natural UV-B radiation on a perennial *Salicornia* salt marsh in Bahía San Sebastián, Tierra del Fuego, Argentina: a 3-year field study. *Journal of Photochemistry and Photobiology B*, **70** (3), 177–85.

BirdLife International (2012). *Important Bird Areas Factsheet: Bao Bolon Wetland Reserve*. http://www.birdlife.org (accessed June 2012).

Black. M. T. and Pezza, A. B. (2013). A universal, broad-environment energy conversion signature of explosive cyclones. *Geophysical Research Letters*, **40**, 452–457.

Blais-Stevens, A. and Clague, J. J. (2001). Paleoseismic signature in late Holocene sediment cores from Saanich Inlet, British Columbia. *Marine Geology*, **175**, 131–148.

Blasco, F. (1977). Outlines of ecology, botany and forestry of the mangals of the Indian subcontinent. In *Ecosystems of the World 1: Wet Coastal Ecosystems*, ed. V. J. Chapman. Amsterdam, Oxford, New York: Elsevier Scientific Publishing Company, pp. 241–260.

Blasco, F. (1984). Mangrove evolution and palynology. In *The Mangrove Ecosystem: Research Methods*, ed. S. C. Snedaker and J. G. Snedaker. Paris: UNESCO, pp. 36–49.

Blasco, F., Janodet, E. and Bellan, M. F. (1994). Natural hazards and mangroves in the Bay of Bengal. *Journal of Coastal Research Special Issue*, **12**, 277–288.

Blum, M. D. and Roberts, H. H. (2009). Drowning of the Mississippi Delta due to insufficient sediment supply and global sea level rise. *Nature Geoscience* **2**, 488–491.

Boaden, P. J. S. (1985). *An Introduction to Coastal Ecology*. Glasgow: Blackie and Son.

Boggs, K., Klein, S. C., Grunblatt, J., Streveler, G. P. and Koltun, B. (2007). *Landcover classes and plant associations of Glacier Bay National Park and Preserve*. Natural Resource Technical Report NPS/GLBA/NRTR-2008/093. Fort Collings, CO: National Park Service. D-147, 255 pp.

Boorman, L. A., Hazelden, J. and Boorman, M. (2001). The effect of rates of sedimentation and tidal submersion regimes on the growth of salt marsh plants. *Continental Shelf Research*, **21** (18), 2155–2165.

Borstad, G. A., Álvarez, M. M. de S., Hines, J. E. and Dufour, J.-F. (2008). Reduction in vegetation cover at the Anderson River delta, Northwest Territories, identified by Landsat imagery, 1972–2003. Technical Report Series No. 496. Yellowknife, Northwest Territories: Canadian Wildlife Service.

Bortolus, A. (2006). The austral cordgrass *Spartina densiflora* Brong.: its taxonomy, biogeography and natural history. *Journal of Biogeography*, **33**, 158–168.

Bortolus, A. and Iribarne, O. (1999). Effects of SW Atlantic burrowing crab *Chasmagnathus granulata* on a *Spartina* salt marsh. *Marine Ecology Progress Series*, **178**, 78–88.

Bortolus, A., Schwindt, E., Bouza, P. J. and Idaszkin, Y. L. (2009). A characterization of Patagonian salt marshes. *Wetlands*, **29** (2), 772–780.

Boto, K. G. and Robertson, A. I. (1990). The relationship between nitrogen fixation and tidal export of nitrogen in a tropical mangrove system. *Estuarine and Coastal Shelf Science*, **31**, 531–540.

Bowyer, P. (2003). The storm surge and waves at Halifax with Hurricane Juan. Environment Canada, Canadian Hurricane Center Archives, 17 October 2003.

Brasington, J. (2001). Monitoring marshland degradation using multispectral remote sensed imagery. In *The Iraqi Marsh Lands: A Human and Environmental Study*, ed. P. Clark and S. Magee. London: Politico's Publishing, pp. 110–132.

Braun-Blanquet, J. (1928). *Pflanzensoziologie: Grundzuge der Vegetationskunde*. Berlin: Springer.

Bromirski, P. D., Miller, A. J. and Flick, R. E. (2012). Understanding North Pacific sea level trends. *EOS Transactions*, **93** (27), 249–250.

Broome, S. W. and Craft, C. B. (2009). Tidal marsh creation. In *Coastal Wetlands: An Integrated Ecosystem Approach*, ed. G. M. Perillo, E. Wolanski, D. R. Cahoon and M. M. Brinson. Amsterdam: Elsevier BV, pp. 715–736.

de Bruxelles, S. (2013). http://www.thetimes.co.uk/tto/science/biology/article3667780.ece (accessed October 2013).

Bryant, E., Young, R. W. and Price, D. M. (1992). Evidence of tsunami sedimentation on the southeastern coast of Australia. *Journal of Geology*, **100**, 753–765.

Bujalesky, G. G. (2007). Coastal geomorphology and evolution of Tierra del Fuego (Southern Argentina). *Geologica Acta*, **5** (4), 337–362.

Byrne, R., Ingram, B. L., Starratt, S., *et al.* (2001). Carbon-isotope, diatom, and pollen evidence for late Holocene salinity change in a brackish Marsh in the San Francisco Estuary. *Quaternary Research*, **55** (1), 66–76.

Caballero, M., Peinalba, M. C., Martinez, M, Ortega-Guerrerol, B. and Vazquez, L. (2005). A Holocene record from a former coastal lagoon in Bahia Kino, Gulf of California, Mexico. *The Holocene*, **15** (8), 1236–1244.

Cahoon, D. R., Hensel, P. F., Spencer, T., *et al.* (2006). Coastal wetland vulnerability to relative sea level rise: wetland elevation trends and process controls. In *Wetlands and Natural Resource Management, Ecological Studies 190*, ed. J. T. A. Verhoeven, B, Beltman, R. Bobbink and D. F. Whigham. Berlin, Heidelberg: Springer-Verlag, pp. 271–292.

Callard, S. L., Gehrels, W. R., Morrison, B. V. and Grenfell, H. R. (2011). Suitability of salt marsh Foraminifera as proxy indicators of sea level in Tasmania. *Marine Micropaleontology*, **79** (3–4), 121–131.

Callaway, J. C. and Zedler, J. B. (2009). Conserving the diverse marshes of the Pacific Coast. In *Human Impacts on Salt Marshes: A Global Perspective*, ed. B. R. Silliman, E. D. Grosholz and M. D. Bertness. Berkeley, CA: University of California Press, pp. 285–307.

Callaway, J. C., Zedler, J. B. and Ross, D. L. (1997). Using tidal marsh mesocosms to aid wetland restoration. *Restoration Ecology*, **5** (2), 135–146.

Campbell, S. J., Rashid, M. A. and Thomas, R. (2001). *Seagrass Habitat of Cairns Harbour and Trinity Inlet*, Queensland Government DPI Information Series QI02059. Cairns: DPI, 25 pp.

Capps, D. M., Wiles, G. C., Clague, J. J. and Luckman, B. H. (2011). Tree-ring dating of the nineteenth-century advance of Brady Glacier and the evolution of two ice-marginal lakes, Alaska. *The Holocene*, **21** (4), 641–649.

Carmack, E. and McLaughlin, F. (2011) Towards recognition of physical and geochemical change in Subarctic and Arctic seas. *Progress in Oceanography*, **90**, 90–104.

Catron, J. E., Ceballos, G. and Felger, R. (Eds) (2005). *Biodiversity, Ecosystems, and Conservation in Northern Maxico*. New York, NY: Oxford University Press, pp. 298–333.

Cerón-Souza, I., Rivera-Ocasio, E., Medina, E., *et al.* (2010). Hybridization and introgression in New World red mangroves, Rhizophora (Rhizophoraceae). *American Journal of Botany*, **97** (6), 945–957.

Chagué-Goff, C., Hamilton, T. S. and Scott, D. B. (2001). Geochemical evidence for the recent changes in a salt marsh, Chezzetcook Inlet, Nova Scotia, Canada. *Proceedings of the Nova Scotian Institute of Science*, **41** (4), 149–159.

Chagué-Goff, C., Schneider, J-L., Goff, J. R., Dominey-Howes, D. and Strotz, L. (2011). Expanding the proxy toolkit to help identify past events: lessons from the 2004 Indian Ocean Tsunami and the 2009 South Pacific Tsunami. *Earth-Science Reviews*, **107**, 107–122.

Chapman, V. J. (1960). *Salt Marshes and Salt Deserts of the World*. London: Leonard Hill Limited, 392 pp.

Chapman, V. J. (1974). *Salt Marshes and Salt Deserts of the World*. Germany: Lubrecht & Cramer, Ltd.

Chapman, V. J. (1976). *Coastal Vegetation*, 2nd edn. Oxford: Pergamon Press Limited.

Chapman, V. J. (editor) (1977). *Ecosystems of the World 1: Wet Coastal Ecosystems*. Amsterdam, Oxford, New York: Elsevier Scientific Publishing Company, pp. 1–29; 261–270.

Chapman, V. J. (1984). Botanical surveys in mangrove communities. In *The Mangrove Ecosystem: Research Methods*, ed. S. C. Snedaker and J. G. Snedaker. Paris: UNESCO, pp. 53–80.

Chen, R. and Twilley, R. R. (1998). A gap dynamic model of mangrove forest development along gradients of soil salinity and nutrient resources. *Journal of Ecology*, **86** (1), 37–51.

Chen, K., Yuan, J. and Yan, C. (2007) Shandong Yellow River Delta National Nature Reserve. Wetlands International-China Programme, Ramsar Handbook 5, 22 October 2007, 13 pp.

Chmura, G. L. (2011). What do we need to assess the sustainability of the tidal salt marsh carbon sink? *Ocean and Coastal Management*, **83**, 25–31.

Chmura, G. L., Stone, P. A. and Ross, M. S. (2006). Non-pollen microfossils in everglades sediments. *Review of Paleobotony and Palynology*, **141** (1–2), 103–119.

Christian, R. R. and Mazzilli, S. (2007). Defining the coast and sentinel ecosystems for coastal observations of global change. In *Lagoons and Coastal Wetlands in the Global Change Context: Impacts and Management Issues*, ed. P. Viaroli, P. Lasserre and P. Campostrini. Reprinted from *Hydrobiologia*, **577**, 55–70.

Chu, Z., Yang, X., Feng, X., *et al.* (2013). Temporal and spatial changes in coastline movement of the Yangtze delta during 1974–2010. *Journal of Asian Earth Sciences*, **66**, 166–174.

Chung, C. H. (2006). Forty years of ecological engineering with *Spartina* plantations in China. *Ecological Engineering*, **27** (1), 49–57.

CIMI: Canada-Iraqi Marshland Initiative. (2010). *Atlas of Iraqi Marshes*. Canadian International Development Agency, 72 pp. Online at https://docs.google.com/file/d/0B0iKVcebg9VQU1hrb0RFNGlrSkk/edit?pli=1 (accessed October 2013).

Clement, J. P., Bengtson, J. L. and Kelly, B. P. (2013). *Managing for the Future in a Rapidly Changing Arctic: A Report to the President*. Washington, DC: Interagency Working Group on Coordination of Domestic Energy Development and Permitting in Alaska (D. J. Hayes, Chair), 59 pp.

Cleveland, C. J. (editor) (2008). Guianan mangroves. In *Encyclopedia of Earth*. Washington, DC: Environmental Information Coalition, National Council for Science and the Environment.

Coch, N. K. (1994). Hurricane hazards along the northeastern Atlantic coast of the United States. In *Coastal Hazards Perception, Susceptibility and Mitigation*, ed. C. W. Finkl Jr. Fort Lauderdale, FL: Coastal Education and Research Foundation, pp. 115–148.

Cohen, M. C. L and Lara, R. J. (2003). Temporal changes of mangrove vegetation boundaries in Amazonia: application of GIS and remote sensing techniques. *Wetlands Ecology and Management*, **11**, 223–231.

Cohen, M. C. L., Behling, H. and Lara, R. J. (2005). Amazonian mangrove dynamics during the last millennium: the relative sea level and the Little Ice Age. *Review of Palaeobotany and Palynology*, **136** (1–2), 93–108.

Cohen, M. C. L., Guimarães, J. T. F.; França, M., Lara, R. J. and Behling, H. (2009). Tannin as an indicator of paleomangrove in sediment cores from Amapá, northern Brazil. *Wetlands Ecology and Management*, **17** (2), 145–155.

Combellick, R. A. and Reger, R. D. (1994). *Sedimentological and radiocarbon-age data for tidal marshes along the eastern and upper Cook Inlet, Alaska: Fairbanks*. Alaska Division of Geological and Geophysical Surveys Professional Report of Investigations 94–6, 60 pp.

CONABIO (2009). *Manglares de México: Extensión y distribución*, 2ª edn. Tlalpan México: Comisión Nacional para el Conocimiento y Uso de la Biodiversidad, 99 pp.

Conn, S. and Gilliham, M. (2010). Comparative physiology of elemental distributions in plants. *Annals of Botany*, **105**, 1081–1102.

Connor, C. and O'Haire, D. (1988). *Roadside Geology of Alaska*. Missoula, MT: Mountain Press Publishing Co., 250 pp.

Connor, R. F., Chmura, G. L. and Beecher, C. B. (2001). Carbon accumulation in Bay of Fundy salt marshes: Implications for restoration of reclaimed marshes. *Global Biogeochemical Cycles*, **15** (4), 943–954.

Connor, C., Streveler, G., Post, A., Monteith, D. and Howell, W. (2009). The Neoglacial landscape and human history of Glacier Bay, Glacier Bay National Park and Preserve, southeast Alaska, USA. *The Holocene*, **19** (3), 381–393.

Cook, E. R., Anchukaitis, K. J., Buckley, B. M., *et al.* (2012). Asian monsoon failure and megadrought during the last millennium. *Science*, **328**, 486–489.

Coops, H., Hanganu, J., Tudor, M. and Oosterberg, W. (1999). Classification of Danube Delta lakes based on aquatic vegetation and turbidity. *Hydrobiologia*, **415**, 187–191.

Copertino, M. and Seeliger, U. (2010). Hábitats de Ruppia maritime e de macroalgas. In *O Estuário da Lagoa dos Patos: Um Seculo de Transformações*, ed. U. Seeliger and C. Odebrecht. Rio Grande, Brazil: FURG, pp. 91–98.

Corcoran, E. Ravilious, C. and Skuja, M. (2007). *Mangroves of Western and Central Africa*. UNEP-Regional Seas Programme/UNEP-WCMC. Online at http://www.unep-wcmc.org/resources/publications/UNEP_WCMC_bio_series/26.htm (accessed October 2013).

COSEWIC (2009). *COSEWIC Assessment and Status Report on the Eskimo Curlew Numenius borealis in Canada*. Ottawa: Committee on the Status of Endangered Wildlife in Canada, 32 pp.

Costa, C. S. B, and Davy, A. J. (1992). Coastal saltmarsh communities and Latin America. In *Coastal Plant Communities of Latin America*, ed. U. Seeliger. San Diego, CA: Academic Press, Inc., pp. 179–199.

Costa, C. S. B., Marangoni, J. C. and Azevedo, A. M. G. (2003). Plant zonation in irregularly flooded salt marshes: relative importance of stress tolerance and biological interactions. *Journal of Ecology*, **91**, 951–965.

Costa, C. S. B., Iribarne, O. O. and Farina, J. M. (2009). Human impacts and threats to the conservation of South American salt marshes. In *Human Impacts on Salt Marshes: A Global Perspective*, ed. B. R. Silliman, E. D. Grosholz and M. D. Bertness. Berkeley, CA: University of California Press, pp. 337–359.

Crain, C. M., Gedan, K. B. and Dionne, M. (2009). Tidal restrictions and mosquito ditching in New England marshes. In *Human Impacts on Salt Marshes: A Global Perspective*, ed. B. R. Silliman, E. D. Grosholz and M. D. Bertness. Berkeley, CA: University of California Press, pp. 149–170.

Craft, C., Megonigal, P., Broome, S., Stevenson, J., Freese, R., Cornell, J., and Sacco, J. (2003). The pace of ecosystem development of constructed *Spartina alterniflora* marshes. *Ecological Applications*, **13** (5), 1417–1432.

Craft, C., Clough, J., Ehman, J., *et al.* (2009). Forecasting the effects of accelerated sea-level rise on tidal marsh ecosystem services. *Frontiers in Ecology and the Environment*, **7** (2), 73–78.

Crow, B. And Carney, J. (2013). Commercializing nature: mangrove conservation and female oyster collectors in The Gambia. *Antipode*, **45** (2), 275–293.

Culleton, B. J., Kennett, D. J. and Jones, T. L. (2009). Oxygen isotope seasonality in a temperate estuarine shell midden: a case study from CA-ALA-17 on the San Francisco Bay, California. *Journal of Archaeological Science*, **36** (7), 1354–1363.

Daborn, G. R., Brylinsky, M. and Van Proosdij, D. (2003). *Ecological Studies of the Windsor Causeway and Pesaquid Lake 2002*. Publication No. 69. Wolfville, NS: Acadia Centre for Estuarine Research.

Daiber, F. C. (1977). Salt marsh animals: distributions related to tidal flooding, salinity and vegetation. In *Ecosystems of the World 1, Wet Coastal Ecosystems*, ed. V. J. Chapman. Amsterdam, Oxford, New York: Elsevier Scientific Publishing Company, pp. 70–108.

Dale, P. E. R. and Knight, J. M. (2008). Wetlands and mosquitoes: a review. *Wetlands Ecology Management*, **16**, 255–276.

Dale, P. E. R. and Knight, J. M. (2012). Destroying mosquitos without destroying wetlands: an eastern Australian approach. *Wetlands Ecology Management*, **20**, 233–242.

Daleo, P. and Iribarne, O. (2009). Beyond competition: the stress-gradient hypothesis tested in plant-herbivore interactions. *Ecology*, **90** (9), 2368–2374.

Daleo, P., Fanjul, E., Casariego, A. M., *et al.* (2007). Ecosystem engineers activate mycorrhizal mutualism in salt marshes. *Ecology Letters*, **10** (10), 902–908.

Darboe, F. S. (2002). *Fish Species Abundance and Distribution in the Gambia Estuary*. United Nations University, Fisheries Training Program, Final Project 2002. Reykjavik, Iceland: United Nations University.

Davy, A. J., Bakker, J. P. and Figuerroa, M. E. (2009). Human modification of European salt marshes. In *Human Impacts on Salt Marshes: A Global Perspective*, ed. B. R. Silliman, E. D. Grosholz and M. D. Bertness. Berkeley, CA: University of California Press, pp. 311–335.

Day Jr., J. W., Rybczyk, J., Scarton, F., Rismondo, A. and Cecconi, G. (1999). Soil accretionary dynamics, sea level rise and the survival of wetlands in Venice Lagoon: a field and modelling approach. *Estuarine, Coastal and Shelf Science*, **49**, 607–628.

Deegan, L. A., Johnson, D. S., Warren, R. S., *et al.* (2012). Coastal eutrophication as a driver of salt marsh loss. *Nature*, **490**, 388–394.

Degeorges, A. and Reilly, B. K. (2007). Eco-Politics of dams on the Gambia River. *International Journal of Water Resources Development*, **23** (4), 641–657.

de Lange, W. P. and de Lange, P. J. (1994). An appraisal of factors controlling the latitudinal distribution of mangrove (*Avicennia marina* var. *resinifera*) in New Zealand. *Journal of Coastal Research*, **10** (3), 539–548.

De Luca, V., Salim, V., Atsumi, S. M. and Yu, F. (2012). Mining the biodiversity of plants: a revolution in the making. *Science*, **336**, 1658–1661.

Demarcq, H. and Demarcq, G. (1992). Le biostrome à *Crassostrea gasar* (Bivalvia) de la Holocène du Sine-Saloum, Sénégal; données nouvelles et interpretation écostratigraphique. *Geobios*, **25** (2), 225–250.

Dennison, W. C., Saxby, T. and Walsh, B. M. (editors) (2012). Responding to major storm impacts: ecological impacts of Hurricane Sandy on Chesapeake and Delmarva coastal bays. University of Maryland Center for Environmental Science. Online at ian.umces.edu (accessed October 2013).

De Selincourt, A. (1972). Translation. *Herodotus: The Histories*. London: Penguin Classics.

Dijkema, K. S. (1987). Geography of the salt marshes in Europe. *Zeitschrift für Geomorphologie*, **31**, 489–499.

Dolan, R. and Davis, R. E. (1994). Coastal storm hazards. In *Coastal Hazards Perception, Susceptibility and Mitigation*, ed. C. W. Finkl Jr. Fort Lauderdale, FL: Coastal Education and Research Foundation, pp. 103–114.

Dominguez, J. M. L., Martin, L. and Bittencourt, A. C. S. P. (1987). Sea level history and Quaternary evolution of river mouth-associated beach-ridge plains along the East-Southeast Brazilian coast: a summary. *Society of Economic Paleontologists and Mineralogists Special Publication*, **41**, 115–127.

Dominguez-Cadena, R., Léon de la Luz, J. L. and Riosmena-Rodriguez, R. (2011). Análisis de la influencia de la condiciones micro-topográphicas del substrato en la estrctura del manglar en el Golfo de California. In *Los Manglares de la Península de Baja California*, ed. E. F. F. Pico, E. S. Zaragosa, R. R. Rodriguez and J. L., Léon de la Luz. La Paz, México: Centro de Investigaciones Biológicas del Noroeste, S.C., p. 29–66.

Doody, J. P. (2008). *Saltmarsh Conservation, Management and Restoration*. New York: Springer.

Dos Santos, V. M., Matheson, F. E., Pilditch, C. A. and Elger, A. (2012). Is black swan grazing a threat to seagrass? Indications from an observational study in New Zealand. *Aquatic Botany*, **100**, 41–50.

Dowty, R. A., Shaffer, G. P., Hester, M. W., *et al.* (2001). Phytoremediation of small-scale oil spills in fresh marsh environments: a mesocosm simulation. *Marine Environmental Research*, **52** (3), 195–211.

Duke, N. C., Ball, M. C. and Ellison, J. C. (1998). Factors influencing biodiversity and distributional gradients in mangroves. *Global Ecology and Biogeography Letters*, **7**, 27–47.

Dunton, K. H., Weingartner, T. and Carmack, E. C. (2006). The nearshore western Beaufort Sea ecosystem: circulation and importance of terrestrial carbon in arctic coastal food webs. *Progress in Oceanography*, **71**, 362–378.

Dutton, A. and Lambeck, K. (2012). Ice volume and sea level during the last interglacial. *Science*, **337**, 216–219.

Eisma, D. (1997). *Intertidal Deposits: River Mouths, Tidal Flats, and Coastal Lagoons*. Boca Raton, FL: CRC Press.

Ellenblum, R. (2012). *The Collapse of the Eastern Mediterranean*. Cambridge: Cambridge University Press.

Ellison, J. C. (2009). Geomorphology and sedimentology of mangrove swamps. In *Coastal Wetlands: An Integrated Ecosystems Approach*, ed. G. M. E. Perillo, E. Wolanski, D. R. Cahoon and M. M. Brison. Amsterdam: Elsevier BV, pp. 565–591.

Ellison, A. M., Farnsworth, J. T. and Merkt, R. E. (1999). Origins of mangrove ecosystems and the mangrove biodiversity anomaly. *Global Ecology and Biogeography*, **8** (2), 95–115.

Ellison, J. C. and Zouh, I. (2012). Vulnerability to climate change of mangroves: assessment from Cameroon, Central Africa. *Biology*, **1**, 617–638.

Elton, C. (1927). *Animal Ecology*, 1st edn. London: Sidgwick and Jackson.

Engelhart, S. E., Horton, B. P., Roberts, D. H., Bryant, C. L. and Corbett, D. R. (2007). Mangrove pollen of Indonesia and its suitability as a sea level indicator. *Marine Geology*, **242**, 65–81.

FAO (2007). *The World's Mangroves 1980–2005*. FAO Forestry Paper 153. Rome: Food and Agriculture Organisation of the United Nations.

FAO (2012). *The State of Food Insecurity in the World 2012*. Rome: Food and Agriculture Organisation of the United Nations.

FAO/UNESCO (1973). *Irrigation, Drainage and Salinity: An International Sourcebook*. Paris: Unesco/Hutchinson and Co. (Publishers) Ltd., 510 pp.

Faraco, L. F. D. and da Cunha Lana. P. (2004). Leaf-consumption in subtropical mangroves of Paranaguá Bay (SE Brazil). *Wetlands Ecology and Management*, **12**, 115–122.

Falk, D., Palmer, M. and Zedler, J. B. (editors) (2006). *Foundations of Restoration Ecology*. Washington, DC: Island Press.

Feddes, F. (2012). *A Millenium of Amsterdam*. Bussum, the Netherlands: THOTH Publishers.

Fernandez, L. D. and Zapata, J. A. (2010). Distribution of benthic Foraminifera (Protozoa: Foraminiferida) in the Quillaipe Inlet (41°32'S; 72°44'W), Chile: Implications for sea level studies. *Revista Chilena de Historia Natural*, **83**: 567–583.

Ficke, A. D., Myrick, C. A. and Hansen L. J. (2007). Potential impacts of global climate change on freshwater fisheries. *Reviews in Fish Biology and Fisheries*, **17**, 581–613.

Field, C. B., Osborn, J. G., Hoffman, L. L., et al. (1998). Mangrove diversity and ecosystem function. *Global Ecology and Biogeography Letters*, **7**, 3–14.

Fischetti, M. (2013). Storm of the century. *Scientific American*, **308**, (6), 38–67.

Flowers, T. J. (2004). Improving crop salt tolerance. *Journal of Experimental Botany*, **55** (396), 307–319.

Flowers, T. J. and Colmer, T. D. (2008). Salinity tolerance in halophytes. *New Phytologist*, **179**, 945–963.

Forbes, D. L. (editor) (2011). *State of the Arctic Coast 2010: Scientific Review and Outlook*. International Arctic Science Committee, Land-Ocean Interactions in the Coastal Zone, Arctic Monitoring and Assessment Programme, International Permafrost Association. Geestacht, Germany: Helmholtz-Zentrum, 178 pp.

Forbes, D. L. and Hansom, J. D. (2011). Polar Coasts. In *Treatise on Estuarine and Coastal Science*, Vol. **3**, ed. E. Wolanski and D. S. McLusky. Waltham: Academic Press, pp. 245–283.

Frail-Gauthier, J. L., Scott, D. B. and Batt, J. H. (2007). Studies of living marsh foraminifera to enhance their usefulness as paleoenvironmental indicators. *2007 Geological Society of America. Abstracts with Programs*, **39** (6), p. 447

Frail-Gauthier, J., Scott, D. B. and Romanuk, T. R. (2011). Ecology and habits of temperate salt marsh foraminifera and associated meiofauna from a laboratory salt marsh system. Annual Meeting of the Geological Society of America, Minneapolis; Cushman Symposium, session no. 231–3, p. 555.

França, M. C., Francisquini, M. I., Cohen, M. C. L., *et al.* (2012). The last mangroves of Marajó Island – Eastern Amazon: Impact of climate and/or relative sea level changes. *Review of Palaeobotany and Palynology*, **187**, 50–65.

Fritz, H. M., Blount, C. D., Thwin, S., Thu, M. K. and Chan, N. (2009). Cyclone Nargis storm surge in Myanmar. *Nature Geoscience*, **2**, 448–449.

Fundy Ocean Research Center for Energy (2013). Online at http://fundyforce.ca (accessed October 2013).

Galván, K., Fleeger, J. W., Peterson, B., *et al.* (2011). Natural abundance stable isotopes and dual isotope tracer additions help to resolve resources supporting a saltmarsh food web. *Journal of Experimental Marine Biology and Ecology*, **410**, 1–11.

Gang, P. O. and Agatsiva, J. L. (1992). The current status of mangroves along the Kenyan coast: a case study of Mida Creek mangroves based on remote sensing. In *The Ecology of Mangrove and Related Ecosystems*, ed. V. Jaccarini and E. Martens. Reprinted from *Hydrobiologia*, **247**, 29–36.

Ganong, W. F. (1903). The vegetation of the Bay of Fundy salt and diked marshes: an ecological study. *Botanical Gazette*, **36**, 161–186.

Garcillán, P. P. and Ezcurra, E. (2003). Biogeographic regions and β-diversity of woody dryland legumes in the Baja California peninsula. *Journal of Vegetation Science*, **14**, 859–868.

Gardner, A. S., Mohloldt, G., Cogley, J. G., *et al.* (2013). A reconciled estimate of glacier contributions to sea level rise: 2003 to 2009. *Science*, **340**, 852–857.

Gehrels, W. R., Kirby, J. R., Prokoph, A., *et al.* (2005). Onset of recent rapid sea level rise in the western Atlantic Ocean. *Quaternary Science Reviews*, **24**, 2083–2100.

Giesen, W., Wulfraat, S., Zieren, M. and Scholten, L. (2007). *Mangrove Guidebook for Southeast Asia*. Bangkok, Thailand: Dharmasarn Co., Ltd. FAO and Wetlands International 2006, Regional Office for Asia and the Pacific.

Giosan, L, Constantinescu, S., Clift, P. D., *et al.* (2006). Recent morphodynamics of the Indus delta shore and shelf. *Continental Shelf Research*, **26**, 1668–1684.

Giosan, L., Coolen, M. J. L., Kaplan, J. O., *et al.* (2012). Early anthropogenic transformation of the Danube-Black Sea system. *Nature Scientific Reports* **2**, 582.

Giri, C., Ochieng, E., Tieszen, L. L., *et al.* (2011). Status and distribution of mangrove forests of the world using earth observation satellite data. *Global Ecology and Biogeography*, **20**, 154–159. Online at http://earthobservatory.nasa.gov/ (accessed October 2013).

Gleason, M. L., Elmer, D. A., Pien, N. C. and Fisher, J. S. (1979). Effects of stem density upon sediment retention by salt marsh cord grass *Spartina alterniflora* Loisel. *Estuaries*, **2** (4), 271–273.

Glenn, E. P., Brown, J. J. and Bluwald, E. (1999). Salt tolerance and crop potential of halophytes. *Critical Reviews in Plant Sciences*, **18** (2), 227–255.

Glenn, E. P., Nagleru, P. A., Brusca, R. C. and Hinojosa-Huerta, O. (2006). Coastal wetlands of the northern Gulf of California: inventory and conservation status. *Aquatic Conservation: Marine and Freshwater Ecosystems*, **l6**, 5–28.

Glooschenko, W. A., Martini, I. P. and Clarke-Whistler, K. (1988). *Salt Marshes of Canada*. Wetlands of Canada, National Wetlands Working Group, Canada Committee on Ecological Land Classification, Ecological Land Classification Series No. 24, p. 349–376.

GNS (2011). *GNS Science Annual Report, 2011: Resilience to Natural Hazards*. Lower Hutt, NZ: GNS Science, pp. 25–26.

Godfrey, P. J. and Godfrey, M. M. (1974). The role of overwash and inlet dynamics in the formation of salt marshes on North Carolina barrier islands. In *Ecology of Halophytes*, ed. R. Reimold and W. H. Queen. New York: Academic Press, pp. 407–428.

Goff, J., Lamarche, G., Pelletier, B., Chagué-Goff, C. and Strotz, L. (2011). Predecessors to the 2009 South Pacific tsunami in the Wallis and Futuna archipelago. *Earth-Science Reviews*, **107**, 91–106.

Gornitz, V. M., Daniels, R. C., White, T. W. and Birdswell, K. R. (1994). The development of a coastal risk assessment database: vulnerability to sea level rise in the U.S. Southeast. *Journal of Coastal Research*, Special Issue No. 12: Coastal Hazards, 327–338.

González-Zamorano, P., Nava-Sánchez, E. H., León de la Luz, J. L. and Díaz-Castro, S. C. (2011). Patrones de distribución y determinants ambientales de los manglares peninsulares. In *Los Manglares de la Península de Baja California*, ed. E. F. F. Pico, E. S. Zaragosa, R. R. Rodriguez and J. L. Léon de la Luz. La Paz, México: Centro de Investigaciones Biológicas del Noroeste, S. C., pp. 67–104.

González-Zamorano, P., Lluch-Cota, S. E. and Nava-Sánchez, E. H. (2012). Relation between the structure of mangrove forests and geomorphic types of lagoons of the Baja California Peninsula. *Journal of Coastal Research*, **29** (1), 173–181.

Gordillo, S., Coronato, A. M. J. and Rabassa, J. O. (2005). Quaternary molluscan faunas from the island of Tierra del Fuego after the last glacial maximum. *Scientia Marina*, **69** (S2), 337–348.

Goto, K., Chagué-Goff, C, Fujino, C. S., *et al.* (2011). New insights of tsunami hazard from the 2011 Tohoku-oki event. *Marine Geology*, **290**, 46–50.

Government of Canada (1996). *The State of Canada's Environment – 1996*. Ottawa, Ontario: Environment Canada.

Graham, S. (2003). Environmental effects of *Exxon Valdez* spill still being felt. *Scientific American*, December 19 issue.

Graf, M-T. and Chmura, G. L. (2006). Reinterpretation of past sea level variation of the Bay of Fundy. *The Holocene*, **20** (1), 7–11.

Greenberg, R., Maldonato, D. S. and Macdonald, M. V. (2006). Tidal marshes: a global perspective on the evolution and conservation of their terrestrial vertebrates. *Bioscience*, **56** (8), 675–685.

Greer, K. and Stow, D. (2003). Vegetation type conversion in Los Penasquitos Lagoon, California: an examination of the role of watershed urbanization. *Environmental Management*, **31**, 489–503.

Gregory, J. M., Bi, D., Collier, M. A., *et al.* (2013). Climate models without pre-industrial volcanic forcing underestimate historical ocean expansion. *Geophysical Research Letters*, **40**, 1800–1604.

Gribsholt, B. and Kristensen, E. (2002). Effects of bioturbation and plant roots on salt marsh biogeochemistry: a mesocosm study. *Marine Ecology Progress Series*, **241**, 71–87.

Griggs, G. B. (1994). California's coastal hazards. *Journal of Coastal Research Special Issue*, **12**, 1–15.

Grosholz, E. D., Levin, L. A., Tyler, A. C. and Neira, C. (2009). Changes in community structure and ecosystem function following *Spartina alterniflora* invasion of Pacific estuaries. In *Human Impacts on Salt Marshes: A Global Perspective*, ed. B. R. Silliman, E. D. Grosholz and M. D. Bertness. Berkeley, CA: University of California Press, pp. 23–40.

Guillard, J., Albaret, J-J., Simier, M., *et al.* (2004). Spatio-temporal variability of fish assemblages in the Gambia Estuary (West Africa) observed by two vertical hydroacoustic methods: Moored and mobile sampling. *Aquatic Living Resources*, **17**, 47–55.

Guo, Y., Mo, D., Mao, L., Wang, S. S. and Li, S. (2013). Settlement distribution and its relationship with environmental changes from the Neolithic to Shang-Zhou dynasties in northern Shandong, China. *Journal of Geographical Science*, **23** (4), 679–694.

Haacks, M. and Thannheiser, D. (2003). The salt marsh vegetation of New Zealand. *Phytocoenologia*, **33**, 267–288.

Hall, B. L. and Denton, G. H. (1999). New relative sea level curves for the southern Scott Coast, Antarctica: evidence for Holocene deglaciation of the western Ross Sea. *Journal of Quaternary Science*, **14** (7), 641–650.

Hamdan, M. A., Asada, T., Hassan, F. M., Warner, B. G., Douabul, A., Al-Hilli, M. R. A. and Alwan, A. A. (2010). Vegetation response to re-flooding in the Mesopotamian Wetlands, Southern Iraq. *Wetlands*, **30**, 177–188.

Hansen, B. B., Grøtan, V., Aanes, R., *et al.* (2013). Climate events synchronise the dynamics of a resident vertebrate community in the High Arctic. *Science*, **339**, 313–315.

Hartmann-Schröder, G. (1991). Contribution to the Bahia polychaete fauna Quillaipe (southern Chile). *Helgoland Marine Research*, **45** (1–2), 39–58.

Hawkes, A. D., Scott, D. B., Lipps, J. H. and Combellick, R. (2005). Evidence for possible precursor events of megathrust earthquakes on the west coast of North America. *Geological Society of America, Bulletin*, **117**, 996–1008.

Hayes, M. (2010). South Carolina. In *Encyclopedia of the World's Coastal Landforms*, ed. E. C. E. Bird. Dordrecht, the Netherlands: Springer and Business Media, pp. 94–95.

Henry, H. A. L. and Jeffries, R. L. (2009). Opportunistic herbivores, migratory connectivity, and catstrophic shifts in Arctic coastal systems. In *Human Impacts on Salt Marshes: A Global Perspective*, ed. B. R. Silliman, E. D. Grosholz and M. D. Bertness. Berkeley, CA: University of California Press, pp. 85–103.

Hernes, P. J., Benner, R., Cowie, G. L., Goñi, M. A., Bergamaschi, B. A., and Hedges, J. I. (2001). Tannin diagenesis in mangrove leaves from a tropical estuary: A novel molecular approach. *Geochimica et Cosmochimica Acta*, **65** (18), 3109–3122.

Heyvaert, V. M. A. and Baeteman, C. (2008). A middle to late Holocene avulsion history of the Euphrates river: a case study from Telled-Dër, Iraq, Lower Mesopotamia. *Quaternary Science Reviews*, **27**, 2401–2410.

Hillaire-Marcel, C. and de Vernal, A. (editors) (2007). *Proxies in Late Cenozoic Paleoceanography*. Amsterdam: Elsevier BV.

Hines, J. E. and Wiebe Robertson, M. O. (2006). Surveys of geese and swans in the Inuvialuit Settlement Region, Western Canadian Arctic 1989–2001. Occasional Paper Number 112. Yellowknife, Northern Territories, Canada: Canadian Wildlife Service, Environmental Stewardship Branch, Environment Canada.

Holt, B. G., Lessard, J-P., Borregaard, M. K., *et al.* (2013). An update of Wallace's zoogeographic regions of the world. *Science*, **339**, 74–77.

Honig, C. A. and Scott, D. B. (1987). Post-glacial strtigraphy and sea level change in southwestern New Brunswick. *Canadian Journal of Earth Sciences*, **24**, 354–364.

Hoppe-Speer, S. C. L, Adams, J. B., Rajkaran, A. and Bailey, D. (2011). The response of the red mangrove *Rhizophora mucronata* Lam. to salinity and inundation in South Africa. *Aquatic Botany*, **95** (2), 71–76.

Hori, K., Saito, Y., Zhao, Q. and Wang, P. (2002). Architecture and evolution of the tide-dominated Changjiang (Yangtze) River delta, China. *Sedimentary Geology*, **146**, 249–264.

Hosokawa, T., Tagawa, H. and Chapman, V. J. (1977). Mangals of Micronesia, Taiwan, Japan, the Philippines and Oceania. In *Ecosystems of the World. 1. Wet Coastal Ecosystems*, ed. V. J. Chapman. Amsterdam, Oxford, New York: Elsevier Scientific Publishing Company, pp. 271–291.

Huang, Y. Zhua, C, Li, X., *et al.* (2012). Differentiated population structure of a genetically depauperate mangrove species *Ceriops tagal* revealed by both Sanger and deep sequencing. *Aquatic Botany*, **101**, 46–54.

Iacovelli, D. and Vasquez, T. (1998) Supertyphoon Tip. *Mariners Weather Log*, **42** (2), 4–9.

Ibáñez, C., Morris, J. T., Mendelssohn, I. A. and Day, J. W. (2013). Coastal Marshes. In *Estuarine Ecology,* 2nd edn, ed. J. W. Day, B. C. Crump, W., M. Kemp and A. Yanez-Arancibia. Hoboken, NJ: Wiley-Blackwell, pp. 129–164.

Idaszkin, Y. L. and Bortolus, A. (2011). Does low temperature prevent *Spartina alterniflora* from expanding toward the austral-most salt marshes? *Plant Ecology*, **212** (4), 553

Idaszkin, Y. L., Bortolus, A. and Bouza, P. J. (2010). Ecological processes shaping Central Patagonian salt marsh landscapes. *Austral Ecology*, **36** (1), 59–67.

Inomata, N., Wang, X-R., Changtragoon, S. and Smzidt, A. E. (2009). Levels and patterns of DNA variation in two sympatric mangrove species, *Rhizophora apiculata* and *R. mucronata* from Thailand. *Genes and Genetic Systems*, **84**, 277–286.

IPCC (2007). Climate change 2007: the physical science basis. In *Contribution of Working Group I to the Fourth Assessment Report of the Intergovernmental Panel on Climate Change*, ed. S. Solomon, D. Qin, M. Manning, Z. Chen, M. Marquis, K. B. Averyt, M. Tignor and H. L. Miller. Cambridge: Cambridge University Press, 996 p.

Isacch, J. P., Costa, C. S. B., Rodrıguez-Gallego, L., *et al.* (2006). Distribution of saltmarsh plant communities associated with environmental factors along a latitudinal gradient on the south-west Atlantic coast. *Journal of Biogeography*, **33**, 888–900.

Isla, F. I. (2013). Coastal lagoons. In *Encyclopedia of Life Support Systems*. Online at http://www.eolss.net/Eolss-sampleAllChapter.aspx (accessed October 2013).

Isla, F. I., Gustavo, G., Bujakesky, G. G., Galasso, M. L. and De Francesco, C. G., (2005). Morphology, grain-size and faunistic composition of the macrotidal beaches of Tierra del Fuego. *Revista de la Asociación Geológica Argentina*, **60** (3), 435–445.

IUCN (2010). The IUCN Red List of Threatened Species. Version 2010.3. Online at http://www.iucnredlist.org (accessed October 2013).

Jacobs, D., Stein, E. D. and Longcore, T. (2010). *Classification of California estuaries based on natural closure patterns: templates for restoration and management.* Southern Californian Coastal Water Research Project Technical Report **619**, 50 pp.

Jahnert, R. J. and Collins, L. B. (2013). Controls on microbial activity and tidal flat evolution in Shark Bay, Western Australia. *Sedimentology*, **60** (4), 1071–1099, doi: 10.1111/sed.12023

Jallow, B. P., Barrow, M. K. A. and Leatherman, S. P. (1996). Vulnerability of the coastal zone of The Gambia to sea level rise and development of response strategies and adaptation options. *Climate Research*, **6**, 166–177.

Jaradat, A. A. (2003). Halophytes for sustainable biosaline farming in the Middle East. In *Desertification in the Third Millenium*, ed. A. S. Alsharhan, W. W. Wood., A. S. Goudie, A. Fowler and E. Abdellatif. Rotterdam: A.A. Balkema/Swetz and Zeitlinger, pp. 1–18.

Javaux, E. J. and Scott, D. B. (2003). Illustration of modern benthic Foraminifera from Bermuda and remarks on distribution in other subtropical/tropical areas. *Palaeontologia Electronica* **6** (1), 1–29.

Jefferies, R. L. (1977). Long-term damage to subarctic coastal ecosystems by geese: ecological indicators and measures of ecosystem dysfunction. In *Disturbance and Recovery in Arctic Lands: An Ecological Perspective*, ed. R. M. M. Crawford. Boston, MA: Kluwer Academic, p. 151–166.

Jefferies, R. L. and Rockwell, R. F. (2002). Foraging geese, vegetation loss and soil degradation in an Arctic salt marsh. *Applied Vegetation Science*, **5**, 7–16.

Jefferies, R. L., Jensen, A. and Abraham, K. F. (1979). Vegetational development and the effect of geese on vegetation at La Pérouse Bay. *Canadian Journal of Botany*, **57**, 1439–1450.

Jefferies, R. L., Svoboda, J., Henry, G., Raillard, M. and Riess, R. (1992). Tundra grazing systems and climatic systems. In *Arctic Ecosystems in a Changing Climate: An Ecophysiological Perspective*, ed. F. S. Chapin III, R. L. Jefferies, G. R. Shaver and J. Svoboda. San Diego, CA: Academic Press Inc., pp. 391–412.

Jenner, K. A. and Hill, P. R. (1998). Recent arctic deltaic sedimentation: Olivier Island, Mackenzie Delta, Northwest Territories, Canada. *Sedimentology*, **45**, 987–2004.

Jennings, J. N. and Bird, E. C. F. (1967). Regional geomorphological characteristics of some Australian estuaries. In *Estuaries*, ed. G. H. Luff. Washington, DC: American Association for the Advancement of Science, pp. 121–128.

Jennings, S. C., Carter, R. G. W. and Orford, J. D. (1993). Late Holocene salt marsh development under a regime of rapid relative-sea level rise: Chezzetcook Inlet, Nova Scotia. Implications for the interpretation of palaeomarsh sequences. *Canadian Journal of Earth Sciences*, **30**, 1374–1384.

Jennings, A. E., Nelson, A. R., Scott, D. B. and Aravena, J. C. (1995). Marsh foraminiferal assemblages in the Valdivia Estuary, south-central Chile, relative to vascular plants and sea level. *Journal of Coastal Research*, **11** (1), 107–123.

Jing, K., Ma, Zh., Li, B., Li, J. and Chen, J. (2007). Foraging strategies involved in habitat use of shorebirds at the intertidal area of Chongming Dongtan, China. *Ecological Research*, **22** (4), 559–570.

Johnson, D. S. and Fleeger, J. W. (2009). Weak response of saltmarsh infauna to ecosystem-wide nutrient enrichment and fish predator reduction: a four-year study. *Journal of Experimental Marine Biology and Ecology*, **373**, 35–44.

Johnson, M. E., Ledesma-Vázquez, J., Backus, D. H. and González, M. R. (2012). Lagoon microbialites on Isla Angel de la Guarda and associated peninsular shores, Gulf of California (Mexico). *Sedimentary Geology*, **263–264**, 76–84.

Johnstone, J. F. and Kokelj, S. V. (2008). Environmental conditions and vegetation recovery at abandoned drilling mud sumps in the Mackenzie Delta Region, Northwest Territories, Canada. *Arctic*, **61** (2), 199–211.

Jorgenson, M. T. (editor) (2011). *Coastal Region of Northern Alaska: Guidebook to Permafrost and Related Features*. Fairbanks, AK: State of Alaska Department of Natural Resources, Department of Natural Resources, Division of Geological and Geophysical Surveys, 188 pp. Free download from www.dggs.alaska.gov (accessed October 2013).

Jorgenson, M. T. and Brown, J. (2005). Classification of the Alaskan Beaufort Sea Coast and estimation of carbon and sediment inputs from coastal erosion. *Geo-Marine Letters*, **25**, 69–80.

Jorgensen, M. T., Frost, G. and Miller, A. (2009). Salt marsh monitoring in Lake Clark and Katmai National Parks and Preserves. *Arctic Park Science*, **9** (1), 47–49.

Joye, S. B., de Beer, D. and Cook, P. L. M. (2009). Biogeochemical dynamics of coastal tidal flats. In *Coastal Wetlands: An Integrated Ecosystems Approach*, ed. G. M. E. Perillo, E. Wolanski, D. R. Cahoon and M. M. Brison. Amsterdam: Elsevier BV, pp. 345–373.

Kangas, P. and Adey, W. (1996). Mesocosms and ecological engineering. *Ecological Engineering*, **6**, 1–5.

Kassens, H., Bauch, H. A., Dmitrenko, I. A., Eickem, H., Hubberten, H.-W., Melles, M., Thiede, J. and Tinokhov, L. A. (editors) (1999). *Land-Ocean Systems in the Siberian Arctic*. Berlin, New York, Heidelberg: Springer-Verlag, 711 pp.

Keddy, P. A. (2010). *Wetland Ecology Principles and Conservation*. Cambridge: Cambridge University Press.

Keddy, P. A., Gough, L., Nyman, J. A., *et al.* (2009). Alligator hunters, pelt traders, and runaway consumption of Gulf Coast marshes. In *Human Impacts on Salt Marshes: A Global Perspective*, ed. B. R. Silliman, E. D. Grosholz and M. D. Bertness. Berkeley, CA: University of California Press, pp. 115–133.

Kemp, A. C., Horton, B. P., Corbett, D. R., *et al.* (2009). The relative utility of foraminifera and diatoms for reconstructing late Holocene sea level change in North Carolina, USA. *Quaternary Research*, **71**, 9–21.

Kemp, A. C., Vane, C. H., Horton, B. P. and Culver, S. J. (2010). Stable carbon isotopes as potential sea level indicators in salt marshes, North Carolina, USA. *The Holocene*, **20** (4), 623–636.

Kemp, A. C., Engelhart, A. E., Culver, S. J., Nelson, A., Briggs, R. W. and Haeussler, P. J., (2013). Modern salt marsh and tidal-flat foraminifera from Sitkinak and Simeonof islands, Southwestern Alaska. *Journal of Foraminiferal Research*, **43**, 88–98.

Khoa, L. V. and Roth-Nelson, W. (1994). Sustainable wetland use for agriculture in the Mekong River delta of Vietnam. In *Global Wetlands: Old World and New*, ed. W. J. Mitsch. Amsterdam: Elsevier BV, pp. 737–748.

Kholeif, S. E. A. (2007). Palynology of mangrove sediments in the Hamata Area, Red Sea Coast, Egypt: vegetation and restoration overview. In *Restoration of Coastal Ecosystems: An Introduction*, ed. M. Isermann and K. Kiehl. Coastline Reports 7, Leiden, the Netherlands: EUCC, pp. 5–16.

Kholeif, S. E. A. and Mudie, P. J. (2009). Palynomorph and amorphous organic matter records of climate and oceanic conditions in late Pleistocene and Holocene sediments of the Nilecone, southeastern Mediterranean. *Palynology*, **32**, 1–24.

Kingsbury, R. W., Radlow, A., Mudie, P. J., Rutherford, J. and Radlow, R. (1976). Salt stress responses in *Lasthenia glabrata*, a winter annual composite endemic to saline soils. *Canadian Journal of Botany*, **54**, 1377–1385.

Kirwan, M. L., Langley, J. A., Guntenspergen, G. R. and Megonigal, J. P. (2013). The impact of sea level rise on organic matter decay rates in Chesapeake Bay brackish tidal marshes. *Biogeosciences*, **10**, 1869–1876.

Koch, B. P., Souza Filho, P. W. M., Behling, H., *et al.* (2011). Triterpenols in mangrove sediments as a proxy for organic matter derived fron the red mangrove (*Rhizophora mangle*). *Organic Geochemistry*, **42**, 62–73.

Kokelj, S. V., Lantz, T. C., Solomon, S., Pisaric, M. F. J., Keith, D., Morse, P., Thienpont, J. R., Smol, J. P., and Esagok, D. (2012). Utilizing multiple sources of knowledge to investigate northern environmental change: Regional ecological impacts of a storm surge in the outer Mackenzie Delta, N.W.T. *Arctic*, **65** (3), 257–272.

Kraft, J. C., Rapp, G., Kayan, I. and Luce, J. V. (2003). Harbor areas at ancient Troy: sedimentology and geomorphology complement Homer's Iliad, *Geology*, **31** (2), 163–166.

Kristensen, E. and Alongi, D. M. (2006). Control by fiddler crabs (*Uca vocans*) and plant roots (*Avicennia marina*) on carbon, iron, and sulfur biogeochemistry in mangrove sediment. *Limnology and Oceanography*, **51** (4), 1557–1571.

Kuipers, B. R., de Wilde, P. A. W. J., and Creutzberg, F. (1981). Energy flow in a tidal flat ecosystem. *Marine Ecology Progress Series*, **5**, 215–221.

Kumar, A. and Nistor, I. (2012). Guest editorial: paleotsunami. *Natural Hazards*, **63**, 1–3.

Kupferschmidt, K. (2013). A worm vaccine, coming at a snail's pace. *Science*, **339**, 502–503.

Langley, J. A., Mozdzer, T. J., Shepard, K. A., Hagerty, S. B. and Megonigal, J. P. (2013). Tidal marsh plant responses to elevated CO_2, nitrogen fertilization, and sea level rise. *Global Change Biology*, **19** (5), 1495–1503.

Lantuit, H., Overduin, P. P., Couture, N., *et al.* (2012). The Arctic Coastal Dynamics Database: a new classification scheme and statistics on Arctic permafrost coastlines. *Estuaries and Coasts*, **35**, 383–400.

Laprida, C., Chapori, N. G., Violante, R. A. and Compagnucci, R. H. (2007). Mid-Holocene evolution and paleoenvironments of the shoreface–offshore transition, north-eastern Argentina: new evidence based on benthic microfauna. *Marine Geology*, **240**, 43–56.

Lara, R. J. and Cohen, M. C. L. (2006). Sediment porewater salinity, inundation frequency and mangrove vegetation height in Braganca, North Brazil: an ecohydrology-based empirical model. *Wetlands Ecology and Management*, **14**, 349–358.

Lara, R. J., Szlafsztein, C. F., Cohen, M. C. L., *et al.* (2009). Geomorphology and sedimentology of mangroves and salt marshes: the formation of geobotanical units. In *Coastal Wetlands: An Integrated Ecosystems Approach*, ed. G. M. E. Perillo, E. Wolanski, D. R. Cahoon and M. M. Brison. Amsterdam: Elsevier BV, pp. 539–624.

Larsen, C. F., Motyka, R. J., Freymueller, J. T., Echelmeyer, K. A. and Ivins, E. R. (2005). Rapid viscoelastic uplift in southeast Alaska caused by post-Little Ice Age glacial retreat. *Earth and Planetary Science Letters*, **237**, 548–560.

Latham, R. (1958). *Translation of The Travels of Marco Polo*. London: Penguin Books, Ltd, 379 pp.

Latorre, F., Claudio F. Pérez, C. F., Stutz, S. and Pastorino, S. (2010). Pollen deposition in tauber traps and surface soil samples in the Mar Chiquita coastal lagoon area, pampa grasslands (Argentina). *Boletín de la Sociedad Argentina de Botánica*, **45** (3–4), 321–332.

Läuchi, A. and Epstein, E. (1984). How plants adapt to salinity. *California Agriculture*, **38** (10), 18–20.

Lawton, J. H. (1995). Ecological experiments with model ecosystems. *Science*, **269**, 328–331.

Lees, G. M. and Falcon N. L. (1952). The geographical history of the Mesopotamian plains. *Geographical Journal*, **118**, 24–39.

Leroy, S., Kazanci, N., Íteri, Ő., *et al.* (2002). Abrupt environmental changes within a late Holocene laustrine sequence south of the Marmara Sea (Lake Manyas, N-W Turkey): possible links with seismic events. *Marine Geology*, **190**, 531–552.

Letey, J. (2000). Soil salinity poses challenges for sustainable agriculture and wildlife. *California Agriculture*, **54** (2), 43–48.

Lewis, R. R. III (2005). Ecological engineering for successful management and restoration of mangrove forests. *Ecological Engineering*, **24** (4), 403–418.

Lewis, R. R. III (2009). Methods and criteria for successful mangrove forest restoration. In *Coastal Wetlands: An Integrated Ecosystems Approach*, ed. G. M. E. Perillo, E. Wolanski, D. R. Cahoon and M. M. Brison. Amsterdam: Elsevier BV, pp. 787–800.

Lézine, A-M. (1997). Evolution of the West African mangrove during the Late Quaternary: a review. *Géographie physique et Quaternaire*, **51** (3), 405–414.

Li, W.-H. (1992). Late Holocene paleoseismology in the lower Eel River valley, Northern California. MSc thesis, Arcata, CA: Humboldt State University, 78 pp.

Limaye, R. B. and Kumaran, K. P. N. (2012). Mangrove vegetation responses to Holocene climate change along Konkan coast of south-western India. *Quaternary International*, **263**, 114–128.

Lin, Q. and Mendelssohn, I. A. (2012). Impacts and recovery of the Deepwater Horizon oil spill on vegetation structure and function of coastal salt marshes in the Northern Gulf of Mexico. *Environmental Science and Technology*, **46** (7), 3737–3743.

Liu, J. P., Milliman, J. D., Gao, S. and Cheng, P. (2004). Holocene development of the Yellow River's subaqueous delta, North Yellow Sea. *Marine Geology*, **209**, 45–67.

Liu, J., Zhuang Z. H. and Cai X. M. (2005). Artificial techniques for the control of *Spartina alterniflora*. *Plant Protection*, **31**, 70–72.

Liu, K. (2004). Paleotempestology: principles, methods, and examples from Gulf coast lake-sediments. In *Hurricanes and Typhoons: Past, Present, and Future*, ed. R. Murnane and Liu, K. New York: Colombia University Press, pp. 13–57.

Liu, K. (2007). Uncovering prehistoric hurricane activity. *American Scientist*, **95**, 126–133.

Lindeman, R. L. (1942). The trophic-dynamic aspect of ecology. *Ecology*, **23**, 399–418.

Lipps J. H. and Valentine, J. W. (1970). The role of Foraminifera in the trophic structure of marine communities. *Lethaia*, **3** (3), 279–286.

Long, S. P. and Mason, C. F. (1983). *Saltmarsh Ecology*. Glasgow: Blackie.

López-Medellín, X., Ezcurra, E. Charlotte González-Abraham, C., Hak, J., Santiago, L. S. and Sickman, J. O. (2011). Oceanographic anomalies and sea level rise drive mangroves inland in the Pacific coast of Mexico. *Journal of Vegetation Science*, **22**, 143–151.

Luecke, D. F., Pitt, J., Congdon, C., et al. (1999). *A Delta Once More: Restoring Riparian and Wetland Habitat in the Colorado River Delta*. Washington, DC: Environmental Defense Fund, EDF Publications.

Luo,. X. X., Yang, S. L. and Zhang, J. (2012). Sediment dispersal in the East China Sea. *Geomorphology*, **179**, 126–140.

Macdonald, K. B. (1969). Quantitative studies of salt marsh faunas from the North American Pacific coast. *Ecological Monographs*, **39**, 33–69.

MacDonald, K. B. (1977). Plant and animal communities of Pacific North American Salt Marshes. In *Ecosystems of the World. 1, Wet Coastal Ecosystems*, ed. V. J. Chapman. Amsterdam, Oxford, New York: Elsevier Scientific Publishing Company, pp. 167–191.

Macdonald, K. B. and Barbour, M. (1974). Beach and salt marsh vegetation of the North American Pacific Coast. In *Ecology of Halophytes*, ed. R. J. Reimold and W. H. Queen. New York: Academic Press, pp. 175–233.

Macnae, W. (1963). Mangrove swamps in South Africa. *Journal of Ecology*, **51** (1), 1–25.

Maltby, E. (editor) (1994). *An Environmental and Ecological Study of the Marshlands of Mesopotamia*. Draft Consultative Bulletin. Wetland Ecosystems Research Group, University of Exeter. London: The AMAR Appeal Trust.

Malygina, N., Vlasova, E. and Bogdanova, V. (2013). Wild reindeer (*Rangifer tarandus* L.) resources use in the Taimyr peninsula: aspects of the principle of ecological law. *Czech Polar Reports*, **3** (1), 69–73.

Mann, K. H. (2000). *Ecology of Coastal Waters: With Implications For Management*, 2nd edn. Hoboken, NJ: John Wiley & Sons, Inc.

Manson, G. K., Solomon, S. M., Forbes, D. L., Atkinson, D. E. and Craymer, M. (2005). Spatial variability of factors influencing coastal change in the Western Canadian Arctic. *Geo-Marine Letters*, **25**, 138–145.

Mao, L., Zhang, Y. and Bi, H. (2006). Modern pollen deposits in coastal mangrove swamps from Northern Hainan Island, China. *Journal of Coastal Research*, **22** (6), 1423–1436.

Mao, L., Batten, D. J., Fujiki, T., *et al.* (2012). Key to mangrove pollen and spores of southern China: an aid to palynological interpretation of Quaternary deposits in the South China Sea. *Review of Palaeobotany and Palynology*, **176–177**, 41–67.

Marangoni, J. C. and Costa, C. S. B. (2009). Natural and anthropogenic effects on salt marsh over five decades in the Patos Lagoon (Southern Brazil). *Brazilian Journal of Oceanography*, **57** (4), 345–350.

Marín-Guirao, L., Sandoval-Gil, J. M., Ruíz, J. M. and Sánchez-Lizaso, J. L. (2011). Photosynthesis, growth and survival of the Mediterranean seagrass *Posidonia oceanica* in response to simulated salinity increases in a laboratory mesocosm system. *Estuarine, Coastal and Shelf Science*, **92** (2), 286–296.

Markovskya, E., Schmakova, N. and Sergienko, L. (2012). Ecophysiological characteristics of the coastal plants in the conditions of the tidal zone on the coasts of Svalbard. *Czech Polar Reports*, **2** (2), 103–108.

Marshall, W. A., Gehrels, W. R., Garnett, M. H., *et al.* (2007). The use of 'bomb spike' calibration and high-precision AMS C-14 analyses to date salt marsh sediments deposited during the past three centuries. *Quaternary Research*, **68**, 325–337.

Martell, A. M. and Pearson, A. M. (1978). The small mammals of the Mackenzie Delta Region, Northwest Territories, Canada. *Arctic*, **31** (4), 475–488.

Martini, I. P., Morrison, R. I. G., Glooschenko, W. A. and Protz, R. (1980). Coastal studies in James Bay, Ontario. *Geoscience Canada*, **7** (1), 11–20. See also http://www.uoguelph.ca/geology/hudsonbay/ for details (accessed October 2013).

Martini, I. P., Jeffries, R. L., Morrison, R. I. G., and Abraham, K. F. (2009). Polar coastal wetlands: development, structure, and land use. In *Coastal Wetlands: An Integrated Ecosystems Approach*, ed. G. M. E. Perillo, E. Wolanski, D. R. Cahoon and M. M. Brison. Amsterdam: Elsevier BV, pp. 119–115.

McAlpin, J. P. (editor) (2009). *Paleoseismology*, 2nd edn. International Geophysics Series, 95. New York: Elsevier.

McFadden, L., Spencer, T. and Nichols, R. J. (2007). Broad-scale modelling of coastal wetlands: what is required? In *Lagoons and Coastal Wetlands in the Global Change Context: Impacts and Management Issues*, ed. P. Viaroli, P. Lasserre and P. Campostrini. Reprinted from *Hydrobiologia*, **577**, 5–15.

McFarlin, C. R., Brewer, J. S., Buck, T. L. and Pennings, S. C. (2008). Impact of fertilization on a salt marsh food web in Georgia. *Estuaries and Coasts*, **31**, 313–325.

McGann, M. (2008). High resolution foraminiferal, isotopic, and trace element records from Holocene estuarine deposits of San Francisco Bay, California. *Journal of Coastal Research*, **24** (5), 1092–1109.

Mckenna, M. (2007). *The Leschenault Estuarine System, South-Western Australia*. Perth, Western Australia: Department of Water, Government of Western Australia, 172 pp.

McKillop, H. (2005). Finds in Belize document Late Classic Maya salt making and canoe transport. *Proceedings of the National Academy Sciences*, **205** (15), 5630–5634.

McKnight, A., Sullivan, K. M., Irons, D. B., Stephensen, S. W. and Howlin, S. (2006). Marine bird and sea otter population abundance of Prince William Sound, Alaska: trends following the T/V Exxon Valdez Oil Spill, 1989–2005. Exxon Valdez Oil Spill Restoration Project Final Report. Anchorage, AK: US Fish and Wildlife Service.

Mcleod, E., Chmura, G. L., Bouillon, S., *et al.* (2011). A blueprint for blue carbon: toward an improved understanding of the role of vegetated coastal habitats in sequestering CO_2. *Frontiers in Ecology and the Environment*, **9**, 552–560.

McWilliam, J. R. (1986). The national and international importance of drought and salinity effects on agricultural production. *Australian Journal of Plant Physiology*, **13**, 1–13.

Mendelssohn, I. A. and McKee, K. L. (2000). Saltmarshes and Mangroves. In *North American Vegetation*, ed. M. G. Barbour and W. D. Billings. Cambridge, New York: University Press, pp. 501–536.

Mendez, M. M., Schwindt, E. and Bortolus, A. (2013). Patterns of substrata use by the invasive acorn barnacle *Balanus glandula* in Patagonian salt marshes. *Hydrobiologia*, **700** (1), 99–107.

Mendoza-Salgado, R. A., Lechuga-Devéze, C. H., Amador, E. and Pedrin-Avilés, S. (2011). La calidad ambiental de manglares de B.C.S. In *Los Manglares de la Península de Baja*

California, ed. E. F. F. Pico, E. S. Zaragosa, R. R. Rodriguez and J. L. Léon de la Luz. La Paz, México: Centro de Investigaciones Biológicas del Noroeste, S.C., pp. 9–28.

Mildenhall, D. C. and Brown, L. J. (1987). An early Holocene occurrence of the mangrove *Avicennia marina* in Poverty Bay, North Island, New Zealand; its climatic and geological interpretations. *New Zealand Journal of Botany*, **25**, 281–294.

Millennium Ecosystem Assessment, 2005. *Ecosystems and Human Well-being: Wetlands and Water Synthesis*. Washington, DC: World Resources Institute.

Milo, R. and Last, R. R. (2012). Achieving diversity in the face of constraints: lessons from metabolism. *Science*, **336**, 1663–1667.

Mitsch, W. J. (2000). Self-design applied to coastal restoration. In: *Concepts and Controversies in Tidal Marsh Ecology*, ed. M. P. Weinstein and D. A. Kreeger. Boston, MA: Kluwer Academic Publishers, pp. 554–564.

Mitsch, W. J. and Day Jr, J. W. (2004). Thinking big with whole-ecosystem studies and ecosystem restoration: a legacy of HT Odum. *Ecological Modelling*, **178** (1), 133–155.

Mitsch, W. J. and Gosselink, J. G. (2007). *Wetlands*. Hoboken, NJ: John Wiley & Sons, Inc., 600 pp.

Mitsch, W. J., Gosselink, J. G., Anderson, C. J. and Zhang, L. (2009). *Wetland Ecosystems*. Hoboken, NJ: John Wiley & Sons, Inc.

Moslow, T. (1984). Depositional models of Shelf and Shoreline Sandstones. In *Continuing Education Course Note Series No. 27*. 1983 American Association of Petroleum Geologists, Fall Education Course Conference, 102 pp.

Morozova, L. M. and Ektova, S. N. (2013). Salt marsh vegetation of the southern tundra subzone of Western Siberia: An example of the Baydaratskaya Bay coasts in the Kara Sea. *Czech Polar Reports*, **3** (1), 58–68.

Morris, J. T. (2007). Ecological engineering in intertidial saltmarshes. In *Lagoons and Coastal Wetlands in the Global Change Context: Impacts and Management Issues*. Dordrecht, the Netherlands: Springer, pp. 161–168.

Morrisey, D. J., Swales, A, Dittmann, S., *et al.* (2010). The ecology and management of temperate mangroves. *Oceanography and Marine Biology: An Annual Review* **48**, 43–60.

Mudie, P. J. (1970). A survey of the coastal wetland vegetation of north San Diego County. California Department Fish and Game, Wildlife Management Administration Report 70–4, 18 pp.

Mudie, P. (1974). The potential economic uses of halophytes. In *Ecology of Halophytes*, ed. R. Reimold and W. H. Queen. New York: Academic Press, pp. 565–598.

Mudie, P. J. (1992). Circum-Arctic Quaternary and Neogene marine palynofloras: paleo-ecology and statistical analysis. In *Neogene and Quaternary Dinoflagellate Cysts and Acritarchs*, ed. M. J. Head and J. H. Wrenn. American Association of Stratigraphic Palynologists Foundation, pp. 347–390.

Mudie, P. and Byrne, R. (1980). Pollen evidence for historic sedimentation rates in California coastal marshes. *Estuarine Coastal Marine Science*, **10**, 305–316.

Mudie, P. J. and Lelièvre, M. A. (2013). Palynological study of a Mi'kmaw shell midden, northeast Nova Scotia, Canada. *Journal of Archaeological Science*, **40** (4), 2161–2175.

Mudie, P., Browning, B. and Speth, J. (1974). *The Natural Resources of Los Peñasquitos Lagoon: Recommendations for Use and Development*. Coastal Wetlands Series, no. 7. Sacramento, CA: California Department of Fish and Game, 75 pp.

Mudie, P., Browning, B. and Speth, J. (1976). *The Natural Resources of San Dieguito and Batiquitos Lagoons*. Coastal Wetland Series, no. 12. Sacramento, CA, California Department of Fish and Game, 311 pp.

Mudie, P. J., Keen, C. E., Hardy, I. A. and Vilks, G. (1984). Multivariate analysis and quantitative paleocology of benthic Foraminifera in surface and late Quarternary shelf sediments, northern Canada. *Marine Micropaleontology*, **8**, 283–313.

Mudie, P., Rochon, A. and Levac, E. (2002). Palynological records of red tide-producing species in Canada: past trends and implications for the future. *Palaeogeography, Palaeoclimatology, Palaeoecology*, **180**, 159–186.

Mudie, P., Rochon, A. and Levac, E. (2005a). Decadal-scale sea ice changes in the Canadian Arctic and their impacts on humans during the past 4,000 years. *Environmental Archaeology*, **10**, 113–126.

Mudie, P. J., Greer, S., Brakel, J., *et al.* (2005b). Forensic palynology and ethnobotany of *Salicornia* species (Chenopodiaceae) in northwest Canada and Alaska. *Canadian Journal of Botany*, **83** (1), 111–123.

Mudie, P. J., Leroy, S. A. G., Marret, F., *et al.* (2011). Non-Pollen Palynomorphs (NPP): Indicators of salinity and environmental change in the Caspian-Black Sea-Mediterranean Corridor. In *Geology and Geoarchaeology of the Black Sea Region: Beyond the Flood Hypothesis*, ed. I. Buynevich, V. Yanko-Hombach, O. Smyntyna and R. Martin, Geological Society of America, Special Publication 473, ed. D. Siegel. Syracuse, NY: Syracuse University, pp. 89–115.

Mudie, P. J., Dickson, J., Hebda, R. and Thomas, F. C. (2014). Environmental scanning electron microscopy: a modern tool for unlocking ancient secrets about the last journey of the Kwäday Dän Ts'inchi Man. In *Teachings From Long Ago Person Found: Highlights from the Kwäday Dän Ts'inchi Project*, ed. R. J. Hebda, S. Greer and A. Mackie. Victoria, British Columbia, Canada: Royal British Columbia Museum Press. Online at http://www. royalbcmuseum.bc.ca/KDT/default.aspx (accessed October 2013).

Munns, R. and Tester, M. (2008). Mechanisms of salt tolerance. *Annual Review Plant Biology*, **59**, 651–681.

Munns, R., James, R. A., Xu, B., *et al.* (2012). Wheat grain yield on saline soils is improved by an ancestral Na+ transporter gene. *Nature Biotechnology* **30**: 360–364.

Murty, T. S. and Flather, R. A. (1994). Impact of storm surges in the Bay of Bengal. *Journal of Coastal Research Special Issue*, **12**, 149–161.

Nagelkerken, I., Blaber, S. J. M., Bouillon, S., *et al.* (2008). The habitat function of mangroves for terrestrial and marine fauna: a review. *Aquatic Botany*, **89**, 155–185.

Nehring, S., Christian Boestfleisch, C., Buhmann, A. and Papenbrock, J. (2012). The North American toxic fungal pathogen G3 *Claviceps purpurea* (Fries) Tulasne is established in the German Wadden Sea. *BioInvasions Records*, **1** (1), 5–10.

Nelson, A. R., Jennings, A. E. and Kashima, K. (1996). An earthquake history derived from stratigraphic and microfossil evidence of relative sea level change at Coos Bay, southern coastal Oregon. *GSA Bulletin*, **108** (2), 141–154.

Nixon, F. C., Reinhardt, E. G. and Rothaus, R. (2009). Foraminifera and tidal notches: dating neotectonic events at Korphos, Greece. *Marine Geology*, **257**, 41–53.

Njie, M. and Drammah, O. (2011). *Value Chain of the Artisanal Oyster Harvesting Fishery of The Gambia*. Kingston, RI: Coastal Resources Center, University of Rhode Island, 74 pp.

NOAA (2013). Online at http://chesapeakebay.noaa.gov/fish-facts/oysters (accessed October 2013).

Novitsky, P. (2010). Analysis of mangrove structure and latitudinal relationships on the Gulf Coast of peninsular Florida. MA Dissertation. Tampa, FL: Department of Geography, University of South Florida, 73 pp.

Nyugen, H. N. (2007). Flooding in the Mekong River Delta, Viet Nam. Human Development Report 2007/2008, Occasional Paper, 23 pp.

Odum, E. P. (1970). *Fundamentals of Ecology*. Philadelphia, PA: Saunders.

Odum, E. P. (1984). The mesocosm. *BioScience*, **34**, 558–562.

Odum, E. P. and de la Cruz, A. (1967). Particulate organic detritus in a Georgia salt marsh-estuarine system. In *Estuaries*, ed. G. H. Lauff. AAAS Publication 83. Washington DC: AAAS, p. 383–385.

Oliver, T. S., Rogers, K., Chafe, C. J. and Woodroffe, C. D. (2012). Measuring, mapping and modelling: an integrated approach to the management of mangrove and saltmarsh in the Minnamurra River estuary, southeast Australia. *Wetlands Ecology Management*, **20**, 353–371.

Oo, N. W. (2004). Changes in Habitat Conditions and Conservation of Mangrove Ecosystem in Myanmar: A Case Study of Pyindaye Forest Reserve, Ayeyarwady Delta. Status Report for MAB Young Scientist Award, Yangon, Myanmar: Yangon University of Distance Education.

Oppenheimer, S. (1998). *Eden in the East: The Drowned Continent of Southeast Asia*. London: Phoenix.

Osgood, D. T. and Silliman, B. R. (2009). From climate change to snails: potential causes of salt marsh dieback along the US eastern seaboard and Gulf coasts. In *Human Impacts on Salt Marshes: A Global Perspective*, ed. B. R. Silliman, E. D. Grosholz and M. D. Bertness. Berkeley, CA: University of California Press, p. 237–252.

Osterman, L. E. and Smith, C. G. (2012). Over 100 years of environmental change recorded by foraminifers and sediments in Mobile Bay, Alabama, Gulf of Mexico, USA. *Estuarine, Coastal and Shelf Science*, **115**, 345–358.

Osterman, L. E., Kelly, W. S. and Ricardo, J. P. (2009). Benthic foraminiferal census data from Louisiana continental shelf cores, Gulf of Mexico. Reston: US Geological Survey Open-File Report 2008–1348, 16 pp.

Overpeck, J. T., Francis, J. A., Perovich, D. K., *et al.* (2005). Arctic system on trajectory to new seasonally ice-free state. *EOS, Transactions of the American Geophysical Union* **86** (34), 309, 312–313.

Padgett, D. E. and Brown, J. L. (1999). Effects of soil drainage and soil organic content on growth of *Spartina alterniflora* (Poaceae) in an artificial salt marsh mesocosm. *American Journal of Botany*, **86** (5), 697–702.

Packham, J. R. and Willis, A. J. (1997). *Ecology of Dunes, Salt Marsh and Shingle*. London: Chapman and Hall.

Pala, C. (2013). Detective work uncovers under-reported overfishing. *Nature*, **496**, 18.

Paling, E. I., Fonseca, M., van Katwijk, M. M. and van Keulen, M. (2009). Seagrass restoration. In *Coastal Wetlands: An Integrated Ecosystems Approach*, ed. G. M. E. Perillo, E. Wolanski, D. R. Cahoon and M. M. Brison. Amsterdam: Elsevier BV, pp. 687–714.

Paris, R., Lavigne, F., Wassmer, P. and Sartohadi, J. (2007). Coastal sedimentation associated with the December 26, 2004 tsunami in Lhok Nga, west Banda Aceh (Sumatra, Indonesia). *Marine Geology*, **238**, 93–106.

Parker J. D., Montoya J. P. and Hay, M. E. (2008). A specialist detritivore links *Spartina alterniflora* to salt marsh food webs. *Marine Ecology Progress Series*, **364**, 87–95.

Paterson, D. M., Aspden, R. J. and Black, K. S. (2009). Intertidal flats: ecosystem functioning of soft sediment systems. In *Coastal Wetlands: An Integrated Ecosystems Approach*, ed. G. M. E. Perillo, E. Wolanski, D. R. Cahoon and M. M. Brison. Amsterdam: Elsevier BV, pp. 317–343.

Pattanaik, C., Reddt, C. S., Dahl, N. K. and Das, R. (2008). Utilisation of mangrove forests in Bhitarkarnika wildlife sanctuary, Orissa. *Indian Journal of Traditional Knowledge*, **7** (4), 598–603.

Paull, C. K., Ussler III, W., Dallimore, S. R., *et al.* (2007). Origin of pingo-like features on the Beaufort Sea shelf and their possible relationship to decomposing methane gas hydrates. *Geophysical Research Letters*, **34**, L01603.

Peinado, M., Alcaraz, F., Delgadillo, J., *et al.* (1994). The coastal salt marshes of California and Baja California. *Vegetatio*, **110** (1), 55–66.

Peinado, M., Alcares, F. and Delgadillo, J. (1995). Syntaxonomy of some halophilous communities of North and Central America. *Phytocoenologia*, **25**, 23–31.

Peinado, M., Aguirre, J. L., Delgadillo, J. and Macias, M. A. M. (2007). Zonobiomes, zonoecotones and azonal vegetation along the Pacific Coast of North America. *Plant Ecology*, **191** (2), 221–252.

Peinado, M., Aguirre, J. L., Delgadillo, J. and Macías, M. A. (2008). A phytosociological and phytogeographical survey of the coastal vegetation of western North America. Part I: plant communities of Baja California, Mexico. *Plant Ecology*, **196**, 27–60.

Peinado, M., Macias, M. A. M., Aguirre, J. L. and Delgadillo, J. (2009). A phytogeographical classification of the North American Pacific Coast based on climate, vegetation and a floristic analysis of vascular plants. *Journal of Botany*, Article 389414, 30 pp.

Peinado, M., Macías, M. Á., Ocaña-Peinado, F. M., Aguirre, J. L. and Delgadillo, J. (2011). Bioclimates and vegetation along the Pacific basin of Northwestern Mexico. *Plant Ecology*, **212**, 263–281.

Pennings, S. C. and Bertness, M. D. (2001). Salt Marsh Communities. In *Marine Community Ecology*, ed. M. D. Bertness, S. D. Gaines and M. Hay. Sunderland, MA: Sinauer Associates, Inc., pp. 299–315.

Pérez-Ruzafa, A., Mompeán, M. C. and Concepción, M. (2007). Hydrogeographic, geomorphologic and fish assemblage relationships in coastal lagoons. In *Lagoons and Coastal Wetlands in the Global Change Context: Impacts and Management Issues*, ed. P. Viaroli, P. Lasserre and P. Campostrini. Reprinted from *Hydrobiologia*, **577**, 107–125.

Perillo, G. M., Wolanski, E., Cahoon, D. R. and Brinson, M. M. (editors) (2009). *Coastal Wetlands: An Integrated Ecosystem Approach*. Amsterdam: Elsevier BV.

Perry, J. E., Barnard, T. A., Bradshaw, J. G., *et al.* (2001). Creating tidal salt marshes in the Chesapeake Bay. *Journal of Coastal Research, Special Issue*, **27**, 170–191.

Pessenda, L. C. R., Vidotto, E., De Oliveira, P. E., *et al.* (2012). Late Quaternary vegetation and coastal environmental changes at Ilha do Cardoso mangrove, southeastern Brazil. *Palaeogeography, Palaeoclimatology, Palaeoecology*, **363**–364, 57–68.

Pezeshki, S. R., Hester, M. W., Lin, Q. and Nyman, J. C. (2000). The effects of oil spill and clean-up on dominant US Gulf coast marsh macrophytes: a review. *Environmental Pollution*, **108**, 129–139.

Plafker, G. (1969). *Tectonics of the March 27, 1964, Alaska Earthquake*. US Geological Survey Professional Paper 543-I, pp. 1–74.

Phleger, F. B and Bradshaw, J. S. (1966). Sedimentary environments in a marine marsh. *Science*, **154**, 1551–1553.

Phleger, F. and Ayala-Castañares, A. (1969). Marine geology of Topolobampo lagoons, Sinaloa, Mexico. In *Coastal Lagoons: A Symposium*, ed. A. Ayala-Castañares and F. B. Phleger. Ciudad Universitaria, Mexico: Universidad Nacional Autónoma de México, pp. 101–136.

Pico, E. F. F., Zaragosa, E. S., Rodriguez, R. R. and Léon de la Luz, J. L. (editors) (2011). *Los Manglares de la Península de Baja California*. La Paz, México: Centro de Investigaciones Biológicas del Noroeste, S.C., 326 pp.

Pilkey, O. H. and Theiler, E. R. (1992). *Coastal Erosion*. Tulsa, OK: Society of Economic Paleontologists and Mineralogists, Slide set no. 6.

Pilkey, O. H. and Young, R. (2009). *The Rising Sea*. Washington, DC: Island Press, 203 pp.

Pillay, T. V. R. (1969). Estuarine fisheries of West Africa. In *Estuaries*, ed. G. H. Luff. Washington, DC: American Association for the Advancement of Science, pp. 639–646.

Ping, C-L., Michaelson, G. J., Joregenson, M. T., *et al.* (2008). High stocks of soil organic carbon in the North American Arctic region. *Nature Geoscience*, **1**, 615–619.

Piou, C., Feller, I. C., Berger, U. and Chi, F. (2006). Zonation patterns of Belizean offshore mangrove forests 41 years after a catastrophic hurricane. *Biotropica*, **38** (3), 365–374.

Pisaric, M. F. J., Thienpont, J., Kokelj, S. V., *et al.* (2011). Impacts of a recent storm surge on an Arctic ecosystem examined within a millennial timescale. *Proceedings of the National Academy of Sciences*, **108** (22), 8960–8965.

Pitcher A. M., Ollerhead, J., Kellman, L., Risk, D. and Campbell, D. A. (2005). Evidence for subsurface pooling of CH_4 in saltmarsh sediments in the Musquash Estuary, New Brunswick. *Canadian Coastal Conference 2005*, 7 pp.

Plaziat, J. C. (1984). Mollusc distribution in the mangal. In *Hydrobiology of the Mangal*, ed. F. D. Por and I. Dor. The Hague: W. Junk Publishers, pp. 111–154.

Pokhrel, Y. N., Hanasaki, N., Yeh, P. J-F., *et al.* (2012). Model estimates of sea level change due to anthropogenic impacts on terrestrial water storage. *Nature Geoscience*, **5**, 389–392.

Polidoro, B. A., Carpenter, K. E., Collins, L., *et al.* (2010). The loss of species: mangrove extinction risk and geographic areas of global concern. *PLoS ONE*, **5** (4). www.plosone.org

Por, F. D. (1984). The ecosystem of the mangal: general considerations. In *Hydrobiology of the Mangal*, ed. F. D. Por and I. Dor. The Hague: W. Junk Publishers, pp. 1–14.

Por, F. D. and Dor, I. (editors) (1984). *Hydrobiology of the Mangal: the Hydrobiology of the Mangrove Forests*. The Hague: W. Junk Publishers, 260 pp.

Por, M. S., Prado Por, A. and Oliviera, E. C. (1984). The mangal of the estuary and lagoon system of Cananeia (Brazil). In *Hydrobiology of the Mangal*, ed. F. D. Por and I. Dor. The Hague: W. Junk Publishers, pp. 212–228.

Pratolongo, P. D., Kirby, J. R., Plater, A., and Brinson, M. M. (2009). Temperate coastal wetlands: morphology, sediment processes, and plant communities. In *Coastal Wetlands: An Integrated Ecosystems Approach*, ed. G. M. E. Perillo, E. Wolanski, D. R. Cahoon and M. M. Brison. Amsterdam: Elsevier B.V., pp. 89–118.

Pringle, C., Velidis, G., Heliotis, F., Bandacu, D. and Cristofor, F. (1993). Environmental problems of the Danube Delta. *American Scientist*, **81** (4), 350–361.

Pugh, D. (2004). *Changing Sea Levels: Effects of Tides, Weather and Climate*. Cambridge: Cambridge University Press.

Punwong, P., Marchant, R. and Selby, K. (2013). Holocene mangrove dynamics from Unguja Ukuu, Zanzibar. *Quaternary International*, **298**, 4–19.

Qui, J. (2013). Monsoon melee. *Science*, **340**, 1400–1401.

Rains, D. W. and Epstein, E. (1967). Sodium absorption by barley roots: Role of dual mechanisms of alkali cation transport. *Plant Physiology*, **42**, 314–318.

Rajkaran, A. (2011). A status assessment of mangrove forests in South Africa and the utilization of mangroves in Mngazana Estuary. PhD Thesis. Port Elizabeth, South Africa: Nelson Mandela Metropolitan University, 155 pp.

Ramsar Convention Secretariat (2007) *Wise Use of Wetlands: A Conceptual Framework for the Wise Use of Wetlands*. Ramsar Handbooks for the Wise Use of Wetlands, 3rd edn, Vol. 1. Gland, Switzerland: Ramsar Convention Secretariat.

Ramsar (2013). *Ramsar's Liquid Assets: 40 years of the Convention on Wetlands*. Gland, Switzerland: Convention Secretariat.

Ramsar database (2013). www.wetlands.org/rsis/ (accessed October 2013).

Ravishankar, T., Gnanappazham L., Ramasubramanian, R., *et al.* (2004). *Atlas of Mangrove Wetlands of India*, Vol. **II**. Chennai, India: Andhra Pradesh.

Rechinger, K. (1964). *Flora of Lowland Iraq*. New York: Hafner Publishing. Ltd.

Redondo-Gómez, S., Mateos-Naranjo, E., Figueroa, M. E. and Davy, A. J. (2010). Salt stimulation of growth and photosynthesis in an extreme halophyte, *Arthrocnemum macrostachyum*. *Plant Biology*, **12** (1), 79–87.

Regev-Rudzki, N., Wilson, D. W., Carvalho, T. G., *et al.* (2013). Cell-Cell communication between malaria-infected red blood cells via exosome-like vesicles. *Cell*, **153** (5), 1120–1133.

Reinhardt, E. G., Nairn, R. B. and Lopez, G. (2010). Recovery estimates for the Río Cruces after the May 1960 Chilean earthquake. *Marine Geology*, **269**, 18–33.

Reise, K. (1991). Macrofauna in mud and sand of tropical and temperate mudflats. In *Estuaries and Coasts: Spatial and Temporal Intercomparison*, ed. M. Elliott and J.-P. Ducrotoy. Caen, France: Olsen and Olsen, pp. 211–216.

Ren, H., Lu, H, Shen, W., Huang, C., Guo, Q., Li, Z. and Jian, C. (2009). *Sonneratia apetala* Buch. Ham in the mangrove ecosystems of China: an invasive species or restoration species? *Ecological Engineering*, **35**, 1243–1248.

Richardson, C. and Hussain, N. (2006). Restoring the Garden of Eden: an ecological assessment of the marshes of Iraq (pdf). *BioScience*, **56** (6), 477–489.

Riehl, S., Zeidi, M. and Conard, N. J. (2013). Emergence of agriculture in the foothills of the Zagros mountains of Iran. *Science*, **341**, 65–67.

Robert E. M. R., Schmitz N., Okello J. A., *et al.* (2011). Mangrove growth rings: fact or fiction? *Trees – Structure and Function*, **25**, 49–58.

Roberts, N. (1998). *The Holocene*, 2nd edn. Hoboken, NJ: Wiley-Blackwell, 316 pp.

Rodríguez-Salinas, P., Riosmena-Rodríguez, R., Hinojosa-Arango, G. and Muñiz-Salazar, R. (2010). Restoration experiment of *Zostera marina* L. in a subtropical coastal lagoon. *Ecological Engineering*, **36** (1), 12–18.

Rodwell, J. S. (2000). *British Plant Communities*. Vol. 5: Maritime Cliffs, Sand Dunes, Saltmarshes and Other Vegetation. Cambridge: Cambridge Univesity Press.

Rose, R. I. (2001). Pesticides and public health: integrated methods of mosquito management. *Emerging Infectious Diseases*, **7** (1), 17.

Rovai, A. S., Menghini, R. P., Schaeffer-Novelli, Y. Molera, G. C. and Coelho, C. (2012). Protecting Brazil's coastal wetlands. *Science*, **335**, 1571–1572.

Rowley, D. B., Forte, A. M., Moucha, R., *et al.* (2013). Dynamic topography change of the eastern United States since 3 million years ago. *Science*, **340**, 1560–1563.

Ruiz, F., Abad, M., Cáceres, L. M., *et al.* (2010). Ostracods as tsunami tracers in Holocene sequences. *Quaternary Research*, **73**, 130–135.

Russell, T. L., Kay, B. H. and Skilleter, G. A. (2009). Environmental effects of mosquito insecticides on saltmarsh invertebrate fauna. *Aquatic Biology*, **6**, 77–90.

Rydin, H. and Jeglum, J. K. (2006). *The Biology of Peatland*. New York: Oxford University Press Inc.

Saatcioglu, M., Ghobarah, A. and Nistor, I. (2005). *Reconnaissance Report on The December 26, 2004 Sumatra Earthquake and Tsunami*. Ottawa, Canada: Canadian Association of Earthquake Engineering, 21 pp.

Saenger, P., Specht, M. M., Specht, R. L. and Chapman, V. L. (1977). Mangal and coastal salt marsh communities in Australia. In *Ecosystems of the World. Vol. I. Wet coastal ecosystems*, ed. V. J. Chapman. Amsterdam: Elsevier BV, pp. 293–346.

Saenger, P. and Bellan, M. F. (1995). *The mangrove vegetation of the Atlantic Coast of Africa: a review*. Université de Toulouse, Toulouse, France.

Saintilan, N. (editor) 2009. *Australian Salt Marsh Ecology*. Collingwood, Victoria, Australia: CSIRO Publishing.

Saintilan, N. and Hashimoto, T. R. (1999). Mangrove-saltmarsh dynamics on a bay-head delta in the Hawkesbury River estuary, New South Wales, Australia. *Hydrobiologia*, **413**: 95–102.

Saintilan, N., Rogers, K. and McKee, K. (2009). Salt marsh–mangrove interactions in Australasia and the Americas. In *Coastal Wetlands: An Integrated Ecosystems Approach*, ed. G. M. E. Perillo, E. Wolanski, D. R. Cahoon and M. M. Brison. Amsterdam: Elsevier BV, pp. 855–884.

Salkin, R. M. and Ring, T. (1996). Asia and Oceania. In *International Dictionary of Historic Places* ed. P. E. Schellinger and R. M. Salkin. London: Routledge, Taylor and Francis, pp. 353–234.

San Martin, C., Contreras, D., San Martin, J. and Ramirez, C. (1992). Vegetation of salt marshes in south-central Chile. *Revista Chile de Historia Natural*, **65** (3), 327–342.

Sanchez, H. (2009). *Saving Colombia's Mangrove Forests* [online]. Available from: http://www.youtube.com/watch?v=AECq3Y39GLY (accessed October 2013).

Sanders, R. (2011). Marine ice and other issues during harvesting of tidal electricity from Nova Scotia's Minas Passage in 2011. *Journal of Ocean Technology, Subsea Oil and Gas*, **6** (1), 34–55.

Santamaria-Gallegos, N. A, Danemann, G. D. and Escurra, E. (2011). Conservación y manejo de los manglares de la Península de Baja California. In *Los Manglares de la Península de Baja California*, ed. E. F. F. Pico, E. S. Zaragosa, R. R. Rodriguez and J. L. Léon de la Luz. La Paz, México: Centro de Investigaciones Biológicas del Noroeste, S.C., pp. 273–294.

Sârbu, A., Janauer, G, Schmidt-Mumm, U., *et al.* (2011). Characterisation of the potamal Danube River and the Delta: connectivity determines indicative macrophyte assemblages. *Hydrobiologia*, **671**, 75–93.

Sasekumar, A. and Chong, V. C. (1998). Faunal diversity in Malaysian mangroves. *Global Ecology and Biogeography Letters*, **7**, 57–60.

Sasser, C. E., Gosselink, J. G., Swenson, E. M. and Evers, D. E. (1995). Hydrologic, substrate and ecological characteristics of floating marshes in sediment-rich peatlands of the Mississippi delta plain, Louisiana, USA. *Wetlands Ecology*, **3** (3), 171–187.

Scarton, F., Cecconi, G., Cerasuolo, C. and Valle, R. (2013). The importance of dredge islands for breeding waterbirds. A three-year study in the Venice Lagoon (Italy). *Ecological Engineering* **54**, 39–48.

Schaefer, K., Zhang, T., Bruhwiler, L. and Barrett, A. P. (2011). Amount and timing of permafrost carbon release in response to climate warming. *Tellus*, **63B**, 165–180.

Schrijvers, J., Okondo, J. and Vincx, M. (1993). Ecological study of the benthos of the mangroves and surrounding beaches. In *Dynamics and Assessment of Kenyan Mangrove Ecosystems* No. TS2-0240-C (GDF), Final Report (April 1993), ed. A. F. Woitchik, P. Polk and E. Okemwa. Brussels: Vrije Universiteit Brussel, 239 pp.

Schuerch, M., Vafeidis, A., Slawig, T. and Temmerman, S. (2013). Modeling the influence of changing storm patterns on the ability of a salt marsh to keep pace with sea level. *Journal of Geophysical Research: Earth Surface*, **118**, 1–13.

Scott, D. (editor) (1995). *A Directory of Wetlands in the Middle East*. London: Earthscan Publications Ltd.

Scott, D. and Evans, M. (1994). Wildlife of the Mesopotamian Marshlands: Unpublished report. Wetland Ecosystems Research Group, University of Exeter.

Scott, D. B. (1976a). Quantitative studies of marsh foraminiferal patterns in Southern California and their applications to Holocene stratigraphic problems. In *Foraminifera of the Continental Margins, Part A: Ecology and Biology*, ed. C. T. Schafer and B. R. Pelletier. Maritime Sediments, Special Publication **1**: 153–170.

Scott, D. B. (1976b). Brackish water foraminifera from Southern California and description of *Polysaccammina ipohalina*, n.gen., n. sp. *Journal of Foraminiferal Research*, **6**, 312–321.

Scott, D. B. and Medioli, F. S. (1978). Vertical zonations of marsh foraminifera as accurate indicators of former sea levels. *Nature*, **272**, 528–531.

Scott, D. B. and Medioli, F. S. (1980). Quantitative studies of marsh foraminiferal distributions in Nova Scotia and comparison with those in other parts of the world: implications for sea level studies. *Special Publication of the Cushman Foundation for Foraminiferal Research* 17, 58 pp.

Scott, D. B. and Martini, I. P. (1982). Marsh foraminiferal zonations in western James-Hudson Bay. *Naturaliste Canadien*, **109**, 399–414.

Scott, D. B. and Lipps, J. (2008). Paleo-hazard recognition in coastal settings-microfossil determinations of the history and precursors of major catastrophic events. *Abstracts: Geological Society of America Meeting 2008*, p. 569.

Scott, D. B., Mudie, P. J. and Bradshaw, J. S. (1976). Benthonic foraminifera of three Southern California lagoons: ecology and recent stratigraphy. *Journal of Foraminiferal Research*, **6** (1), 59–75.

Scott, D. B., Piper, D. J. W. and Panagos, A. G. (1979). Recent salt marsh and intertidal marsh foraminifera from the western coast of Greece. *Rivista Italiana Paleontologica*, **85** (1), 243–266.

Scott, D. B., Schnack, E. J., Ferrero, L., Espinosa, M. and Barbosa, C. F. (1990). Recent marsh foraminifera from the east coast of South America: comparison to the northern hemisphere. In *Paleoecology, Biostratigraphy, Paleoceanography and Taxononomy of Agglutinated Foraminifera. NATO ASI Series C*, **327**: 717–738.

Scott, D. B., Suter, J. R., and Kosters, E. C. (1991). Marsh foraminifera and and arcellaceans of the lower Mississippi Delta: controls on spatial distribution. *Micropaleontology*, **37** (4), 373–392.

Scott, D. B., Collins, E. S., Duggan, J., *et al.* (1996). Pacific Rim marsh foraminiferal distributions. *Journal of Coastal Research*, **12**(4), 850–93.

Scott, D. B., Medioli, F. S. and Schafer, C. T. (2001). *Monitoring of Coastal Environments Using Foraminifera and Thecamoebian Indicators*. Cambridge: Cambridge University Press.

Scott, D. B., Mudie, P. J. and Bradshaw, J. S. (2011). Coastal evolution of Southern California as interpreted from benthic foraminifera, ostracodes, and pollen. *Journal of Foraminiferal Research*, **41** (3), 285–307.

Scourse, J. T., Marret, F., Versteegh, G. J. M., Jansen, F., Schefug, E. and van der Plicht, J. (2005). High-resolution last deglaciation record from the Congo fan reveals significance of mangrove pollen and biomarkers as indicators of shelf transgression. *Quaternary Research*, **64**, 57–69.

Seabrook, C. (2012). *The World of the Salt Marsh: Appreciating and Protecting the Tidal Marshes of the Southeastern Atlantic Coast*. Atlanta, GA: University of Georgia Press.

Seliskar, D. M. and Gallagher, J. L. (1983). *The Ecology of Tidal Marshes of the Pacific Northwest Coast: A Community Profile*. Washington, DC: US Fish and Wildlife Service, Division of Biological Services, FWS/OBS-82/32, 65 pp.

Senner, N. R. (2007). *Conservation Plan for the Hudsonian Godwit*. Version 1.0. Manomet, MA: Manomet Center for Conservation Science.

Sergienko, L. (2013). Salt marsh flora and vegetation of the Russian Arctic coasts. *Czech Polar Reports*, **3** (1), 30–37.

Sharpe, P. J. and A. H. Baldwin (2012). Tidal marsh plant community response to sea level rise: a mesocosm study. *Aquatic Botany* **101**, 34–40.

Shaw, J. and Ceman, J. (1999). Salt marsh aggradation in response to late-Holocene sea level rise at Amherst Point, Nova Scotia, Canada. *The Holocene* **9**, 439–51.

Shennan, I. A. and Hamilton, S. (2006). Coseismic and pre-seismic subsidence associated with great earthquakes in Alaska. *Quaternary Science Reviews*, **25**, 1–8.

Shennan, I. A., Scott, D. B., Rutherford, M. and Zong, V. (1999). Microfossil analysis of sediments representing the 1964 earthquake, exposed at Girdwood Flats, Alaska, USA. *Quaternary International*, **60**, 55–73.

Shnyukov, E. F., Maslakov, N. and Yank-Hombach, V. (2010). Mud volcanoes of the Azov-Black Sea basin, onshore and offshore. *Abstract volume, INQUA 501- IGCP 521 Sixth Plenary Meeting and Field Trip, Rhodes, Greece, 27 September–5 October 2010*, pp. 190–194.

Shum, C., Kuo, C., and Guo, J. (2008). Role of Antarctic ice mass balances in present-day sea level change. *Polar Science* **2**: 149–161.

Silliman, B. R., Bertness, M. D., Grosholz, E. D. (editors) (2009). *Human Impacts on Salt Marshes: A Global Perspective*. Berkeley, CA: University of California Press.

Silliman, B. R., van de Koppel, J., McCoy, M. W., *et al*. (2012). Degradation and resilience in Louisiana salt marshes after the BP–Deepwater Horizon oil spil. *Proceedings of the National Academy of Science*, **109** (28), 11234–11239.

Silvestri, S, Defina, A. and Marani, M. (2005). Tidal regime, salinity and salt marsh plant zonation. *Estuarine, Coastal and Shelf Science*, **6**, 119–130.

Simier, M., Laurent. C., Ecoutin, J-M. and Albaret, J-J. (2006). The Gambia River estuary: a reference point for estuarine fish assemblages studies in West Africa. *Estuarine, Coastal and Shelf Science*, **69**, 615−628.

Simmonds, I. and Rudeva, I. (2012). The great Arctic cyclone of August 2012. *Geophysical Research Letters*, **39**, L23709.

Smith, C. B., Cohen, M. C. L., Pessenda, L. C. R., *et al*. (2011). Holocene coastal vegetation changes at the mouth of the Amazon River. *Review of Palaeobotany and Palynology*, **68**, 21–30.

Solomon, S. M. (2005). Spatial and temporal variability of shoreline change in the Beaufort-Mackenzie region, northwest territories, Canada. *Geo-Marine Letters*, **25**, 127–137.

Solomon, S. M. and Forbes, D. L. (1999). Coastal hazards and associated management issues on South Pacific Islands. *Oceans and Coastal Management*, **42**, 523–554.

Solomon, S., Mudie, P. J., Cranston, R., *et al*. (2000). Paleohydrology of a drowned thermokarst embayment, Richards Island, Beaufort Sea, Canada. *International Journal Earth Sciences*, **89**, 503–521.

Sorokin, Y. I. (2002). *The Black Sea: Ecology and Oceanography*. Leiden, the Netherlands: Backhuys Publishers.

Sousa, S. H. M., Amaral, P. G. C., Martins, V., *et al*. (2013). Environmental evolution of the Caravelas Estuary (Northeastern Brazilian Coast, 17°S. 39°W) based on multiple proxies in a sedimentary record of the last century. *Journal of Coastal Research*, epublished ahead of print. Online at http://dx.doi.org/10.2112/JCOASTRES-D-12-00051.1 (accessed October 2013).

Southall, K. E., Gehrels, W. R. and Hayward, B. W. (2006). Foraminifera in a New Zealand salt marsh and their suitability as sea level indicators. *Marine Micropaleontology*, **60**, 167–179.

Spalding, M., Kainuma, M. and Collins, L. (2010). *World Atlas of Mangroves*. Abingdon, UK: Routledge.

Sparks, E. L., Cebrian, J., Biber, P. D., Sheehan, K. L. and Tobias, C. R. (2013). Cost-effectiveness of two small-scale marsh restoration designs. *Ecological Engineering*, **53**, 250–256.

Stanley, D. J. and Warne, A. G., (1998). Nile Delta in its destruction phase. *Journal of Coastal Research*, **14** (3), 794–825.

Stanley, R. G. Charpentier, R. R., Cook, T. A., et al. (2011). Assessment of undiscovered oil and gas resources of the Cook Inlet Region, South-Central Alaska, 2011. US Geological Survey Fact Sheet 2011–3068, 2 pp.

Steinke, T. D. (1999). Mangroves in South African estuaries. In *Estuaries of South Africa*, ed. B. R. Allanson and D. Baird. Cambridge: Cambridge University Press, pp. 119–140.

Stern, N. (2007). *The Economics of Climate Change: The Stern Review*. Cambridge: Cambridge University Press.

Stern, G. A., Macdonald, R. W., Outridge, P. M., *et al.* (2012). How does climate change influence arctic mercury? *Science of the Total Environment*, **414**, 22–42.

Stevenson, J. C., Rooth, J. E., Kearney, M. S. and Sundberg, K. L. (2000). The health and long term stability of natural and restored marshes in Chesapeake Bay. In *Concepts and Controversies in Tidal Marsh Ecology*, ed. M. P. Weinstein and D. A. Kreeger. Dordrecht, the Netherlands: Kluwer Academic Press, pp. 709–736.

Stewart, D. B. and Lockhart, W. L. 2005. *An overview of the Hudson Bay marine ecosystem*. Canadian Technical Report of Fisheries and Aquatic Science, 2586, 487 pp.

Stokstad, E. (2013). BP research dollars yield signs of cautious hope. *Science*, **339**, 636–637.

Stone, C. S. (1993). Vegetation of coastal marshes near Juneau, Alaska. *Northwest Science*, **67** (4), 215–230.

Stow, D. (2010). *The Vanished Ocean: How Tethys Reshaped the World*. Oxford: Oxford University Press.

Strong, D. R. and Ayres, D. R. (2009). *Spartina* introductions and consequences in salt marshes: arrive, survive, thrive, and sometimes hybridize. In: *Human Impacts on Salt Marshes: A Global Perspective*, ed. B. R. Silliman, E. D. Grosholz and M. D. Bertness. Berkeley, CA: University of California Press, pp. 3–22.

Stover, C. W. and Coffman, J. L. (1993). *Seismicity of the United States, 1568–1989 (Revised)*, US Geological Survey Professional Paper 1527. Washington, DC: United States Government Printing Office.

Su, G-H., Huang. Y-L., Tan, F-X., *et al.* (2006). Genetic variation in *Lumnitzera racemosa*, a mangrove species from the Indo-West Pacific. *Aquatic Botany*, **84**, 341–6.

Syvitski, J. P. M., Kettner, A. J., Overeem, I., *et al.* (2009). Sinking deltas due to human activities. *Nature Geoscience*, **2**, 681–686.

Szeliga, W. (2013). 2012 Haida Gwaii quake: insight into Cascadia's subduction extent. *EOS, Transactions, American Geophysical Union*, **94** (9), 85–86.

Teal, J. M. (1962). Energy flow in the salt marsh ecosystem of Georgia. *Ecology*, **43**, 614–624.

Teal, J. and M. Teal (1969). *Life and Death of the Salt Marsh*. New York: Little and Brown Publishers, 278 pp.

Tian, B., Zhou, Y., Zhang, L. and Yuan, L. (2008). Analyzing the habitat suitability for migratory birds at the Chongming Dongtan Nature Reserve in Shanghai, China. *Estuarine, Coastal and Shelf Science*, **80**, 296–302.

Thannheiser, D. and Haacks, M. (2004). Plant sociological studies on the salt marshes of Tasmania. In *Glimpses of a Gaian World*, ed. G. Kearsley and B. Fitzharris. Dunedin, New Zealand: University of Otago, pp. 81–96.

Thilenius, J. F. (1990). Woody plant succession on earthquake-uplifted coastal wetlands of the Copper River Delta, Alaska. *Forest Ecology and Management*, **33**/34, 439–462.

Thompsen, M. S., Adam, P. and Silliman, B. S. (2009). Anthropogenic threats to Australasian coastal salt marshes. In *Human Impacts on Salt Marshes: A Global Perspective*, ed. B. R. Silliman, E. D. Grosholz and M. D. Bertness. Berkeley, CA: University of California Press, pp. 367–390.

Thu, P. M. and Populus, J. (2007). Status and changes of mangrove forest in Mekong Delta: case study in Tra Vinh, Vietnam. *Estuarine, Coastal and Shelf Science*, **71**, 98–109.

TNC (2011). *Coastal Habitat Mapping Program, South East Alaska Data Summary Report*, prepared by Coastal & Ocean Resources Inc. and Archipelago Marine Research Inc. for NOAA National Marine Fishereries Service, Alaska Region, The Nature Conservancy (TNC), Project 10–12; mapping data accessible at www.shorezone.org (accessed October 2013).

Tobias, C. and Neubauer, S. C. (2009). Salt marsh biogeochemistry: an overview. In *Coastal Wetlands: An Integrated Ecosystems Approach*, ed. G. M. E. Perillo, E. Wolanski, D. R. Cahoon and M. M. Brison. Amsterdam: Elsevier BV, pp. 445–492.

Toledo, G., Rojas, A. and Bashan, Y. (2001). Monitoring of black mangrove restoration with nursery-reared seedlings on an arid coastal lagoon. *Hydrobiologia*, **444** (1–3), 101–109.

Tomlinson, P. A. (1986). *The Botany of Mangroves*. Cambridge: Cambridge University Press, 413 pp.

Tonev, R., Lysenko, T., Gussev, C. and Zhelev, P. (2008). The halophytic vegetation in South-East Bulgaria and along the Black Sea coast. *Hacquetia*, **7** (2), 95–121.

Turner, R. E. (1976). Geographic variations in salt marsh macrophyte production: a review. *Contributions in Marine Science*, **20**, 47–68.

Tuttle, M. P., Ruffman, A., Anderson, T. and Jeter, H. (2004). Distinguishing tsunami from storm deposits in eastern North America: the 1929 Grand Banks tsunami versus the 1991 Halloween storm. *Seismological Research Letters*, **75** (1), 117–131.

Twilley, R. R. (1985). *An Analysis of Mangrove Forests along the Gambia River Estuary: Implications for the Management of Estuarine Resources*. Ann Arbor, MI: Great Lakes and Marine Waters Center, University of Michigan.

Twilley, R. R. and Day, J. W., Jr. (2013). Mangrove wetlands. In *Estuarine Ecology*, 2nd edn, ed. J. W. Day, Jr., B. C. Crump, W. M. Kemp and A. Yáñez-Arancibia. Hoboken, NJ: John Wiley & Sons, Inc.

Twilley, R. R. and Rivera-Monroy, V. H. (2009). Ecogeomorphic models of nutrient biogeo-chemistry for mangrove wetlands. In *Coastal Wetlands: An Integrated Ecosystems Approach*, ed. G. M. E. Perillo, E. Wolanski, D. R. Cahoon and M. M. Brison. Amsterdam: Elsevier BV, pp. 641–683.

UNEP (2001). Partow, H. *The Mesopotamian Marshlands: Demise of an Ecosystem.* Nairobi, Kenya: Division of Early Warning and Assessment, United Nations Environment Programme.

UNEP/FAO/PAP (2002). Integrated Coastal and Marine Areas Management in the Gambia: Southern Coastal Region Coastal Profile and Management Strategy. West African Regional Seas Technical Reports Series No.1. Split, Croatia, UNEP/FAO/PAP, 1998.

UNESCO/FAO (1973). *Irrigation, Drainage and Salinity. An International Source Book.* London: PARIS/UNESCO, HUTCHINSON & CO. Ltd., 510 pp.

University of Washington (2007). *Ruesink Lab* [online]. Available from: http://depts.washington.edu/jlrlab/aboutthebay.php (accessed October 2013).

Vaiphasa, C., de Boer, W. F., Skidmore, A. K., *et al.* (2007). Impact of solid shrimp pond waste materials on mangrove growth and mortality: a case study from Pak Phanang, Thailand. *Hydrobiologia* **591**: 47–57.

Valiela, I., Kinney, E., Culbertson, J., Peacock, E. and Smith, S. (2009). Global losses of mangroves and salt marshes, In *Global Loss of Coastal Habitats: Rates, Causes and Consequences*, ed. C. M. Duarte. Madrid, Spain: Fundacion BBVA, pp. 108–138.

Van der Graaf, A. J., Stahl., J., Klimkowska, A., Bakker, J. P. and Drent, R. H. (2006). Surfing on a green wave: how plant growth drives spring migration in the barnacle goose *Branta leucopsis*. *Ardea*, **94** (3), 567–577.

Verkerk, M. P. and van Rens, C. P. M. (2005). *Saline Intrusion in Gambia River After Dam Construction*. Enshede, the Netherlands: University of Twente, Civil Engineering and Management.

Vermaire, J. C., Pisaric, M. F. J., Thienpont, J. R., *et al.* (2013). Arctic climate warming and sea ice declines lead to increased storm surge activity. *Geophysical Research Letters*, **40**, 1386–1390.

Vince, S. W. and Snow, A. (1984). Plant zonation in an Alaskan salt marsh 1. Distribution, abundance and environmental factors. *Journal of Ecology*, **72**, 651–667.

Von Caemmerer, S., Quick, W. P. and Furbank, R. T. (2012). The development of C4 rice: current progress and future challenges. *Science*, **336**, 1671–1672.

Waisel, Y. (1972). *The Biology of Halophytes*. New York: Academic Press.

Walsh, G. E. (1974). Mangroves: a review. In *Ecology of Halophytes*, ed. R. J. Reimold and W. H. Queen. New York: Academic Press, pp. 51–174.

Walsh, G. E. (1977). Exploitation of mangal. In *Ecosystems of the World 1, Wet Coastal Ecosystems*, ed. V. J. Chapman. Amsterdam, Oxford, New York: Elsevier Scientific Publishing Company, pp. 347–362.

Walters, B. B., Rönnbäck, P., Kovacs, J. M., *et al.* (2008). Ethnobiology, socio-economics and management of mangrove forests: a review. *Aquatic Botany*, **89**, 220–236.

Walton, M. E., Samonte-Tan, G. P., Primavera, J. H., Edwards-Jones, G., and Le Vay, L. (2006). Are mangroves worth replanting? The direct economic benefits of a community-based reforestation project. *Environmental Conservation*, **33** (4), 335–343.

Warme, J. E. (1969). Mugu Lagoon, coastal Southern California: origin, sediments and productivity. In *Coastal Lagoons. A Symposium*, ed. A. Ayales-Castañares and F. B. Phleger. Ciudad Universitaria, Mexico: Universidad Nacional Autónoma de México, pp. 137–169.

Watcham, E. P., Shennan, I. and Barlow, N. L. (2013). Scale considerations in using diatoms as indicators of sea level change: lessons from Alaska. *Journal of Quaternary Science*, **28** (2), 165–179.

Watkeys, M. K., Mason, T. R. and Goodman, P. S. (1993). The role of geology on the development of Maputaland, South Africa. *Journal of African Earth Sciences* **16** (1): 1–16.

Watson, E. B. and Byrne, R. (2009). Abundance and diversity of tidal marsh plants along the salinity gradient of the San Francisco Estuary: implications for global change ecology. *Plant Ecology*, **205**, 113–128.

Weis, J. S. and Butler, C. A. (2009). *Salt Marshes: A Natural and Unnatural History.* New Brunswick, NJ: Rutgers University Press.

West, R. C. (1977). Tidal salt marsh and mangal formations of Middle and South America. In *Ecosystems of the World 1, Wet Coastal Ecosystems*, ed. V. J. Chapman. Amsterdam, Oxford, New York: Elsevier Scientific Publishing Company, pp. 193–213.

Whigham, D. F., Baldwin, A. H. and Barendregt, A. (2009). Tidal freshwater wetlands. In *Coastal Wetlands: An Integrated Ecosystems Approach*, ed. G. M. E. Perillo, E. Wolanski, D. R. Cahoon and M. M. Brison. Amsterdam: Elsevier B.V., Amsterdam, pp. 515–533.

White, D. S. and Howes, B. L. (1994). Long-term ^{13}N-nitrogen retention in the vegetated sediments of a New England salt marsh. *Limnology and Oceanography*, **39**, 1878–1892.

Whitfield, A. K. (1992). A characterisation of southern African estuarine systems. *Southern African Journal of Aquatic Sciences* **12**: 89–103.

Whittaker, R. H. (1975). *Communities and ecosystems.* New York: Macmillan.

Wicander, R. and Monroe J. S. (2004). *Historical Geology*, 4th edn. Belmont, CA: Brooks/Cole-Thompson Learning.

Wiegert, R. G. and Freeman, B. J. (1990). *Tidal Salt Marshes of the Southeast Atlantic Coast: A Community Profile.* US Fish and Wildlife Service Biological Report 85, 70 p.

Willems, B. A., Powell, R. D., Cowan, E. and Jaeger, J. M. (2011). Glacial outburst flood sediments within Disenchantment Bay, Alaska: implications of recognising marine jökulhaup deposits in the stratigraphic record. *Marine Geology*, **284**, 1–12.

Willis, J. M., Hester, M. W. and Shaffer, G. P. (2005). A mesocosm evaluation of processed drill cuttings for wetland restoration. *Ecological Engineering*, **25** (1), 41–50.

Wilmshurst, J. M., Anderson, A. J., Higham, T. F. G. and Worthy, T. H. (2008). Dating the late prehistoric dispersal of Polynesians to New Zealand using the commensal Pacific rat. *Proceedings of the National Academy of Sciences*, **105** (22), 7676–7680.

Wilson, N. C. and Saintilan, N. (2012). Growth of the mangrove species *Rhizophora stylosa* Griff. at its southern latitudinal limit in eastern Australia. *Aquatic Botany*, **101**, 8–17.

Winzeler, E. A. (2008). Malaria research in the post-genomic era. *Nature*, **455** (7214), 751–756.

Woitchik, A. F., Polk, P. and Okemwa, E. (editors) (1993). *Dynamics and Assessment of Kenyan Mangrove Ecosystems*. No. TS2-0240-C (GDF), Final Report (April, 1993). Brussels: Vrije Universiteit Brussel, 239 pp.

Wolanski, E., Brinson, M. M., Cahoon, D. R. and Perillo, G. M. E. (2009). Coastal wetlands: a synthesis. In *Coastal Wetlands: An Integrated Ecosystem Approach*, ed. G. M. E. Perillo, E. Wolanski, D. R. Cahoon and M. M. Brinson. Asterdam: Elsevier BV, pp. 1–62.

Woo, I. and Takekawa, J. Y. (2012). Will inundation and salinity levels associated with projected sea level rise reduce the survival, growth, and reproductive capacity of *Sarcocornia pacifica* (pickleweed)? *Aquatic Botany*, **102**, 8–14.

Woodroffe, C. D., Thom, B. G. and Chappell, J. (1985). Development of widespread mangrove swamps in mid-Holocene times in northern Australia. *Nature*, **317**, 711–713.

Wright, A. L., Weaver, R. W. and Webb, J. W. (1997). Oil bioremediation in salt marsh mesocosms as influenced by N and P fertilization, flooding, and season. *Water, Air, and Soil Pollution*, **95** (1–4), 179–191.

Xue, Z., Liu, J. P., DeMaster, D., Van Nguyen, L. and Ta, T. K. O. (2010). Late Holocene evolution of the Mekong subaqueous delta, Southern Vietnam. *Marine Geology*, **269**, 46–60.

Yang, S-L. (1999). Tidal wetland sedimentation in the Yangtze Delta. *Journal of Coastal Research*, **15** (4), 1091–1099.

Yang, S. L., Milliman, J. D., Li, P. and Xu, K. (2011). 50,000 dams later: erosion of the Yangtze River and its delta. *Global and Planetary Change*, **75**, 14–20.

Yanko-Hombach, V., Mudie, P. J., Kadurin, S. and Larchenchov, E. (2013). Holocene marine transgression in the Black Sea: new evidence from the northwestern Black Sea shelf. *Quaternary International*, epublished ahead of print. doi:10.1016/j.quaint.2013.07.027.

Young, G. (1977). *Return to the Marshes: Life with the Marsh Arabs of Iraq*. London: Collins.

Zahran, M. A. (1977). Africa A. Wet formations of the African Red Sea Coast. In *Ecosystems of the World 1. Wet Coastal Ecosystems*, ed. V. J. Chapman. Amsterdam, Oxford, New York: Elsevier Scientific Publishing Company, pp. 215–232.

Zaitsev, Y. P. and Alexandrov, B. G. (1998). *Black Sea Biological Diversity Ukraine*. New York: United Nations Publications. Sales No. E.98.III.B.19, Black Sea Environmental Series, Vol. 7.

Zedler, J. B. and Nordby, C. S. (1986). *The Ecology of Tijuana Estuary, California: An Estuarine Profile*. US Fish and Wildlife Service Biological Report 85 (7.5), 104 pp.

Zedler, J. B. and West, J. M. (2008). Declining diversity in natural and restored salt marshes: A 30-year study of Tijuana Estuary. *Restoration Ecology*, **16** (2), 249–262.

Zedler, J. B., Nordby, C. S. and Kus, B. E. (1992). *The Ecology of Tijuana Estuary, California: A National Estuarine Research Reserve*. Washington, DC: NOAA Office of Coastal Resource Management, Sanctuaries and Reserves Division, 151 pp.

Zhu, R., Chen, Q., Ding, W. and Xu, H. (2012). Impact of seabird activity on nitrous oxide and methane fluxes from High Arctic tundra in Svalbard, Norway. *Journal of Geophysical Research*, **117**, G04015.

Zong, Y., Chen, Z., Innes J. B., Chen, C., Wang, Z. and Wang, H. (2007). Fire and flood management of coastal swamp enabled first rice paddy cultivation in east China. *Nature*, **449**, 459–462.

Zong, Y., Huang, G., Switzer, A. D., Yu, F. and Yim, W. W-S. (2009). An evolutionary model for the Holocene formation of the Pearl River delta, China. *The Holocene*, **19** (1), 129–142.

Zuo, P., Zhao, S., Liu, C., Wang, C. and Liang, Y. (2012). Distribution of *Spartina* spp. along China's coast. *Ecological Engineering*, **40**, 160–166.

Index

Figures and Tables are indicated in bold typeface

40856665R00213